BC Science 7

Authors

Adrienne Mason
Professional Writer/Educator
Tofino, British Columbia

Karen Charleson
Professional Writer/Educator
Hooksum Outdoor School
Hot Springs Cove, British Columbia

Eric Grace
Professional Writer/Educator
Victoria, British Columbia

Jacqueline Martin
Simon Fraser Junior High School
Calgary, Alberta

Senior Pedagogical Consultant

Sandy Wohl
Hugh Boyd Secondary School
Richmond, British Columbia

Consultants

Herb Johnston
Department of Curriculum Studies
University of British Columbia
Vancouver, British Columbia

Douglas A. Roberts
Professor Emeritus of Education
University of Calgary
Calgary, Alberta

Aboriginal Consultants

Dr. E. R. Atleo (Umeek of Ahousaht)
First Nations Studies Department
Malaspina University-College
Nanaimo, British Columbia

Dr. David Blades
Department of Curriculum and Instruction
University of Victoria
Victoria, British Columbia

Assessment Consultant

Bruce McAskill
Hold Fast Consultants, Inc.
Victoria, British Columbia

Reading Consultant

Shirley Choo
Resource Teacher
Montroyal School
North Vancouver, British Columbia

Literacy Consultant

Gloria Gustafson
Coordinator
District Staff Development Team
Coquitlam School District
Coquitlam, British Columbia

ESL Consultant

Sylvia Helmer
Vancouver School Board
District Learning Services
Vancouver, British Columbia

Technology Consultant

Glen Holmes
Claremont Secondary School
Victoria, British Columbia

Social Considerations Consultant

Nancy Josephson
School of Education
University College of the Cariboo
Kamloops, British Columbia

Special Education/At Risk Consultant

Elizabeth Sparling
Integration Support Teacher
Parkland Secondary School
Sidney, British Columbia

Health & Safety Consultant

Ken P. Waldman, C.R.S.P.
Richmond, British Columbia

Advisory Panel

Elspeth Anderson
Crofton House School
Vancouver, British Columbia

Dr. E. R. Atleo (Umeek of Ahousaht)
First Nations Studies Department
Malaspina University-College
Nanaimo, British Columbia

Dr. Allan MacKinnon
Faculty of Education
Simon Fraser University
Burnaby, British Columbia

Darren Macmillan
Middle School Coordinator
Winslow Centre
Coquitlam, British Columbia

Dr. Jolie Mayer-Smith
Department of Curriculum Studies
University of British Columbia
Vancouver, British Columbia

Steve Williams
Shuswap Junior Secondary School
Salmon Arm, British Columbia

Josie Wilson
Ashcroft Secondary School
Ashcroft, British Columbia

McGraw-Hill Ryerson

Toronto   Montréal   Boston   Burr Ridge, IL   Dubuque, IA   Madison, WI   New York
San Francisco   St. Louis   Bangkok   Bogotá   Caracas   Kuala Lumpur   Lisbon   London
Madrid   Mexico City   Milan   New Delhi   Santiago   Seoul   Singapore   Sydney   Taipei

McGraw-Hill Ryerson Limited
A Subsidiary of The McGraw-Hill Companies

COPIES OF THIS BOOK MAY BE OBTAINED BY CONTACTING:

McGraw-Hill Ryerson Ltd.

WEB SITE:
http://www.mcgrawhill.ca

E-MAIL:
orders@mcgrawhill.ca

TOLL-FREE FAX:
1-800-463-5885

TOLL-FREE CALL:
1-800-565-5758

OR BY MAILING YOUR ORDER TO:
McGraw-Hill Ryerson
Order Department
300 Water Street
Whitby, ON L1N 9B6

Please quote the ISBN and title when placing your order.

Student text ISBN:
0-07-094786-4

McGraw-Hill Ryerson BC Science 7

Copyright © 2004, McGraw-Hill Ryerson Limited, a Subsidiary of The McGraw-Hill Companies. All rights reserved. No part of this publication may be reproduced or transmitted in any form or by any means, or stored in a data base or retrieval system, without the prior written permission of McGraw-Hill Ryerson Limited, or, in the case of photocopying or other reprographic copying, a licence from The Canadian Copyright Licensing Agency (Access Copyright). For an Access Copyright licence, visit www.accesscopyright.ca or call toll free to 1-800-893-5777.

The information and activities in this textbook have been carefully developed and reviewed by professionals to ensure safety and accuracy. However, the publisher shall not be liable for any damages resulting, in whole or in part, from the reader's use of the material. Although appropriate safety procedures are discussed in detail and highlighted throughout the textbook, the safety of students remains the responsibility of the classroom teacher, the principal, and the school board/district.

ISBN 13: 978-0-07-094786-3

ISBN 10: 0-07-094786-4

http://www.mcgrawhill.ca

4 5 6 7 8 9 10 TCP 0 9

Printed and bound in Canada

Care has been taken to trace ownership of copyright material contained in this text. The publishers will gladly accept any information that will enable them to rectify any reference or credit in subsequent printings.

National Library of Canada Cataloguing in Publication Data

Grace, Eric

BC science 7 / authors, Eric Grace, Adrienne Mason, Jacqueline Martin.

Includes index.

ISBN 0-07-094786-4

1. Science--Textbooks. I. Mason, Adrienne II. Martin, Jacqueline

III. Title. IV. Title: BC science seven.

Q161.2.G72 2004   500   C2004-901432-3

SCIENCE PUBLISHER: Jane McNulty

PROJECT MANAGER: Keith Owen Richards

DEVELOPMENTAL EDITORS: Tricia Armstrong, Jonathan Bocknek, Lois Edwards, Sara Goodchild

MANAGER, EDITORIAL SERVICES: Linda Allison

SUPERVISING EDITOR: Crystal Shortt

COPY EDITORS: Paula Pettitt-Townsend, Valerie Ahwee

PERMISSIONS EDITOR: Pronk&Associates Inc.

PROJECT CO-ORDINATOR: Valerie Janicki

EDITORIAL ASSISTANT: Erin Hartley

MANAGER, PRODUCTION SERVICES: Yolanda Pigden

PRODUCTION CO-ORDINATOR: Janie Deneau

SPECIAL FEATURES: Darcy Dobell, Kirsten Craven, Michael Keefer, Laura McCoy

SET-UP PHOTOGRAPHY: Ian Crysler

COVER DESIGN, INTERIOR DESIGN, AND ART DIRECTION: Pronk&Associates Inc.

TECHNICAL ILLUSTRATORS: Pronk&Associates Inc.

COVER IMAGE: Grambo Photography/firstlight.ca

OUR COVER: The cougar is Canada's largest wild cat. Cougars were once found all across Canada. Today, they are found mostly on Canada's West Coast. What factors have caused the cougar population to decline? What steps can we take so that animals such as the cougar do not become extinct? In Unit 1 of this textbook, you will find answers to these and many other questions. Read the unit to learn all about ecosystems.

Acknowledgements

The authors, editors, project manager, and publisher of *McGraw-Hill Ryerson BC Science 7* extend sincere thanks to the students, elders, educators, consultants, and reviewers who contributed their time, energy, knowledge, and wisdom during the creation of this textbook. We are grateful to the students of Duncan Cran Elementary School in Fort St. John for their advice on readability and textbook design. We thank the following teachers from Duncan Cran Elementary School for providing similar advice: Catherine Dickie, Darren Dickie, Jason Arsenault, Katie Arsenault, Lance Lloyd, and Donna Bulmer. We thank the students of teacher Bobby Samra of the Inter-A Program at Kwantlen Park Secondary School in Surrey who participated in set-up photography sessions, as well as the students of Truellen Sumner in the school's Aboriginal Leadership course. We thank Merle Williams, Assistant Director, Aboriginal Education, British Columbia Teachers' Federation, for her input on our Ask an Elder features. And we extend deep and respectful gratitude to the elders profiled in these features: Trudy Frank of Ahousaht First Nation, Elizabeth Gravelle of Ktunaxa Nation, and Chief Herb Morven of Nisga'a Nation. Thanks to Dr. Richard Atleo for his kindness in facilitating the interview with Trudy Frank. Thanks to the Ktunaxa Kinbasket Treaty Council, to Michael Keefer, and to Laura McCoy for facilitating the interview with Elizabeth Gravelle. The following educators provided excellent general feedback on resource development: John Eby, Kathy Fladager, Randy Gibbons, Sam Hauck, Craig Hemmerich, Samia Khan, Samson Nashon, Rod Retallick, and Karen Stiles. For help in locating reviewers and consultants, we thank Steve Cardwell, Roland Case, Darlene Monkman, and Judy Thompson. Jim Lee and Donna Matovinovic of CTB McGraw-Hill provided guidance in concept development. Last but not least, we thank the other individuals profiled in this textbook who shared their knowledge and expertise: Hilda Ching, Leah De Forest, Herb Dragert, Mackenzie Gier, Hesham Nabih, John Persaud, Linda Söber, and Charles Young.

Pedagogical Reviewers

Elspeth Anderson
Crofton House School
Vancouver, British Columbia

Sara Beardsell
Nisga'a Elementary Secondary School
New Aiyansh, British Columbia

Jeff Bourgeois
Hawthorne Elementary School
Delta, British Columbia

Phyllis Daly
Lord Strathcona Community
Elementary School
Vancouver, British Columbia

Darren L. Dickie
Duncan Cran Elementary School
Fort St. John, British Columbia

Nancy J. Flood
Department of Biological Sciences
University College of the Cariboo
Kamloops, British Columbia

Gail Ford
Lord Strathcona Community
Elementary School
Vancouver, British Columbia

Manfred Hartmann
Erma Stephenson Elementary School
Surrey, British Columbia

Tammy Hartmann
Green Timbers Elementary School
Surrey, British Columbia

Gurprit Hayher
Glenrosa Middle School
Westbank, British Columbia

Cam Hill
Hartley Bay School
Hartley Bay, British Columbia

Pat Jensen
Aberdeen Elementary School
Abbotsford, British Columbia

Wayne H. J. Kelly
Dunsmuir Middle School
Victoria, British Columbia

Arthur Loyie
Sunshine Hills Elementary School
Delta, British Columbia

Emma Marley
Grandview Heights Elementary
School
Surrey, British Columbia

Colin P. McTaggart
Chief Dan George Middle School
Abbotsford, British Columbia

Georgina Oberle
Glendale Elementary School
Williams Lake, British Columbia

Calvin Parsons
Strawberry Vale Elementary School
Victoria, British Columbia

Chantel Parsons
Courtenay Middle School
Courtenay, British Columbia

Caroline Pennelli
St. Paul's Elementary School
Richmond, British Columbia

Bobby Samra
Kwantlen Park Secondary School
Surrey, British Columbia

Dr. Carol E. Scarff
Faculty of Education
Okanagan University College
Kelowna, British Columbia

Jennifer A. Wilson
Delta Manor Elementary School
Delta, British Columbia

Josie Wilson
Ashcroft Secondary School
Ashcroft, British Columbia

Anita Woode
William Konkin Elementary School
Burns Lake, British Columbia

Accuracy Reviewers

Dr. Margot E. Mandy
Department of Chemistry
University of Northern
 British Columbia
Prince George, British Columbia

Dr. Jonathan Shurin
Department of Zoology
University of British Columbia
Vancouver, British Columbia

Dr. Glenn Woodsworth
Research Scientist Emeritus
Geological Survey of Canada
Vancouver, British Columbia

Contents

Safety in Your Science Classroom viii
Science, Technology, and Society xii
Traditional Ecological Knowledge xviii

Unit 1 Ecosystems 2

Chapter 1 Organisms and Their Environment 4

Starting Point Activity 1-A: Chain of Events 5
- **1.1** Abiotic and Biotic Environments 6
  - Find Out Activity 1-B: Salty Seeds 8
  - Conduct an Investigation 1-C: Soil Sleuths 10
- **1.2** How Organisms Interact in Ecosystems 16
  - Find Out Activity 1-D: Designer Habitat 18
  - Conduct an Investigation 1-E:
     Sampling Populations in an Ecosystem 20
- **1.3** Roles of Organisms in Ecosystems 23
  - Think & Link Investigation 1-F:
     What Goes Up Must Come Down 24
  - Conduct an Investigation 1-G: Don't Waste It! 26

Chapter 1 At a Glance 31
Chapter 1 Review 32

Chapter 2 Cycles in Ecosystems 34

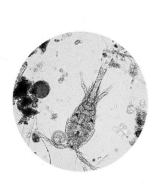

Starting Point Activity 2-A: Ant Alert 35
- **2.1** Food Chains, Food Webs, and Energy Flow 36
  - Design Your Own Investigation 2-B:
     A Mealworm's Food Chain 39
  - Think & Link Investigation 2-C: Dinner at the Copepod Café ... 41
- **2.2** Cycles of Matter 45
  - Find Out Activity 2-D: Modelling the Water Cycle 47
- **2.3** Limiting Factors in Ecosystems 51
  - Conduct an Investigation 2-E: What's the Limit? 53
  - Conduct an Investigation 2-F: Succession in a Bottle 57

Chapter 2 At a Glance 59
Chapter 2 Review 60

Chapter 3 Ecosystems and People 62

Starting Point Activity 3-A: Learning, Harming, Helping 63
- **3.1** Learning about Ecosystems 64
  - Find Out Activity 3-B: Ecosystem Stories 66
  - Conduct an Investigation 3-C: Counting Caribou 68
  - At Home Activity 3-D: Mapping Home 72
- **3.2** Human Impacts on Ecosystems 75
  - Find Out Activity 3-E: Alien Invaders 77
  - Find Out Activity 3-F: Risky Business 80
- **3.3** Conserving and Protecting Ecosystems 82
  - Conduct an Investigation 3-G: Model Ecosystem in a Bottle 84

Chapter 3 At a Glance 89
Chapter 3 Review 90
Unit 1 Ask an Elder: Trudy Frank 92
Unit 1 Ask a Biologist: Hilda Ching 94
Unit 1 Project: Hands-on Habitat 96

Unit 2 Chemistry 98

Chapter 4 Characteristics and Properties of Matter 100
Starting Point Activity 4-A: Mystery Materials. 101

4.1 Describing Matter. 102
Find Out Activity 4-B: Describing Matter 105

4.2 Measuring Matter 111
Find Out Activity 4-C: Practise Measuring Mass 112
Conduct an Investigation 4-D: Practise Measuring Volume 114
Conduct an Investigation 4-E: Building a Density Tower 117

4.3 Changes in Matter 122
Think & Link Investigation 4-F: Name the Change 123
Find Out Activity 4-G: Using the Particle Model 124

Chapter 4 At a Glance 127
Chapter 4 Review. 128

Chapter 5 Classifying Matter 130
Starting Point Activity 5-A: Mixture or Pure Substance? 131

5.1 Pure Substances and Mixtures 132
At Home Activity 5-B: Making Sugar "Disappear" 134
Conduct an Investigation 5-C: Inspector's Corner 135

5.2 Classifying Mixtures 137
Find Out Activity 5-D: Keep it Together! 138
Conduct an Investigation 5-E: What Kind of Mixture? 141

5.3 Pure Substances 143
Find Out Activity 5-F: An Element and a Compound 146

5.4 Pure Substances from Mixtures 151
Find Out Activity 5-G: Separating Mixtures 151
Find Out Activity 5-H: Panning for "Gold" 153

Chapter 5 At a Glance 155
Chapter 5 Review. 156

Chapter 6 Exploring Solutions 158
Starting Point Activity 6-A: Cold Tea, Hot Tea 159

6.1 Solutes and Solvents 160
Find Out Activity 6-B: Does It Dissolve? 161

6.2 Dissolving. 165
Find Out Activity 6-C: How Much Is Too Much? 165
Think & Link Investigation 6-D:
    How Does Temperature Affect Solubility? 168
Design Your Own Investigation 6-E:
    Changing the Rate of Dissolving 170

  6.3 Acids and Bases. 174
    Conduct an Investigation 6-F: Acid, Base, or Neutral? 176
    Find Out Activity 6-G: Home-Grown Indicators 178
  6.4 Processing Solutions. 180
    Conduct an Investigation 6-H:
     Separating Salt from Salt Water 181

 Chapter 6 At a Glance 185
 Chapter 6 Review. 186

 Unit 2 Ask an Elder: Elizabeth Gravelle. 188
 Unit 2 Ask an Oil Spill Adviser 190
 Unit 2 Project: Purifying Mixtures 192

Unit 3 Earth's Crust 194

Chapter 7 The Composition of Planet Earth 196
 Starting Point Activity 7-A: Examine the Evidence 197
  7.1 The Structure of Earth 198
    Find Out Activity 7-B: A Model Planet 200
    Conduct an Investigation 7-C: Movement in the Mantle 201
    At Home Activity 7-D: Convection Currents at Home 202

  7.2 Minerals and Mineral Resources. 204
    Find Out Activity 7-E: Research the Resource 205
    Think & Link Investigation 7-F: Trade Routes 209
    Conduct an Investigation 7-G: A Geologist's Mystery 210
  7.3 How Rocks Are Formed 213
    Conduct an Investigation 7-H: Cool Crystals, Hot Gems 215
    Find Out Activity 7-I: Mystery Moulds 219
    Find Out Activity 7-J: Find the Family 221

 Chapter 7 At a Glance 223
 Chapter 7 Review. 224

Chapter 8 Earth's Moving Crust 226
 Starting Point Activity 8-A: Sediment Shake-up 227
  8.1 Weathering and the Rock Cycle 228
    Conduct an Investigation 8-B: Rocks that Fizz 231
    At Home Activity 8-C: Chewable Rocks 232
    Conduct an Investigation 8-D: Colouring the Rock Cycle 233
    Find Out Activity 8-E: Real Rock and Roll. 235

8.2 Clues in the Crust 236
  Conduct an Investigation 8-F: Give Me a Clue! 239
  Find Out Activity 8-G: Deep Sea Diorama 240

8.3 The Theory of Plate Tectonics 245
  Design Your Own Investigation 8-H: On a Collision Course! 248
  Think & Link Investigation 8-I:
    Where Do You Draw the Line? 250

Chapter 8 At a Glance 253
Chapter 8 Review 254

Chapter 9 Earth's Changing Surface 256

Starting Point Activity 9-A: Use the Force 257

9.1 Changing LandForms 258
  Find Out Activity 9-B: How Does a Glacier Move? 263

9.2 Earthquakes 265
  Find Out Activity 9-C: Waiting For the Waves 269
  Think & Link Investigation 9-D: In the Zone 271
  Find Out Activity 9-E: Pass the Wave 272

9.3 Volcanoes 274
  Find Out Activity 9-F: Ash in the Area 277
  Find Out Activity 9-G: Blast from the Past 279
  Find Out Activity 9-H: Pardon My Intrusion 281
  Find Out Activity 9-I: Discovering the Past 282

Chapter 9 At a Glance 283
Chapter 9 Review 284

Unit 3 Ask an Elder: Chief Herb Morven 286
Unit 3 Ask a Geophysicist 288
Unit 3 Project: Shake and Quake 290

SkillPOWER 1 Using Graphic Organizers 292

SkillPOWER 2 Metric Conversion and SI Units 294

SkillPOWER 3 Measurement 297

SkillPOWER 4 Developing Models in Science 301

SkillPOWER 5 Organizing and Communicating Scientific Data 303

SkillPOWER 6 Designing and Conducting Experiments 309

Glossary 314

Index 322

Credits 325

Safety in Your Science Classroom

Become familiar with the following safety rules and procedures. It is up to you to use them, and your teacher's instructions, to make your activities and investigations in *BC Science* 7 safe and enjoyable. Your teacher will give you specific information about any other special safety rules and procedures that need to be used in your school.

1. **Working with your teacher . . .**
   - Listen carefully to any instructions your teacher gives you.
   - Inform your teacher if you have any allergies, medical conditions, or other physical problems that could affect your work in the science classroom. Tell your teacher if you wear contact lenses or a hearing aid.
   - Obtain your teacher's approval before beginning any activity you have designed yourself.
   - Know the location and proper use of the nearest fire extinguisher, fire blanket, first-aid kit, and fire alarm.

2. **Starting an activity or investigation . . .**
   - Before starting an activity or investigation, read all of it. If you do not understand how to do any step, ask your teacher for help.
   - Be sure you have checked the safety icons and have read and understood the safety precautions.
   - Begin an activity or investigation only after your teacher tells you to begin.

Wearing protective clothing is important when working with certain materials, such as ammonia.

3. **Dressing for success in science . . .**
   - When you are directed to do so, wear protective clothing, such as a lab apron and safety goggles. Always wear protective clothing when you are using materials that could pose a safety problem, such as unidentified substances, or when you are heating something.
   - Tie back long hair, and do not wear scarves, ties, or jewellery such as necklaces and bracelets.

4. **Acting responsibly . . .**
   - Work carefully with a partner, and make sure that your work area is clear.
   - Handle equipment and materials carefully.
   - Make sure that stools and chairs are resting securely on the floor.
   - If other students are doing something that you consider dangerous, report it to your teacher.

5. **Handling edible substances . . .**
   - Do not chew gum, eat, or drink in your science classroom.

- Do not taste any substances or use your mouth to draw any materials into a tube.

6. Working in a science classroom . . .
- Make sure that you understand all the safety labels on school materials and materials you bring from home. Familiarize yourself with the WHMIS symbols and the special safety symbols used in this book (see page xi).
- When carrying equipment for an activity or investigation, hold it carefully in front of your body. Carry only one object or container at a time.
- Be aware of others during activities and investigations. Make room for students who are carrying equipment to their work stations.

7. Working with sharp objects . . .
- Always cut away from yourself and others when using a knife or razor blade.
- Always keep the pointed end of scissors or any other sharp object facing away from yourself and others if you have to walk with it.
- If you notice sharp or jagged edges on any equipment, take special care with it and report it to your teacher.
- Ask your teacher to dispose of broken glass.

8. Working with electrical equipment . . .
- Make sure that your hands are dry when touching electrical cords, plugs, or sockets.
- Pull the plug, not the cord, when unplugging electrical equipment.
- Report damaged equipment or frayed cords to your teacher immediately.
- Place electrical cords in places where people will not trip over them.

9. Working with heat . . .
- When heating something, wear safety goggles and any other safety equipment that the textbook or your teacher advises.
- Always use heatproof containers.
- Do not use broken or cracked containers.
- Point the open end of a container that is being heated away from yourself and others.
- Do not allow a container to boil dry.
- Handle hot objects carefully. Be especially careful with a hot plate that might look as though it has cooled down.
- Before using a Bunsen burner, make sure that you have received training in how to light it and use it safely.
- If you do receive a burn, inform your teacher and apply cold water to the burned area immediately.

Use this method to smell a substance in the laboratory.

Hold containers away from your face when pouring liquids.

10. **Working with various chemicals . . .**
    - If any part of your body comes in contact with a chemical that might be dangerous, inform your teacher. Wash the area immediately and thoroughly with water. If you get anything in your eyes, do not touch them. Wash them immediately and continuously for 15 min.
    - Always handle carefully. If you are asked to smell a substance, never smell it directly. Hold the container slightly in front of and beneath your nose, and waft the fumes toward your nostrils.
    - Hold containers away from your face when pouring a liquid, as shown below.

11. **Working with living things . . .**

    On a field trip:
    - Try to disturb the area as little as possible.
    - If you move something, do it carefully and always replace it carefully.
    - If you are asked to remove plant material, remove it gently and take as little as possible.

    In the classroom:
    - Treat living creatures with respect.
    - Make sure that living creatures receive humane treatment while they are in your care.
    - If possible, return living creatures to their natural environment when your work is complete.

12. **Cleaning up in the science classroom . . .**
    - Clean up any spills, according to your teacher's instructions.
    - Clean equipment before you put it away.
    - Wash your hands thoroughly after doing an activity.
    - Dispose of materials as directed by your teacher. Do not mix different materials when disposing of them, unless your teacher has instructed you to do so. Never dispose of materials in a sink unless your teacher directs you to do this.

13. **Designing, constructing, and experimenting with structures and mechanisms . . .**
    - Use tools safely to cut, join, and shape objects.
    - Handle modelling clay correctly. Wash your hands after using it.
    - Follow proper procedures when studying mechanical systems and the way they operate.
    - Use special care when observing and working with objects in motion (for example, gears and pulleys, elevated objects, and objects that spin, swing, bounce, or vibrate).
    - Do not use power equipment, such as drills, sanders, saws, and lathes, unless you have specialized training in handling such tools.

SAFETY SYMBOLS

The following safety symbols are used in the *BC Science* 7 program to alert you to possible dangers. Be sure that you understand each symbol you see in an activity or investigation before you begin.

	Disposal Alert This symbol appears when care must be taken to dispose of materials properly.
	Fire Safety This symbol appears when care should be taken around open flames.
	Thermal Safety This symbol appears as a reminder to use caution when handling hot objects.
	Sharp Object Safety This symbol appears when a danger of cuts or punctures caused by the use of sharp objects exists.
	Electrical Safety This symbol appears when care should be taken when using electrical equipment.
	Skin Protection Safety This symbol appears when use of caustic chemicals might irritate the skin or when contact with micro-organisms might transmit infection.
	Clothing Protection Safety A lab apron should be worn when this symbol appears.
	Eye Safety This symbol appears when a danger to the eyes exists. Safety goggles should be worn when this symbol appears.
	Poison Safety This symbol appears when poisonous substances are used.
	Chemical Safety This symbol appears when chemicals that are used can cause burns or are poisonous if absorbed through the skin.
	Animal Safety This symbol appears whenever live animals are studied, and the safety of the animals and the students must be ensured.

WHMIS (Workplace Hazardous Materials Information System)

Look carefully at the WHMIS (Workplace Hazardous Materials Information System) safety symbols that are shown below. The WHMIS symbols are used throughout Canada to identify the dangerous materials that are found in all workplaces, including schools. Make sure that you understand what these symbols mean. When you see these symbols on containers in your classroom, at home, or in a workplace, use safety precautions.

Compressed Gas	Flammable and Combustible Material
Oxidizing Material	Corrosive Material
Poisonous and Infectious Material Causing Immediate and Serious Toxic Effects	Poisonous and Infectious Material Causing Other Toxic Effects
Biohazardous Infectious Material	Dangerously Reactive Material

Instant Practice

1. Find four of the *BC Science* 7 safety symbols in activities or investigations in this textbook. Record the page number and title of the investigation or activity in which you found each symbol. What possible dangers that relate to the symbol are in the activity or investigation?

2. Find two of the WHMIS symbols on containers in your school, or ask your parent or guardian to look for two WHMIS symbols in a workplace. Record the contents of each container and the place where the container is stored. What dangers are associated with the contents of each container?

Science, Technology, and Society

Scientists ask a lot of questions. They want to know how and why things in the universe happen the way they do. Scientists ask questions such as:

- Why does iron seem to rust more quickly in damp places?
- How old are the oldest rocks on Earth?
- How many litres of water are there in the ocean?
- What is the mass of a humpback whale?

Scientists use many methods to find answers to their questions. Over time, these answers form a body of knowledge about the universe. People call this body of knowledge "science." **Science** is a way of explaining the objects and events in the universe.

Scientists ask questions that lead them into nature, into laboratories, and into the farthest reaches below and beyond Earth.

Designing Experiments to Answer Questions

What makes scientific knowledge different from other kinds of knowledge? The difference is in the way that scientists construct their knowledge. One method that they use is designing experiments. The diagram shows some common features of a science experiment. SkillPower 6, "Designing and Conducting Experiments," on page 309, outlines the key features of science experiments.

Different Methods for Different Questions

Scientists do not design experiments and make direct observations for every question they ask. For example, you cannot easily place a humpback whale on a balance to find its mass. You cannot possibly measure the volume of all the water in the ocean.

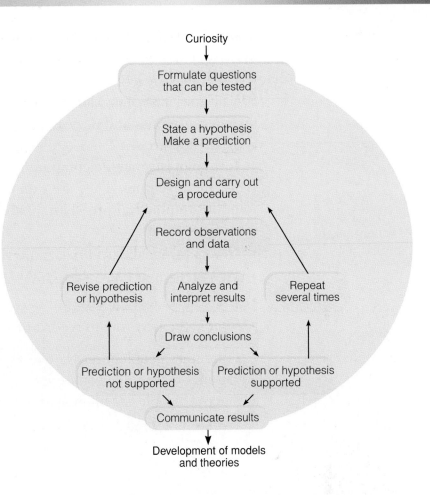

This diagram highlights skills and processes involved in designing and conducting scientific experiments.

Just as scientists do, you will use a variety of methods to find answers to questions. For example, you will:
- infer the meaning of information that you read
- interpret data that you research and read about
- classify objects or events to seek meaningful relationships
- use models to represent matter and the way it behaves

Throughout your Science 7 course this year, you will practise skills and processes that you have learned in previous grades. You will also develop new skills, such as hypothesizing and formulating models. You will continue to build a strong foundation of science knowledge and skills.

Technological Problem Solving

Have you ever used a pencil to flip an object out of a tight spot? That is technology in action! **Technology** is the use of common experience as well as scientific knowledge to solve practical problems. Technology includes machines and devices. It also includes ways to use an object for a variety of purposes.

Technologists include computer programmers, engineers, lab technicians, and inventors. They are experts at using their skills to find solutions to practical problems. Technological problem solvers ask questions such as:

- How can we make iron rust more slowly?
- What is the best type of rock to make a flagstone?
- How can we take the salt out of ocean water so people can drink it?
- What is the best way to record the songs sung by humpback whales?

Sometimes people improve technology by using scientific knowledge. Sometimes technology comes first. Later, scientific knowledge may advance as people try to understand why the technology works or how it could be used.

Each item shown here is an example of technology. What practical problem does each item solve?

Technological Problem Solving in This Course

When technologists design and build a new device, they go through a process of trial and error. Often, their first attempt at building the device is not fully successful. The first attempt still gives useful information for the next attempt, however.

In your Science 7 course, you will use technological problem solving skills to meet a challenge. You will develop and carry out a plan that is based on certain design criteria (standards). You will be asked to evaluate the plan. You will also communicate and reflect on the results. The diagram below outlines one process that you may use to solve technological problems.

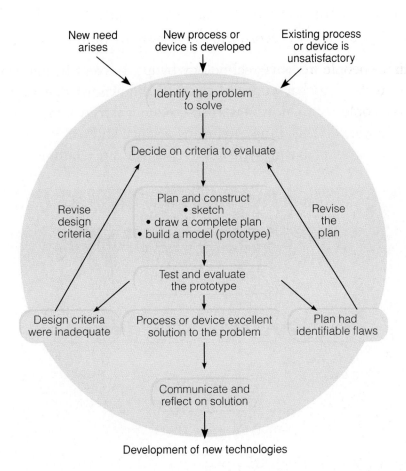

This diagram shows skills and processes involved in solving technological problems.

Societal Decision Making

If you wanted to drink all the water on Earth, how big would you have to be? You would need a stomach that measures about 1360 km across. You would be so tall that your head would be about 14 000 km above your feet. You would stand 1600 times taller than Earth's tallest mountain, Mount Everest. The Earth's atmosphere would be about level with your waistline.

The amount of water on Earth that you could drink is in very short supply. Most of Earth's water — about 97 percent — is salt water. Living things cannot drink it. You need *fresh* water to meet your survival needs. Less than 3 percent of Earth's water is fresh water. Most of it is frozen in large masses of ice such as the ice sheet in the photograph. Barely half of 1 percent of Earth's water is found as liquid fresh water. Believe it or not, most of this occurs below the surface as groundwater. The rest is found in rivers, streams, ponds, lakes, and freshwater wetlands.

Canada has one-tenth of all the world's supply of fresh water. How wisely do we use it? What would we do if we did not have enough water to meet our needs? Should we "harvest" frozen water? We could design the technology to do this. However, using science and technology in this way means we would have many issues to consider. For example:

Icebergs and ice sheets are found mainly in the Arctic and Antarctic regions of Earth.

- How would harvesting water from ice affect the environment?
- Would harvesting water from ice affect the pattern of ocean currents and the weather? If so, should we abandon the idea?
- What alternatives are there?
- What factors do we need to take into account before we reach a decision?

In your study of science this year, you will use processes such as the one in the diagram below to make decisions about issues that involve science, technology, society, and the environment. It is important to think about how science, technology, society, and the environment interact with one another, and how these interactions affect the world we live in. It is important, because no one lives alone. We are all connected. We are connected to each other and to our home planet, Earth. Our actions affect each other, as well as the environment.

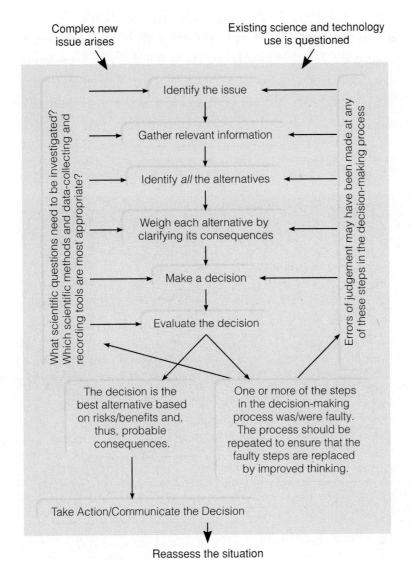

This diagram shows skills and processes involved in making decisions that involve science, technology, society, and the environment.

Traditional Ecological Knowledge

In this textbook, you will learn a great deal about your home province, British Columbia. Nowhere else in Canada do rugged Pacific coastline, plains and plateaus, sub-arctic muskeg, and desert all exist in the same place. Your province has areas with the mildest climate in Canada, and areas with average annual temperatures that barely reach 0°C. In Chapter 1, you will learn that British Columbia is home to a greater variety of plants and animals than any other province in Canada.

British Columbia is also home to a greater diversity of Aboriginal peoples than live anywhere else in Canada. The various Aboriginal peoples of British Columbia each have their own unique cultures and ways of life. There are at least ten major language groups, and dozens of individual languages and language dialects. These languages range from the large Athapaskan language group spanning central and northern British Columbia to the much smaller Ktunaxa language group.

Look at the map on the next page. See which Aboriginal peoples live in your area. Notice that every part of British Columbia is within the traditional territory of one of the Aboriginal peoples that make up this province. In these traditional territories, Aboriginal peoples have ownership rights and responsibilities for care and stewardship of village sites, land and water areas, sacred sites, and fishing, hunting, trapping, and gathering areas.

The Aboriginal peoples of coastal and northern British Columbia divide themselves into Clans. Each Clan is connected to a specific animal and/or plant. This Tlingit hat represents the Nanyaayi (Orca) Clan.

Aboriginal Peoples of British Columbia

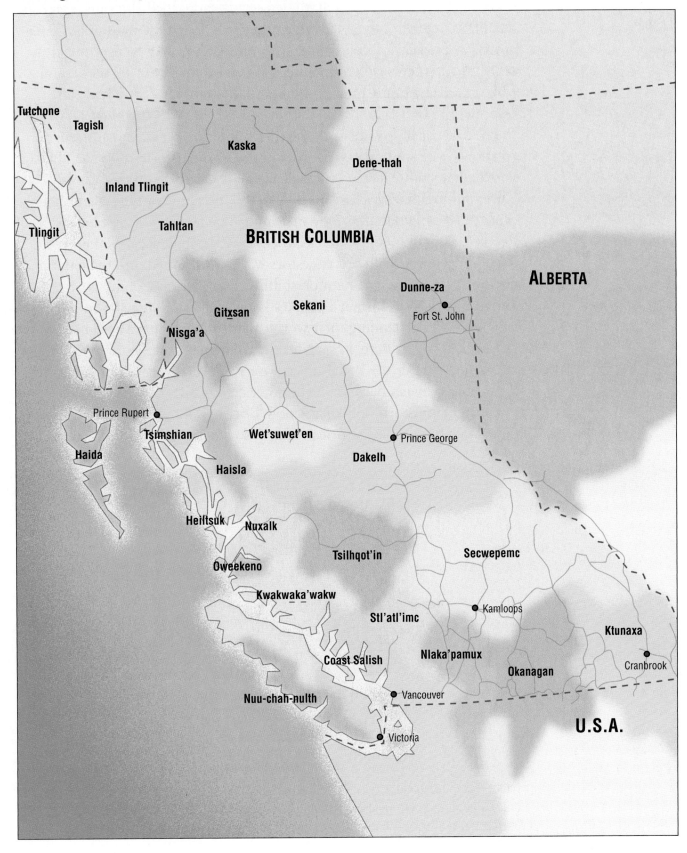

Over thousands of years, Aboriginal peoples have developed complex cultures and societies. They have developed knowledge and skills that suit the different environments in which they live. What the many Aboriginal groups in British Columbia share in common, however, is a spiritual way of looking at the world around them. To Aboriginal peoples, all natural parts of the environment are related and, therefore, to be valued and respected.

The Okanagan People of the southern interior, for example, speak of "plant people" and "animal people" in their origin stories. For the Haida People, human life began on Haida Gwaii (Queen Charlotte Islands). Here, Raven discovered a clam shell on the beach. From this clam shell came the strange sounds of the first human beings. The Coast Salish People recognize plants, animals, rocks, and specific places as their relatives. They know that their ancestors were transformed into the natural features of their environment on south-coastal British Columbia. These examples of what Aboriginal peoples know to be true show us a way of seeing and understanding their worlds that still exists for Aboriginal peoples today. It is here, in specific environments, that Aboriginal peoples know they were created, that life as we have it today began, and that they are meant to live in harmony with other creatures.

Aboriginal histories of what is now British Columbia have been passed down mostly in spoken stories. They also have been passed down in

Histories of Aboriginal peoples have been passed down from generation to generation orally and through art. Bill Reid's sculpture of Raven and the First Man tells the story of Raven discovering the first human beings on Haida Gwaii (Queen Charlotte Islands).

carvings, paintings, songs, and dances, from generation to generation for thousands of years. These histories often agree with more recent scientific discoveries. Aboriginal peoples, for example, tell many stories of a great flood. The Haida speak of an age when their islands were tundra, and they depended on animals such as caribou. At the end of the last Ice Age, about 12 000 years ago, sea levels were much lower than they are today. About 10 000 years ago, the glaciers melted. Sea levels rose well above present-day levels. This huge rise in water levels caused massive flooding along British Columbia's coast. The Aboriginal stories of a great flood and the scientific evidence of sea level and landscape changes related to the last Ice Age tell of the same events in different ways.

The spiritual connections among all living things, combined with the centuries of Aboriginals' experience and observation of their environments, means that vast bodies of knowledge about their environments have been gathered. **Traditional ecological knowledge**, as it is called today, is important to a full understanding of environments, species, and ecosystems. Traditional ecological knowledge is based upon the following ideas.

- The Creator made all things one.
- All things are alive, related, and interconnected.
- All things are sacred and should be respected.
- Balance and harmony are essential among all life forms.

North American Aboriginal peoples discovered that the bark of the willow tree has pain-relieving properties. Willow tree bark contains a substance that is very similar to ASA.

The traditional ecological knowledge of Aboriginal peoples has contributed a great deal to the sciences in North America and worldwide. For example, Aboriginal peoples of North America have given the world cotton, corn, and squash. Aboriginal peoples have also provided more than 3000 varieties of potato, chocolate, beans, and rubber. Their knowledge has led to cures for sicknesses such as scurvy and malaria. Many Aboriginal medicines have found their way into the medicines that we use today. One example is acetylsalicylic acid (ASA), found in pain relievers such as Aspirin™.

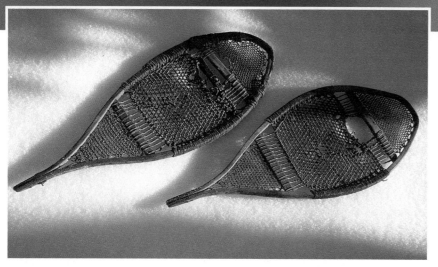

The Nlaka'pamux, Okanagan, Secwepemc, Stl'atl'imx, and Dakelh peoples used Rocky Mountain maple to make snowshoe frames. This basic design is still used today.

In British Columbia, many contributions have come from Aboriginal peoples' in-depth knowledge of and connection to their lands and waters. The environmental and technical knowledge of Aboriginal peoples played an important part in the development of British Columbia's early industries — trapping and fishing. For example, northwest coast canoe designs were ideal for the sea. These designs were adapted into boat designs around the world. Today, Aboriginal methods of selective fish harvesting, grassland and forest management, and wildlife stewardship are all being adopted as tools of sustainable resource management.

Annual numbers and migration routes of wildlife in northeastern British Columbia, for example, can vary greatly from year to year. Such variations can be studied accurately only from many seasons of observation. Aboriginal hunters, who travel regularly over the same areas as wildlife do, are important sources of traditional ecological knowledge. They observe

Much of the practical winter clothing that Canadians wear today has its origins in Aboriginal clothing design.

Traditional-style dugout cedar canoes are still popular today. Their design has been adapted into boat designs around the world.

The Clayoquot Sound UNESCO Biosphere Reserve was created in 2000. The traditional ecological knowledge of Aboriginal peoples played an important part in the creation of the Biosphere Reserve, and plays an important part in its future goals. These peoples include the Nuu-chah-nulth Peoples of Hesquiaht, Ahousaht, and Tla-o-qui-aht, whose traditional territories make up the Biosphere Reserve.

the long-term changes in wildlife behaviour and distribution, as well as environmental change. Along the British Columbia coast, Aboriginal methods of forest management are being investigated in attempts at "new" methods of sustainable forestry. These include the selective harvesting of bark and planks from living cedar trees, as one example.

The histories, cultures, and identities of the Aboriginal peoples of British Columbia are deeply connected to their environments. Over thousands of years here, they have developed complex knowledge systems. These systems regard the environment as a whole. All natural parts are deeply interrelated and ideally live in respectful harmony with one another. Traditional ecological knowledge can help us find ways to ensure the health and well-being of our environment for future generations.

You will learn more about Aboriginal peoples and their interrelationships with your beautiful home province as you read this textbook.

Karen Charleson
Hesquiaht First Nation, Hot Springs Cove, British Columbia

UNIT 1

Ecosystems

These mountain goats live in one of the harshest environments in British Columbia. The steep, rugged mountainside is cold and windy. The goats' 15 cm long wool, however, provides a thick, warm coat and gives them a shaggy appearance. Most animals would have difficulty moving about on this treacherous terrain, but the mountain goats are superbly suited for life here. They have special hooves with hard, sharp edges and a soft flexible inner pad. These unique hooves act almost like suction cups. They help the mountain goats get an incredible grip on the rocks and allow them to climb and leap nimbly on the steep terrain. High in the mountains, the mountain goats are safe from most of their enemies, yet they are still able to find plenty of grasses, mosses, and other plants to eat.

The mountain goat, like all other living things, interacts with its environment. To interact means that the mountain goat and the environment act on each other or have an effect on each other. For example, the mountain goat needs food, water, and a suitable place in which to live. When a heavy snowfall forces it to move down the mountain, it may become food for another animal, such as a cougar, wolverine, or grizzly bear.

Why do organisms live where they do? How do they interact with one another and with their environment? How do natural events, such as fires, affect the environment? How do human activities, such as building roads or towns, affect the environment? These are some of the questions you will explore in this unit.

Unit Contents

Chapter 1
Organisms and
Their Environment 4

Chapter 2
Cycles in
Ecosystems 34

Chapter 3
Ecosystems and
People 62

Getting Ready...

- How do living organisms interact with non-living parts of the environment?
- How does the environment continuously supply energy, food, and water to living organisms?
- How can people use the environment to meet their needs without harming it?

CHAPTER 1
Organisms and

Getting Ready...

- How would you describe the environment in which you live?
- How does the environment affect plants and animals?
- Why are dead plants and animals important to the environment?

Porcupines feed on leaves, bark, and twigs, as well as on discarded antlers and bones.

All living things, from microscopic bacteria to tall spruce trees, are affected by other living things in the environment and by the non-living parts of the environment. This porcupine lives in the northern forests of British Columbia. It feeds on the bark, twigs, and needles of pine and hemlock trees, and on discarded antlers and bones. As well, it chews on axe handles, boat oars, cabins, and wooden buildings! The porcupine is always on the lookout for food. It is also on the lookout for predators, such as wolverines and bobcats. If threatened, it will raise its sharp, barbed quills in defence. The porcupine interacts with plants and other animals that live in the same environment. It also depends on the non-living parts of its environment. The soil, for example, nourishes trees and the porcupine eats the bark and the leaves. The porcupine drinks the water and breathes the air.

In this chapter, you will study the interactions among living things, as well as the interactions between living things and the non-living parts of their environment. You will learn about the interactions and relationships among organisms and the places where they live.

Their Environment

What You Will Learn

In this chapter, you will learn
- how the living and non-living parts of an environment interact
- how populations, communities, and ecosystems are different
- what roles plants and animals play in ecosystems
- why many different organisms and ecosystems are found in British Columbia
- how people interact with ecosystems

Why It Is Important

- You are part of the environment. You have basic needs that can only be met by the environment around you. If you understand your environment, you are better prepared to care for it.
- British Columbia has a rich and varied natural environment. It is important for you to learn about the natural resources of your province.

Skills You Will Use

In this chapter, you will
- model an organism's habitat (home)
- measure and observe populations in local ecosystems
- interpret data on the population cycles of a predator (hunter) and its prey (what it hunts)
- investigate the living and non-living parts of a soil environment

This pine forest is home to porcupines and many other animals that live in northern British Columbia.

Starting Point ACTIVITY 1-A

Chain of Events

Have you ever thought about the impact of one small event on an entire chain of events? Sometimes the results can be surprising.

What to Do

1. With a partner, read the following poem, written by Benjamin Franklin. Discuss the meaning of the poem. If you like, invite other pairs of students to share their ideas.

   For want of a nail, the shoe was lost;
   For want of a shoe, the horse was lost;
   For want of a horse, the rider was lost;
   For want of the rider, the battle was lost;
   For want of the battle, the kingdom was lost.
   And all from the want of a horseshoe nail.

2. This poem is an example of how the loss of a very small object can have a very serious consequence. How are the ideas in the poem similar to events in the world around you? How do living and non-living things depend on each other? Are seemingly small things in the environment critical for survival of living organisms?

3. Think about your "needs." These are the basic things that you, as a living creature, must have in order to survive. What are your needs—the things that are essential for your survival?

Extension

4. Make up your own chain of events, starting with one small event. Include at least six events in your chain. For example, you could start with the following event: You forgot to set your alarm clock, so...

5. Write a poem about how different kinds of living and non-living things depend upon each another.

Section 1.1 Abiotic and Biotic Environments

Figure 1.1 The lynx depends on the hare for food. The hare's white coat is hard for the lynx to see in the snow.

The lynx and the hare shown in Figure 1.1 live in an area of British Columbia where there is a lot of snow. Their thick coats help them keep warm in the frigid weather. The large feet of both the lynx and the hare help them run across the snow without sinking. As well, the hare's white fur helps it blend in with the snow. How these animals look and how they behave tells you something about the area in which they live.

Living organisms are affected by other living things and by the non-living parts of the environment around them. Could you imagine the lynx chasing after the snowshoe hare in a hot, dry desert? Do you think this chase would even happen in the desert?

The Abiotic Parts of the Environment

To understand living things, you need to learn about the abiotic [AE-bih-o-tik] parts of an organism's environment. **Abiotic** means non-living. The abiotic parts of an environment are the non-living things, such as temperature, light, air, water, soil, and climate. The abiotic parts of an environment often determine which organisms can live and survive in that environment. You just learned how the lynx and the hare are well-suited for a cold, snowy climate. The life cycle of a salmon, like the one in Figure 1.2, is another example of the way that the abiotic parts of an environment affect living things. The salmon will not lay its eggs in a muddy pond. They will lay eggs only in certain places in streams.

Figure 1.2 Salmon lay their eggs on beds of small stones in clean, fast-running streams.

Some examples of how the abiotic parts of an environment affect living things are described here.

- ***Temperature and Light*** Temperature often determines where organisms live. Sunlight provides warmth for many animals, such as the alligator lizard shown in Figure 1.3. Green plants need light to make food. The number of hours of daylight triggers changes, such as the flowering of plants and the migration of birds.

- ***Air*** The air contains oxygen gas, which animals breathe. The air also contains carbon dioxide gas, which plants use to make their own food.

- ***Water*** Plants combine water with carbon dioxide to make food and to grow. Animals need water to digest food and move food particles throughout their bodies. Some organisms (such as trout, whales, and British Columbia's largest water insect, the giant water bug) live in water.

- ***Soil*** The soil contains minerals as well as pieces of organisms that were once living. For example, pieces of dead insects and roots from dead plants are found in the soil. The soil provides a home for many animals such as the earthworms shown in Figure 1.4. They burrow into the soil and overturn it. These tunnels allow air and water to mix with the soil. The earthworms eat decaying leaves and other natural materials in the soil and leave the remains behind, close to the surface. In this way, they bring valuable nutrients to the surface. Thus, the soil provides nutrients, such as minerals, for plants. Plants, in turn, hold the soil in place. Their roots prevent wind from blowing the topsoil away and prevent rain from washing it away.

- ***Climate*** The term **climate** means average weather pattern of a region over a long period of time. For example, the climate of northern Canada is very cold, with long, harsh winters and short, cool summers. Climate affects where and how plants and animals live. For example, the bunchgrasses shown in Figure 1.5 on the next page look as if they are dead during the hot summer in the Okanagan. In fact, the roots are alive. As soon as the soil receives some moisture, usually in the fall or early spring, the bunchgrasses begin to grow again.

Figure 1.3 A northern alligator lizard basks in the sunlight each morning until it is warm enough to begin to feed.

Figure 1.4 The action of earthworms help water and oxygen move into the soil.

Chapter 1 Organisms and Their Environment • MHR **7**

There are many other abiotic parts of the environment, in addition to those that were just listed. In the next Find Out Activity, you will learn about one of these abiotic part of the environment. You will see how salt in water can affect plants.

Figure 1.5 Bunchgrasses are able to survive long periods without water.

Find Out ACTIVITY 1-B

Salty Seeds

Salt and other minerals are often found in soil. How does salt water affect the germinating (sprouting) of certain seeds?

What You Need

2 plastic drinking cups (or similar containers)
30 mL salt
water
10 bean seeds
2 paper towels
2 plastic self-sealing bags
masking tape
permanent marker

What to Do

1. Add water to the cups, until each cup is half full. Dissolve the salt in one cup.

2. Add five bean seeds to each cup. Leave the seeds to soak overnight.

3. The next day, wrap each set of seeds in moist paper towel. Place the towels in separate self-sealing bags. Use the masking tape and the marker to label the bags "fresh water" and "salt water."

4. After two days, count the number of seeds in each bag that show signs of root growth (sprouting). **Record** this number.

What Did You Find Out?

1. How did the amount of sprouting differ for the seeds that had been soaked in fresh water compared to the seeds that had been soaked in salt water? Describe any differences in the appearances of the seeds from the two groups.

2. What abiotic factor was tested in this activity?

3. What conditions were kept the same (controlled) for the two groups of seeds?

4. Do you think that seeds from all types of plants respond to salt in a similar way? How could you find out?

5. How does this activity model the way that an abiotic part of an environment can affect a biotic part of the environment?

The Biotic Parts of the Environment

Biotic means living. Living things are called the biotic parts of an environment. For example, seaweeds, crabs, and octopi are all biotic parts of the ocean environment off British Columbia's west coast. All living things affect and interact with other living things. For example, the octopus in Figure 1.6 depends on crabs and other living things for food. When the octopus was smaller, it might have become food for some other living organism. The octopus needs contact with other members of its species in order to reproduce. (A **species** is a group of organisms that can successfully mate with each other and reproduce.) To meet these needs, the octopus might compete with members of the same species for a mate. The octopus might compete with members of a different species for food or a home.

READING check

What are the main differences between the abiotic parts and the biotic parts of an environment?

Figure 1.6 This giant octopus is eating a clam. Two biotic parts of the environment are interacting.

The Needs of Living Things

Since you are a living thing, you are a biotic part of the environment. Think about the basic things that you need to stay alive. Your basic survival needs are very similar to the basic survival needs of other animals. All living organisms need:

- *oxygen gas or carbon dioxide gas from the air* Animals need oxygen gas because it helps them break down food for energy. Plants need carbon dioxide gas because they use it along with energy from the Sun to make their own food.
- *water* Animals need water to dissolve their food and carry oxygen and food throughout their bodies.
- *food* Animals need food for energy and to grow and repair their tissues. Plants make their own food.
- *a suitable place to live* Many creatures find shelter in the environment. Some living things, such as humans, beavers, and wasps, build protective shelters.

When you take a breath, put on a warm coat, wave to a friend, or avoid a buzzing bee, you are interacting with the biotic and abiotic parts of your environment. Organisms that live in the soil also interact with their environment. In the following investigation, you will take a closer look at the abiotic and biotic parts of a soil sample.

READING check

What are the basic needs of plants and animals? Create two lists, one for the needs of plants and one for the needs of animals.

CONDUCT AN INVESTIGATION 1-C

SKILLCHECK
- Observing
- Controlling Variables
- Inferring
- Communicating

Soil Sleuths

In this investigation, you will examine a soil sample.

Question
What are the abiotic parts and the biotic parts of soil?

Safety Precautions

- Do not handle any organisms with bare hands.
- Return all organisms to the places where you found them.
- Handle the glass jar with care so that it does not break.
- Handle the scissors carefully.
- Keep electrical connections for the lamp away from water or moisture.

Apparatus
scissors
large wide-mouthed jar
large plastic funnel
desk lamp with flexible arm
pie plate
hand lens
microscope (optional)

Materials
2 damp paper towels
fine-mesh plastic pot scrubber
garden soil or leaf litter (partially decomposed leaves)
paper towel
paper for sketching

Procedure Group Work

1. Place the damp paper towel in the bottom of the jar. Place the funnel in the mouth of the jar.

2. To make a screen, use the scissors to cut a 3 cm by 3 cm square from one layer of the pot scrubber. Place the screen over the small opening in the neck of the funnel.

3. Fill the funnel with garden soil or leaf litter.

4. Place the lamp over the funnel, and turn it on. Turn out all the other lights in the room, and leave the lamp on overnight.

5. The next day, remove the funnel from the jar. Empty the contents of the jar onto a damp paper towel in the pie plate. (If there are insects that can crawl quickly, you may need to cover the pie plate with a clear plastic wrap to prevent them from escaping.) **Observe** the organisms. Use a hand lens or a microscope if possible.

6. Sketch at least one organism. If possible, everyone in your group should sketch a different organism.

7. **Observe** a small sample of soil. Use a hand lens or a microscope if possible. Describe your soil sample. Draw any interesting objects that you find.

8. When you have completed this investigation, ensure that any organisms are returned to the place where they were collected.

Analyze

1. Describe the organism you sketched. Include details such as size, colour, and number of legs or antennae.

2. Be prepared to provide a short presentation about your organisms to your class.

Conclude and Apply

3. Why do you think organisms in the soil ended up in the jar?

4. Does the soil provide all the things that organisms need to survive?

5. Describe the biotic and abiotic parts of your soil sample.

6. Describe the interactions between the biotic and abiotic parts of the soil. Use information you and your classmates learned in this investigation.

7. Describe two ways in which you could change the abiotic parts of the soil. How do you think these changes might affect the biotic parts of the soil?

Extend Your Skills

8. Use the Internet or reference books on insects, spiders, and other small animals to try to identify the organism you drew in step 6 of the Procedure.

Career CONNECT

Wildlife Biologist

The person in this photograph is Leah De Forest, a wildlife biologist. She is placing bands on the legs of the young chicks that have hatched in the nests on the cliffs. Leah plans to keep track of the chicks' movements.

Other wildlife biologists study animals such as beavers, grizzlies, sea otters, cougars, whales, owls, wolves, and caribou in their natural habitats. These biologists study such things as nutrition, parasites, diseases, and migration patterns. They observe the animals carefully and ask questions about what they observe. Then they find ways to try to answer their questions.

Develop your own question about the murre chicks that Leah De Forest is banding. See if you can find an answer to your question by searching the Internet or visiting a library.

Leah De Forest, a wildlife biologist, is studying murre [MUHR] chicks.

Interactions Between Abiotic and Biotic Parts of the Environment

Figure 1.7 This beaver dam is part of an ecosystem that includes biotic and abiotic factors.

Ecology is the study of the interactions among organisms, as well as the interactions between organisms and their environment. An **ecologist** is someone who observes and studies these relationships. An ecologist who is studying beavers, for example, might investigate what beavers eat, how changes in climate affect beavers, or where beavers build their dams (see Figure 1.7) or have their young.

Ecosystems

Ecologists study ecosystems. An **ecosystem** is all the interacting organisms that live in an environment, as well as the abiotic parts of the environment that affect the organisms. The pond ecosystem in Figure 1.7 is filled with different types of organisms that interact with one another. Organisms eat other organisms, defend themselves, reproduce, and compete for food and space. These are some of the interactions between two biotic parts of the ecosystem. For example, a beaver, one biotic part of the ecosystem, might eat the bark of a poplar tree, another biotic part of the ecosystem.

Its abiotic parts also affect the pond ecosystem. These abiotic parts might include the amount of sunlight and rain the pond receives, and the temperature of the air above it. When a heavy rain lowers the temperature of the water in the pond and raises the height of the water, abiotic parts of the pond ecosystem are interacting. Figure 1.8 shows some of the many different interactions that happen in an ecosystem.

> **READING check**
>
> What is one way in which a beaver might interact with an abiotic part of its environment?

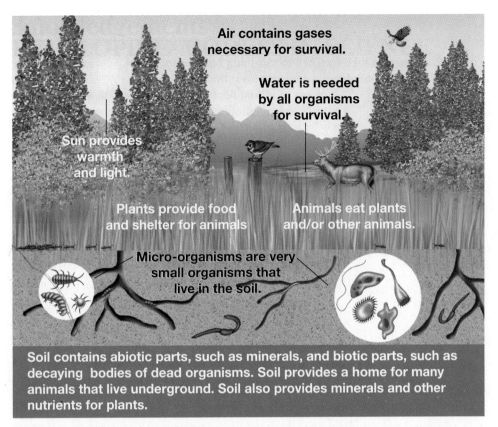

Figure 1.8 This ecosystem includes living things that interact with each other and with the abiotic parts of their environment.

Biomes

Many ecosystems have broadly similar temperatures and have similar amounts of rain every year. Therefore, the plants and animals that are found in these ecosystems could also be similar. Ecologists have a name for large regions that have about the same temperature and amount of rain or snow. These regions are **biomes.** Canada has four *major* biomes as shown in Figure 1.9. The *tundra* is very cold. Very little rain or snow falls in the tundra so there are no trees, just shrubs and grass. Nearly all of British Columbia is considered *boreal forest*. The most common trees in boreal forests are evergreen trees. The *temperate forest* biome receives more rain than the boreal forest. There are many deciduous trees. The fourth biome in Canada is the *grassland* biome. There are large amounts of rain with long dry periods in between. These conditions prevent trees from growing but shrubs and grass grow well.

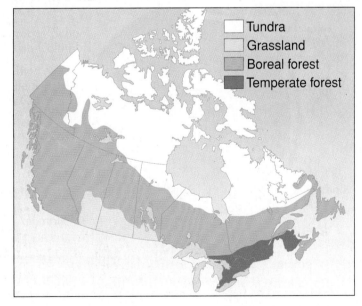

Figure 1.9 Canada has mostly tundra and boreal forest biomes. There are smaller amounts of grassland and temperate forest biomes.

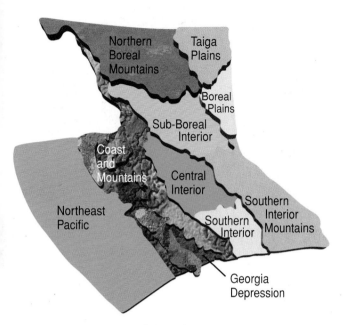

Figure 1.10 British Columbia has ten ecoprovinces. Each ecoprovince has a distinct combination of climate, geography, landforms, and organisms.

British Columbia's Ecosystems

British Columbia has more species of living things than any other province in Canada. In fact, 72 percent of Canada's land mammals, 60 percent of Canada's plant species, 67 percent of Canada's fungus species, and 79 percent of Canada's bird species are found in British Columbia. Why does our province have such a variety of organisms? The answer lies in the diversity of British Columbia's geography and climate. Ecosystems in British Columbia include high mountains, dry grasslands, wet coastal forests, and vast plains. Biologists have described and mapped ten different natural regions, which they call **ecoprovinces,** in British Columbia. Look at the map of ecoprovinces in Figure 1.10. Can you locate your ecoprovince?

Aboriginal Peoples and Ecology

Do you have a favourite outdoor place? Perhaps it is a tree you climb or a site where your family goes camping every year. The more time you spend in this place, the more you learn and care about it. Aboriginal peoples in British Columbia have lived in the same regions of British Columbia for thousands of years. They have gathered a vast store of information and knowledge about these regions. Aboriginal peoples can be considered British Columbia's first ecologists.

In Aboriginal teachings, an ecosystem is a whole, rather than a collection of separate parts. All parts of the environment — biotic and abiotic — are alive, related, and sacred. Origin stories tell of living beings transformed into rock formations and mountains, of animals transformed into people, and of the first people emerging from the ocean. For Aboriginal peoples, all plants, animals, water bodies, land forms, and natural forces such as weather are interconnected and should be respected. Their understanding of ecosystems guides Aboriginal peoples today when they consider how their actions may affect ecosystems.

Section 1.1 Summary

Organisms interact with both the abiotic (non-living) and biotic (living) parts of their environment.

- The abiotic parts of an environment include temperature, light, air, water, soil, and climate.
- The biotic parts of an environment include plants, animals, and all other living things. The biotic parts of an environment interact with each other.
- An ecosystem is all the organisms in an area and all the non-living parts of the environment that affect them.
- People are part of ecosystems. Humans have the same basic needs as all other living things.
- Aboriginal peoples were the first ecologists in Canada. Ecologists study the environment and the relationships among living things and their surroundings.

INTERNET CONNECT
www.mcgrawhill.ca/links/BCscience7
Learn more about British Columbia's plants and animals, by going to the web site above. Click on **Web Links** to find out where to go next.

Check Your Understanding

1. (a) What are the basic needs of all living things?
   (b) Imagine that you are a black bear. You live in a coastal rainforest ecosystem. What are some different ways in which you meet your needs?

2. What is one abiotic part and one biotic part of a polar bear's environment?

3. What are three biotic parts of a marsh community?

4. A lizard is warming itself in the morning sunlight.
   (a) What abiotic parts of the environment affect the lizard? Describe them.
   (b) What biotic parts of the environment affect the lizard? Describe them.

5. **Apply** Describe the ecosystem near your school. What are the most common plants and animals in the area? What is the nearest body of fresh water? How much rainfall and snowfall does the area receive? What kinds of plants and animals lived in the location of your school 100 years ago? How did building a town or city change the ecosystem near your school?

6. **Thinking Critically** The light and fluffy seeds of dandelions are spread by the wind. How are these seeds adapted to the dandelion's ecosystem?

Key Terms
abiotic
climate
biotic
species
ecology
ecologist
ecosystem
biome
ecoprovinces

INTERNET CONNECT
www.mcgrawhill.ca/links/BCscience7
Ecoprovinces are further divided into smaller areas called *biogeoclimatic zones*. Go to the web site above to learn more about the biogeoclimatic zone in which you live. Click on **Web Links** to find out where to go next.

Section 1.2 How Organisms Interact in Ecosystems

Figure 1.11 This bull elk will have to compete with other bulls to find a mate.

A scientific study of the elk in Figure 1.11 might reveal what kind of food he prefers, how often he eats, and how far he roams in search of food or shelter. Although the elk may spend a great deal of time alone, he also interacts with other individuals of his own species. For example, during the mating season, this male elk will probably battle with other male elk as they compete for mates. All organisms depend on other organisms for food, shelter, reproduction, or protection. Thus, it is important to study an organism's relationships with other organisms to better understand its life.

Levels of Organization

In order to organize and communicate their studies, ecologists define the interactions among organisms at different levels. As shown in Figure 1.12, in addition to studying individuals, ecologists study populations, communities, and ecosystems.

Figure 1.12 Study the illustrations to see the relationships among the four levels of biological organization.

Level 1: individual

Level 2: population

Level 3: community

Level 4: ecosystem

Populations

A **population** is a group of individuals of the same species, living together in one ecosystem. For example, all the blue grouse in an alpine meadow are a population. Biological populations vary in size depending on factors such as the species, time of year, weather, or abundance of food. For example, a population of woodland caribou in northern British Columbia might contain thousands of individuals. A population of rare spotted bats in the Okanagan or Cariboo region of British Columbia might have only a few individuals.

Individuals in a population often compete with each other for habitat. Food, water, and shelter make up an organism's **habitat**, the place in which it lives.

How the organisms in a population share the resources of their habitat determines how far apart the organisms must live and how large the population can become. Some species have behaviours that help to reduce competition. For example, the adult and juvenile stages of the rough-skinned newt in Figure 1.13 have different habitat needs. Newt larvae live in ponds. Adult newts live in or under logs in the forest near the pond. The larvae eat small aquatic animals. The adults feed on slugs and worms. Since the larvae and adults live in different places and eat different foods, they can share the same space without competing with each other. You will learn more about the habitat of one population of animals in the Find Out Activity on the next page.

Is a herd of elk a population or a community? Explain your answer.

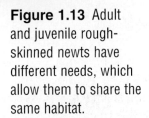

Figure 1.13 Adult and juvenile rough-skinned newts have different needs, which allow them to share the same habitat.

Communities

Populations share their environment and interact with populations of other species. A **community** is made up of all the interacting populations that live in an area. For example, the pine forest community in Figure 1.14 includes animals such as mule deer, hummingbirds, and spotted bats as well as the pine trees. Plants such as Saskatoon berry and balsamroot are also part of the community. In every community, each species has a particular habitat.

Figure 1.14 The animals and plants in the ponderosa pine forest make up a community.

Find Out ACTIVITY 1-D

Designer Habitat

What does an organism's habitat look like? Consider a river otter, for example. A river otter lives near a stream, a lake, or a river. It eats fish, frogs, and sometimes mice and birds. It makes a den on land in a burrow, a hollow tree, or an abandoned beaver lodge. Find out how another British Columbia animal meets its basic needs.

What You Need

sheet of paper
pencil
dip net (optional)
jar (optional)
small cardboard box (such as a shoe box)
art supplies

Safety Precautions

- If you disturb an animal's habitat (for example, if you turn over a stone or dig a hole in the sand), always return the habitat to the condition in which you found it.

What to Do

1. Work with a partner. Choose either (a) or (b).
   (a) Look for animals in your schoolyard or in another outdoor area nearby. Carefully turn over rocks and logs. Look for animals in ponds and ditches. Watch for birds flying overhead. Choose one animal, and **observe** it closely without disturbing it.
   (b) At a library or on the Internet, research the habitat needs of one animal that lives in your region of British Columbia.

2. Draw four large circles on a sheet of paper. Label the circles "food," "water," "shelter," and "space." Describe the habitat of the animal you chose to study by filling in the circles. Consider what your animal needs for shelter. For example, some birds nest in holes in dead trees that are still standing. Also consider where your animal lives and how much space it needs. For example, grey whales live in the Pacific Ocean. They travel the ocean from Mexico to Alaska.

3. Use the cardboard box and the art supplies to make a **model** that shows the habitat of the animal you chose to study. Display your model in the classroom.

What Did You Find Out?

1. Compare the habitat you studied with the habitat studied by another pair of students. How are the habitats the same? How are they different?

2. What are five changes that might affect the survival of the animal you studied? Think about small changes, such as someone riding a bicycle through its habitat. Also think about large changes, such as a flood or a drought.

Extension

3. All organisms need a certain amount of space. If there are too many organisms in a habitat, space can become limited. What might happen if your classroom space was decreased to half its current size?

Ecosystems

Communities are affected by abiotic conditions such as the average amount of sunlight and rain, and the average temperature of the region. An ecosystem includes a biological community as well as the abiotic parts of the environment that affect this community.

The intertidal marine ecosystem shown in Figure 1.15 is on the rocky shoreline of British Columbia's Pacific coast. Tides, weather, the type of sand or rock on which the plants and animals live, and the amount of sunlight and rain that reach the communities are just a few of the abiotic parts of this ecosystem.

Abiotic factors such as the amount of rain or sunshine are almost the same across an ecosystem. You can consider a rotting log with all the organisms that live in or on it to be an ecosystem. You could also consider a forest to be an ecosystem.

Figure 1.15 Intertidal marine life survives in an area that is covered by the ocean for part of the day. During the remainder of the day, it is exposed to sunlight, wind, or rain.

Sampling Populations

Ecologists study ecosystems to learn about relationships between organisms and any changes in populations that take place over a long period of time. For example, ecologists might want to know if a population is declining. For very small ecosystems, you could count all the plants and animals. For large ecosystems, this would be impossible. You could, however, estimate the numbers of plants and animals. Ecologists estimate population sizes in ecosystems by **sampling**. One of the most popular sampling methods involves a **quadrat**, a square that marks off a specific area (see Figure 1.16). The students in Figure 1.17 are conducting a quadrat study on the rocky shoreline of Vancouver Island. You will have an opportunity to use a quadrat in the next investigation.

Describe an individual, a population, a community, and an ecosystem in a freshwater pond or tidal pool. A tidal pool is a small pond that remains on a beach when the tide is out.

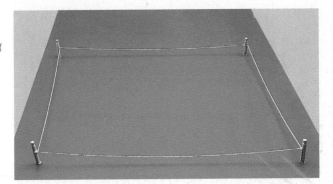

Figure 1.16 You can make a quadrat by joining wooden pegs with string.

Figure 1.17 These students are sampling the intertidal zone using a quadrat.

CONDUCT AN INVESTIGATION 1-E

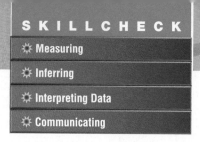

SKILLCHECK
- Measuring
- Inferring
- Interpreting Data
- Communicating

Sampling Populations in an Ecosystem

In this investigation, you will sample the populations of several organisms in an ecosystem. You will use a quadrat to count the individuals in a small area of the ecosystem. The ecosystem you study could be your schoolyard, a park, a meadow, a field, or any another ecosystem.

Question

How can you sample the populations of an ecosystem and examine the interactions among these populations?

Safety Precautions

- Do not harm any animals in your sample area.
- If you disturb a habitat be sure to return it to its original condition.

Apparatus
4 pegs (such as tent pegs)
thermometer
ruler
hand lens
clipboard

Materials
paper or notebook
string
plant, insect, and bird field guides (optional)

Skill POWER

For tips on making a data table, turn to SkillPower 5.

Procedure

① Before you go to your study site, create a data table similar to the one shown here, for each quadrat you will study. Take your tables to your study site.

Name of species	Average # individuals per square metre	Estimated total population

② When you arrive at your study site, sit quietly and **observe** it. **Record** everything that you can see, hear, and even smell to give an overall description of your study site.

③ Toss your pencil at random in your study site. Build a quadrat with one corner where the pencil lands.

④ **Measure** the temperature on the surface of the ground within the quadrat and in the air above it. **Record** these temperatures and any other abiotic factors within the quadrat. For example, is the ground soil, mud, or rock?

⑤ Choose at least five types of plants and five kinds of animals that are inside the quadrat. Count the number of each of these organisms within the quadrat. **Record** the numbers in your data tables. Your teacher will give you a list of tips to follow while counting.

⑥ Repeat step 5 for four more quadrat samples.

⑦ **Measure** or estimate the total size of the ecosystem that you are studying. **Record** your results.

20 MHR • Interactions and Ecosystems

Analyze

1. Calculate the average number of each organism per square metre. (See Sample Calculation) Record your results in your data table.

2. Multiply the average number of each organism per square metre by the number of square metres in the whole ecosystem. See the example in the Sample Calculation. Record your results.

Sample Calculation

The numbers of blue beetles counted in five quadrats (1 m^2) are 13, 15, 16, 14, and 13.

The total number of blue beetles in five quadrats is 71.

$$\text{Average} = \frac{\text{total number of blue beetles}}{\text{total area of quadrats}}$$

$$= \frac{71 \text{ blue beetles}}{5 \text{ quadrats} \times 1 \text{ m}^2}$$

$$= 14.2 \frac{\text{blue beetles}}{\text{m}^2}$$

$$\text{Total ecosystem area} = \text{length} \times \text{width of ecosystem}$$

$$= 50 \text{ m} \times 20 \text{ m}$$

$$= 1000 \text{ m}^2$$

$$\text{Estimated population of blue beetles} = 14.2 \frac{\text{blue beetles}}{\text{m}^2} \times 1000 \text{ m}^2$$

$$= 14\ 200 \text{ blue beetles}$$

Conclude and Apply

3. Write a description of the ecosystem you studied. Include the different types and numbers of plants and animals. Also include the abiotic factors (non-living things) that affect this ecosystem.

4. Why did you sample more than one area in this investigation?

5. Why do you think you were asked to choose your sample areas by chance (by tossing your pencil)?

Extend Your Knowledge

6. Try to identify any unknown organisms by matching them with illustrations in plant and animal guides or by working with other students in your class.

Extend Your Skills

7. Wait for weather conditions to change, and repeat the study. How are the results of your second study different from the results of your first study? How are they similar? What do you think accounts for any differences?

Pause & Reflect

Clams, crabs, sea cucumbers, seaweeds, and over a dozen different kinds of sea stars live in intertidal areas (see Figure 1.17). Local people say, "When the tide is out, the table is set." In your notebook, explain what this expression might mean.

Key Terms

population
habitat
community
sampling
quadrat

Section 1.2 Summary

Organisms and their interactions with the abiotic parts of their environment can be organized into different levels: individuals, populations, communities, and ecosystems.

- A population is a group of individuals of the same species that live together in one place at one time.
- A community is made up of all of the interacting populations that live in one area.
- An ecosystem is a community as well as the abiotic parts of the environment with which the populations in the community interact.
- All individuals need a habitat (a place to live). A habitat includes food, water, shelter, and a suitable amount of space for survival.
- A quadrat study is one way to sample an ecosystem. Ecologists can use sampling to learn about the relationships among organisms in a community or to monitor changes in a community over time.

Check Your Understanding

1. **(a)** How would you use a quadrat to estimate population sizes?
   (b) Why is the result only an estimate, not an exact number?

2. What are the differences and similarities between populations and communities? Write your answers in a chart.

3. Why would it be useful to do at least two population studies of the same ecosystem, with a period of time between the two studies?

4. **Apply** Imagine that you are a biologist. The company for which you work predicts the effects of building large subdivisions or other building projects. There is a plan to build a new luxury resort on the shore of a large bay. Builders need to know how the project will affect the environment in a few particular ecosystems. Your job is to estimate the number of organisms in these ecosystems. Explain how you could sample
   (a) the numbers of different insects in a large tree
   (b) the numbers of different fish in the bay
   (c) the number of groundhogs in a local golf course

5. **Thinking Critically** A home aquarium contains water, an air pump, a light, algae, a goldfish, and algae-eating snails. What are the abiotic parts of this environment? Which parts of this environment would you consider to be a population? Which parts would you consider to be a community?

Section 1.3 Roles of Organisms in Ecosystems

Like all other members of human communities, you play several different roles in your daily life. At school, you are a student. You might also be a member of a sports team. Outside of school, you might be a volunteer at a food bank. Similarly, the organisms in a community of plants and animals play different roles. A **niche** [NEESH] is both the space where an organism lives and the role it plays within its ecosystem. To determine an organism's niche, you must look at what it eats, where it lives, and how it interacts with other organisms in its ecosystem.

Figure 1.18
All these organisms have different roles, or niches, in this pond ecosystem.

Producers and Consumers

Plants are able to grow using energy from the Sun, carbon dioxide in the air, and water and nutrients in the soil. They fill the niche called producers. **Producers** are organisms that create their own food rather than eating other organisms to obtain food. The algae and water lilies in Figure 1.18 are producers. Producers make life possible for all other organisms on Earth.

Organisms that eat, or consume, food are called **consumers.** They cannot create their own food so they must eat producers or other consumers. All animals are consumers.

Types of Consumers

Consumers can be divided into three different groups: herbivores, carnivores, and omnivores. **Herbivores** are animals such as cows and herring that eat plants. **Carnivores** are animals that eat other consumers. Carnivores such as lynxes and dolphins eat meat. **Omnivores** are animals that eat both producers and consumers. Thus, they eat both plants and animals.

How are the niches of different organisms connected? How do they affect one another? Complete the next investigation to explore these questions.

Are you a consumer or a producer? Explain your answer. Are you a carnivore, a herbivore, or an omnivore? Explain your answer.

THINK & LINK
INVESTIGATION 1-F

SKILLCHECK
- Predicting
- Inferring
- Modelling
- Communicating

What Goes Up Must Come Down

Think About It

The niches of lynxes and snowshoe hares are linked together. Lynxes feed mainly on snowshoe hares. Snowshoe hares are herbivores. When there are plenty of plants for snowshoe hares to eat, more of them survive and reproduce. As a result, the lynxes, which feed on the snowshoe hares, will have more food as well. Therefore, more lynxes will survive and reproduce.

After several years, however, there may be so many lynxes killing snowshoe hares that the hare population starts to decline. Then the lynxes would not have enough food, and *their* numbers would decline. More plants would be able to grow because there are fewer snowshoe hares to eat them. As new generations of snowshoe hares are born, there will be plenty of food for them. Since there are fewer lynxes to hunt the hares, the hare population begins to increase. There is more food for the lynxes, so their numbers increase, too. Thus, the whole cycle, which lasts about ten years, will begin again.

The line graph below shows how the numbers of lynxes and hares that were harvested by trappers changed over a period of 90 years.

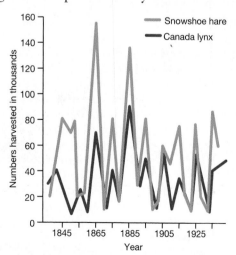

Skill POWER
For tips on writing a hypothesis, turn to SkillPower 6.

What to Do

Use the data in the graph to answer the following questions.
(a) In 1845, approximately how many lynxes and how many hares were harvested by trappers?
(b) How many lynxes and how many hares were harvested in 1855?
(c) In 1865, how did the two populations compare? What do you think led to this change in the numbers of the two populations?

Analyze

1. (a) How can prey control a predator's population?
   (b) How can predators control a prey's population?

2. Examine the following table. Notice that the human population in British Columbia rose steadily from the late 1800s. In what ways do you think the activities of humans might have affected the populations of lynxes and hares?

Year	Number of People
1870	36 247
1921	524 852
1961	1 629 082
2003	4 158 649

3. The graph ends at the year 1935.
   (a) Estimate the numbers of harvested lynxes and hares in 1940, based on the data in the graph.
   (b) Infer what might have happened to these populations by 1945.

5. The last few years that are shown on the graph are the first few years of the Great Depression (1929–1939) a time of mass unemployment. How might unemployment in these years have affected the populations of lynxes and hares?

Scavengers and Decomposers

Have you ever wondered why you seldom see a dead animal in a natural environment? In every community, there are "clean-up squads" that get rid of garbage and sewage. In a biological community, the clean-up squads are consumers called scavengers and decomposers.

Scavengers are organisms that eat decaying plants and animals. The turkey vulture in Figure 1.19 feeds on decaying animal carcasses. The vulture's bald head is easy to keep clean. It uses its excellent sense of smell to find its next meal. The wolverine in Figure 1.20 is another important scavenger in British Columbia with an excellent sense of smell. Other scavengers include the larvae of houseflies, carrion beetles, crows, and some species of gulls.

READING check

What is the main difference between a decomposer and a scavenger?

Figure 1.19 Turkey vultures are found in southern British Columbia and on Vancouver Island. Vultures eat only dead animals.

Figure 1.20 Wolverines are scavengers. They also hunt animals, such as caribou, and eat plant roots and berries.

Have you ever seen food in your refrigerator become mouldy? If so, you have witnessed decomposers at work. **Decomposers** break down (decompose) dead or waste materials, such as rotting wood, dead animals, or animal waste. Many bacteria and fungi, such as the bracket fungi in Figure 1.21, are decomposers. A decomposer does not bite and chew its meal. Instead, a decomposer releases a chemical onto the dead plant or animal that breaks down the tissue. Then the decomposer absorbs the nutrients into its own cells. These nutrients return to the environment when the decomposer dies.

Do different materials decompose as different rates? You will explore this question in the next investigation.

Figure 1.21 Fungi are important decomposers.

Chapter 1 Organisms and Their Environment

CONDUCT AN INVESTIGATION 1-G

SKILLCHECK
- Hypothesizing
- Controlling Variables
- Interpreting Data
- Modelling

Don't Waste It!

Imagine cooking dinner at your home. You may peel potatoes, chop lettuce, and crack an egg. Each of these actions leaves you with waste material to throw away. Kitchen "waste" does not need to be garbage, however. Under the right conditions, it can be composted. When waste is composted, it is broken down so the nutrients can be released. The composted material can then be recycled, for example, as fertilizer in your garden. What kinds of materials break down well? What kinds of materials never break down at all? This investigation will allow you to explore the process of composting.

Skill POWER

For tips on how to write a hypothesis, turn to SkillPower 6.

Question

What kinds of materials decompose, and how long does decomposition take?

Safety Precautions

Apparatus

4 identical large plastic pots with drainage holes (1 per test material)

saucers to go under pots

pieces of window screen or similar material

magnifying glass

Materials

small stones

labels for pots

garden soil (not sterilized)

water

approximately 250 mL of some or all of the test materials in List A and List B

List A: banana peels, cabbage leaves, grass clippings, orange peels, coffee grounds, potato peels, carrot peels, egg shells

List B: aluminium foil, small pieces of plastic, shredded wax paper, shredded paper

Procedure

① Before starting this investigation, formulate a **hypothesis** about the kinds of materials that can decompose. Choose four materials to test: two from List A and two from List B. Based on your hypothesis, **predict** what will happen to each of the materials you are going to test.

② Set each pot on a saucer.
 (a) Put a few small stones over the drainage holes in each pot.
 (b) Add garden soil to each pot until the pot is about half full.

③ Put one test material in each pot. Label the pot to show what material is in it.
 (a) Cover the materials in the pots with equal amounts of soil.

R • Interactions and Ecosystems

(b) Estimate the amount of water that you can add to each pot so that a little water will come out the bottom of the pot into the saucer. Add that amount of water to each pot. If no water drains out of the bottom of the pots, add small but equal amounts of water to each pot until some water drains out of the bottom of the pots.

(c) Cover the open top of each pot with a piece of window screen.

(d) Put the pots in a permanent location for a few weeks. Moisten the soil every few days. Be sure to add the same amount of water to each pot.

4 After a week, remove the uppermost layer of soil. Check that the soil underneath is moist. **Observe** the amount of decomposition. If possible, use a magnifying glass. **Record** your observations. Replace the soil, and continue the process until you can see a difference in the condition of the test materials.

5 Clean up your work area as your teacher directs. Wash your hands thoroughly after completing each part of this investigation.

Skill
POWER
For tips on designing an investigation, turn to SkillPower 6.

Analyze

1. Which test materials decomposed rapidly? Why do you think these test materials decompose rapidly?

2. Which test materials decomposed slowly? Why do you think these test materials decompose slowly?

3. Which materials did not decompose over the course of the investigation? Why do you think these materials not decompose?

4. Did your observations support your hypothesis? Explain.

Conclude and Apply

5. Considering the health of the environment, what should be done to recycle or dispose of the materials you listed in questions 2 and 3? Name several steps your community could take to ensure that wastes are disposed of properly.

6. What factors might speed up the decomposition of the materials you listed in question 2?

Extend Your Skills

7. Design an investigation to determine what effects, if any, temperature has on the rate of decomposition. Have your teacher approve your procedure before you carry out your investigation.

8. Design an investigation to test what effect, if any, using sterilized soil (such as potting soil) has on the rate of decomposition. Have your teacher approve your procedure before you carry out your investigation.

9. Find out about red wrigglers. What are they? What are their habitat requirements? How can they be used to help recycle waste? With your teacher's permission, obtain some red wrigglers to observe and use in your classroom.

There are more individual organisms in a decaying log than there are people on Earth!

Living Relationships

Imagine having a back full of barnacles, each the size of a small orange! Grey whales that travel along the British Columbia coast (like the one in Figure 1.22) know how this feels. The barnacles are a species that attaches only to whales. The barnacle gets a free ride, dining on small animals as it cruises along with the whale. The relationship between barnacles and grey whales is an example of symbiosis. **Symbiosis** is a biological relationship in which two species live closely together in a relationship that lasts over time. The two species are said to have a **symbiotic** relationship. There are three main types of symbiotic relationships: parasitism, mutualism, and commensalism.

Figure 1.22 Grey whales can be covered with clusters of barnacles.

Parasitism

Parasitism is a symbiotic relationship between organisms, in which one partner benefits and the other partner is harmed. The partner that benefits from the relationship is the **parasite**. The partner that is harmed is called the **host**. The whale lice grow on the backs of grey whales, amongst the barnacles. They feed on the grey whale's skin and any damaged tissue. Lice are external parasites which hide in the hair, fur, and feathers of many animals.

The mistletoe in Figure 1.23 is a tiny, yellow-green leafless plant. It grows as a parasite on trees in British Columbia. The roots of the mistletoe burrow into the branch and rob the host tree of sugar, water, and other nutrients.

Figure 1.23 Mistletoe is a parasitic plant. It grows on trees such as hemlock, pine, juniper, and fir in British Columbia.

Some parasites, such as tapeworms, live inside an organism. The tapeworm in Figure 1.24, for example, can live in the small intestine of a human. It may grow as long as 10 m! Tapeworms benefit by absorbing the nutrients from the hosts' food. The hosts are harmed because they do not get the nutrients from the food they eat.

Figure 1.24 Tapeworms are parasites that live inside other animals' intestines.

Figure 1.25 The bacteria, the protozoa, and the termite all benefit from the relationship called mutualism.

Mutualism

Mutualism [MYOO-choo-al-is-uhm] is a relationship between two different organisms, in which each partner benefits from the relationship. The termites in Figure 1.25, for example, are decomposers of wood. They are incapable of breaking down the wood on their own, however. A termite has a mutualistic relationship with bacteria and protozoa that live in its digestive tract. The bacteria and protozoa digest the wood. The termite lives on the waste products of the bacteria and protozoa.

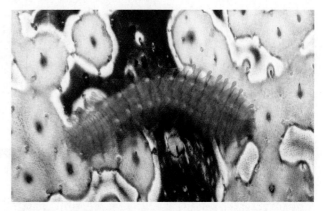

Figure 1.26 The relationship between scale worms and organisms such as sea stars, limpets, worms, and other marine animals is called commensalism.

Commensalism

Commensalism is a symbiotic relationship between organisms, in which one partner benefits and the other partner does not appear to lose or gain from the relationship. The relationship between the barnacles and the grey whale is an example of commensalism. The barnacles gain a habitat (the whale's back) and access to food, but the whale does not appear to be harmed. Similarly, a scale worm (like the one in Figure 1.26) lives on the surface of a sea star, inside the tube of a marine tubeworm, or in a small marine animal called a limpet. The scale worm gets a free ride and "free" food, but it does not harm its host.

READING check

A fish called a topsmelt feeds on whale lice and flakes of old skin on a grey whale's back. What kind of relationship does a topsmelt have with a grey whale? Explain your answer.

Section 1.3 Summary

A niche is where an organism lives and what role it plays within its ecosystem.

- Plants are producers and animals are consumers.
- A consumer can be a herbivore, a carnivore, or an omnivore.
- Scavengers are consumers that eat dead organisms.
- Decomposers are consumers that break down dead organisms.

Symbiosis is a relationship between two species that live in close association over time. The three main types of symbiosis are:

- parasitism (one partner benefits and the other is harmed)
- mutualism (both partners benefit)
- commensalism (one partner benefits and the other does not lose or gain from the relationship)

How can huge humpback whales survive by eating some of the smallest organisms in the ocean? What happens when fire devastates an ecosystem? These are some of the topics you will examine in the next chapter.

Key Terms

niche
producers
consumers
herbivores
carnivores
omnivores
scavengers
decomposers
symbiosis
symbiotic
parasitism
parasite
host
mutualism
commensalism

Check Your Understanding

1. Define and give an example for each term.
   symbiosis   mutualism   omnivore
   producer   parasite

2. What are the effects of predators on a community? What are the effects of parasites? How are the effects of predators and parasites similar or different?

3. Define symbiosis. Give one example of symbiosis from this chapter and one example from your own experience.

4. Classify each organism as a producer, a herbivore, an omnivore, or a carnivore.
   cow   grass   human   green alga
   deer   rabbit   wolf   oak tree

5. **Apply** Hagfish live in the Pacific Ocean. They look something like an eel with a large, sucking mouth. Hagfish enter the mouth of a dead animal then eat it from the inside out. What niche is filled by hagfish?

6. **Thinking Critically** Choose and observe an ecosystem in your neighbourhood. For example, you could observe a park, a pond, or a dead tree. What are the biotic and abiotic parts of this ecosystem? What niche does each organism occupy in this ecosystem?

CHAPTER at a glance

Now that you have completed this chapter, try to do the following. If you cannot, go back to the sections indicated in brackets.

(a) Make a chart that lists the needs of living things. (1.1)

(b) Explain the difference between an ecosystem and a biome. Describe the biome(s) in British Columbia.

(c) What are the key differences between a community and an ecosystem? Give examples of a community and of an ecosystem to help explain the differences. (1.2)

(d) Explain the relationships among individuals, populations, and communities. (1.2)

(e) Give examples of different areas that could be sampled. Explain why sampling is necessary. (1.2)

(f) What is the niche of each of the organisms shown here?

(g) Explain the difference between scavengers and decomposers. What roles do decomposers play in everyday life? (1.3)

(h) Explain why termites cannot function alone as decomposers. (1.3)

(i) Describe one way in which you used a model to explain science ideas in this chapter. (1.1, 1.2, 1.3)

Prepare Your Own Summary

Summarize this chapter by doing one of the following. Use a graphic organizer (such as a concept map), produce a poster, or write a summary to include the key chapter ideas. Here are a few ideas to use as a guide:

- Create a series of pictures to illustrate levels of organization that ecologists use to communicate their observations. Write two or three sentences to explain each of your pictures.
- Identify the producers and consumers in this picture. Do some research to determine whether each animal is a herbivore, a carnivore, or an omnivore.
- Do some research to learn a few ways in which early Aboriginal peoples in British Columbia used the resources in their environment.

- Choose key scientific terms from this chapter. Create a crossword puzzle to review their meanings.
- Create a poster to represent an organism that lives in British Columbia. Describe the biotic and abiotic parts of the organism's environment.

CHAPTER 1 Review

Key Terms

abiotic
climate
biotic
species
ecology
ecologist
ecosystem
biome
ecoprovinces
biogeoclimatic zones
population
habitat
community
sampling
quadrat
niche
producers
consumers
herbivores
carnivores
omnivores
scavengers
decomposers
symbiosis
symbiotic
parasitism
parasite
host
mutualism
commensalism

Reviewing Key Terms

If you need to review, the section numbers show you where these terms were introduced.

1. For each of the following, what is the *difference* between the two terms?
   (a) biotic and abiotic (1.1)
   (b) community and population (1.2)
   (c) ecologist and ecosystem (1.1)
   (d) mutualism and parasitism (1.3)

2. Complete the following table in your notebook. If possible, give examples of organisms that are not described or listed in this chapter. (1.2)

Type of organism	Examples of organism
producer	1. 2.
consumer: herbivore	1. 2.
consumer: carnivore	1. 2.
scavenger	1. 2.
decomposer	1. 2.
parasite	1. 2.
omnivore	1. 2.

3. Define the term "niche." List five different niches in an ecosystem. Give examples of each niche in your list. (1.3)

4. In your notebook, match the description in column A with the correct term in column B.

A	B
(a) long-lasting relationship between two organisms	• adaptation (1.1)
(b) all the interacting living and non-living parts in an area	• ecologist (1.1)
(c) relationship between two organisms, in which one organism benefits and the other organism is harmed	• symbiosis (1.3)
(d) scientist who studies interactions within an environment	• parasitism (1.3)
(e) a characteristic that helps an organism survive in its environment	• ecosystem (1.1)

Understanding Key Ideas

Section numbers are provided if you need to review.

5. Why are quadrats useful for carrying out studies on an ecosystem? (1.2)

6. What are some differences and similarities between communities and ecosystems? (1.2)

7. What are the main parts of an ecosystem? Give examples of each part. (1.2)

8. If you found hawks, field mice, and corn in the same ecosystem, what niche would each organism fill? Explain your answer. (1.3)

9. Why is the study of ecology best done outside a classroom or laboratory? Explain your answer. (1.2)

10. Why are Aboriginal peoples often considered to be the first ecologists in Canada? Explain your answer. (1.1)

11. (a) Why are scavengers and decomposers important in an ecosystem?
    (b) How do scavengers and decomposers differ? (1.3)

12. Choose an organism. Explain how it is affected by the abiotic parts of its environment. (1.1)

Developing Skills

13. Think of an ecosystem, and observe it (if possible). Make a chart, a poster, or another representation to show abiotic-biotic interactions in the ecosystem.

14. Make a graphic organizer showing what you have learned about Canada's four major biomes.

15. How would you create a scientific model of an ecosystem? What would you include in your model? Why?

Problem Solving

16. Do you have a fish, a hamster, or another living creature in your classroom, school, or home? What are some of its adaptations? (For example, what kind of teeth does it have? What kind of feet does it have? Can it run quickly? Is it active during the day or during the night?) After observing the animal, what do you think its original (wild) habitat was?

17. Choose at least four pairs of living organisms in the figure below. What are the interactions between each pair? What niche does each organism fill?

18. What are some examples of relationships that are similar to mutualism in a human community?

Critical Thinking

19. Think about each of the following pairs of organisms, and name the type of symbiotic relationship the partners might have. What are the gains or losses for each partner?
    (a) a flowering plant and a bee
    (b) a dog and a flea
    (c) a nectar-eating bat and a flowering cactus
    (d) a bird and a water buffalo

Pause & Reflect

Go back to the beginning of this chapter on page 4, and check your original answers to the Getting Ready questions. How has your thinking changed? How would you answer those questions now that you have investigated the topics in this chapter?

CHAPTER 2
Cycles in

Getting Ready...

- How can plants grow when no one feeds them?
- What prevents dandelions from covering the surface of Earth?
- Why do "vacant" lots never remain vacant?

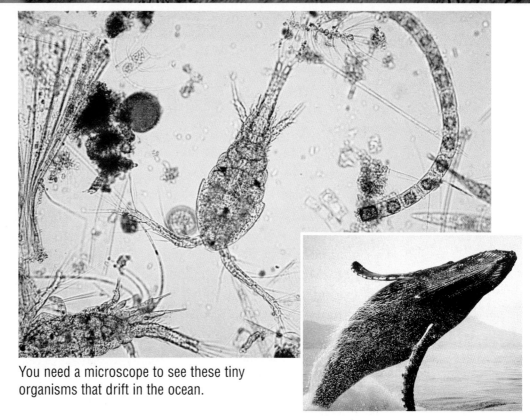

You need a microscope to see these tiny organisms that drift in the ocean.

This 50 t whale lives on small fish and plankton.

Swimming through the ocean is like swimming through a bowl of vegetable and meat soup. The ocean is filled with tiny floating organisms, shown above, called phytoplankton [fih-toh-PLANK-tuhn] and zooplankton. Without these tiny organisms, there would be no life at all in the ocean. In fact, life on Earth would be quite different. Plankton are food for many of the animals in the ocean. Even giant whales, such as this humpback whale, eat plankton.

Humpback whales were once hunted off British Columbia's west coast. When humpback whales were finally protected in 1966 and the hunting stopped, there were only a few thousand humpback whales left. Today, there are over 125 000. Since people no longer hunt these whales, their population has recovered. Their population will not continue to grow and grow, however. There are only so many humpback whales that an ocean ecosystem can support.

In this chapter, you will explore connections between organisms and what they eat. You will also learn about some of the reasons that the sizes of populations are limited.

Ecosystems

What You Will Learn

In this chapter, you will learn
- how plants and animals are interconnected in ecosystems
- how energy is transferred through ecosystems
- what abiotic and biotic cycles occur in ecosystems
- what factors limit the sizes of populations
- how ecosystems change over time

Why It Is Important

- You need living things in order to survive. It is important to learn about the connections between you and other living things.
- Every day, you make choices that affect the environment in which you live. When you understand the connections among organisms in your environment, you can make choices that will not harm your environment.

Skills You Will Use

In this chapter, you will
- infer connections between producers and consumers
- observe the water cycle
- observe and record the food chain of a mealworm
- control variables to determine conditions that affect the growth of plants
- model changes in an ecosystem over time

Every organism in an ecosystem affects the other organisms.

Starting Point  ACTIVITY 2-A

Ant Alert

Ants are found all over the world. They have adapted to many different environments. How do ants live? How do they obtain food?

What to Do

1. With a partner, look outside for some ants on the ground or on a sidewalk. Watch the ants for a little while. Look for patterns in their activities.
    - Do the ants ever bump into each other?
    - What happens when they meet a different kind of insect?
    - Do any ants seem to be carrying food?

2. Scatter a few grains of sugar in front of the ants, and **observe** what they do.

3. Now scatter a few pieces of grass, and **observe** the ants.

4. Try different types of food, such as a small piece of meat or cheese, or birdseed.

5. Make some notes that will help you remember what you observed.

What Did You Find Out?

1. Describe the foods that the ants seem to prefer.

2. Did the ants seem to follow a particular path? Explain your answer by referring to your observations.

3. How did the ants behave when they met an obstacle, such as a piece of grass? Explain your answer by referring to your observations.

4. Compare your observations with the observations of other students in your class.

Chapter 2 Cycles in Ecosystems • MHR **35**

Section 2.1 Food Chains, Food Webs, and Energy Flow

Did you ride your bicycle to school today? Did you play a sport in gym class? Have you ever mowed a lawn or shovelled snow? Do you sleep, breathe, and grow? All of these activities require energy. To obtain energy, you consume food. In this section, you will trace the path of food energy as it is passed from one living organism to another.

Food Chains

A **food chain** is a model that shows how food energy passes (flows) from organism to organism. A food chain begins with a source of energy, which is usually the Sun. Plants trap energy from the Sun. They convert the energy into a form that can be stored in food. Animals eat the plants to obtain this energy. Some animals eat other animals to obtain this energy. Microscopic organisms called phytoplankton (shown on page 34) also trap energy from the Sun and convert it into food. Many of the fish and other organisms that live in the sea depend on phytoplankton for food. Figure 2.1 shows three examples of food chains.

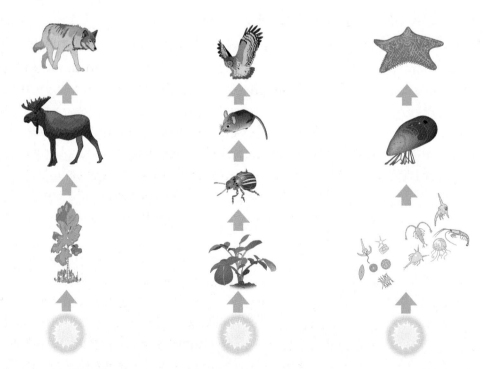

Figure 2.1 In a food chain, the arrows show the direction in which the energy flows from organism to organism. These three food chains exist in ecosystems in British Columbia.

Producers in Food Chains

The plants and phytoplankton in food chains are producers. Producers can make their own food because they contain chlorophyll [KLOHR-uh-fil]. **Chlorophyll** is a green chemical that traps the energy of the Sun. Producers use the trapped energy to make food, in a process called **photosynthesis** [foh-toh-SIN-thuh-sis].

During photosynthesis, the trapped energy is used to convert carbon dioxide gas and water into foods such as sugar and starch. As a result of photosynthesis, oxygen gas is released back into the air. The overall process of photosynthesis is shown in Figure 2.2.

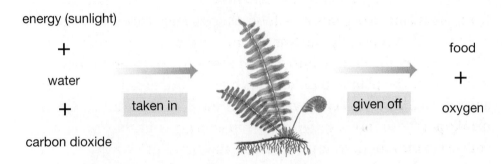

Figure 2.2 Photosynthesis occurs in organisms that contain chlorophyll.

Consumers in Food Chains

Consumers follow producers in a food chain. The carnivore at the end of the food chain is known as the **top consumer**. In the food chains in Figure 2.1, the wolf, the owl, and the sea star are the top consumers.

A consumer's only source of energy is producers or other consumers. A consumer obtains the energy it needs by breaking down high-energy foods, such as sugar and starch, in a process called **cellular respiration**. Cellular respiration occurs inside all living cells. Even plant cells need the energy from cellular respiration when the Sun is not shining.

During cellular respiration, oxygen is used to help break down the high-energy foods. Carbon dioxide, water, and energy are released. The energy is used to carry out all the functions of life, such as growing, repairing tissues, eliminating wastes, breathing, and digestion. The process of cellular respiration is shown in Figure 2.3 on the next page. The relationship between photosynthesis and cellular respiration is shown in Figure 2.4 on the next page.

READING Check

How is every bite of food you eat like eating sunshine? Explain your answer.

Pause & Reflect

Aboriginal peoples honour the importance of the Sun. The Nisga'a tell the story of Txeemsim [KLEE-suhm] bringing the Sun to the world. In the village of Hesquiaht, on the west coast of Vancouver Island, one person was responsible for observing the exact spot where the Sun rose on the horizon each morning.

INTERNET CONNECT

www.mcgrawhill.ca/links/BCscience7

Read more Aboriginal stories about the Sun by visiting the web site above. Click on **Web Links** to find out where to go next.

READING check

Where do humans fit in a food chain? Write a food chain that includes you.

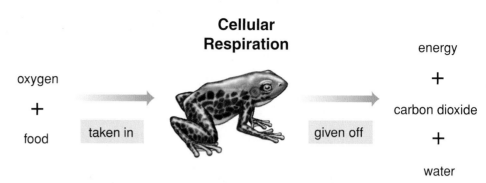

Figure 2.3 Cellular respiration occurs in all living cells. The oxygen that cells use was produced during photosynthesis.

INTERNET CONNECT

www.mcgrawhill.ca/links/BCscience7

Learn more about food chains by trying a quiz, playing a game, or watching a movie. Go to the web site above, and click on **Web Links** to find out where to go next.

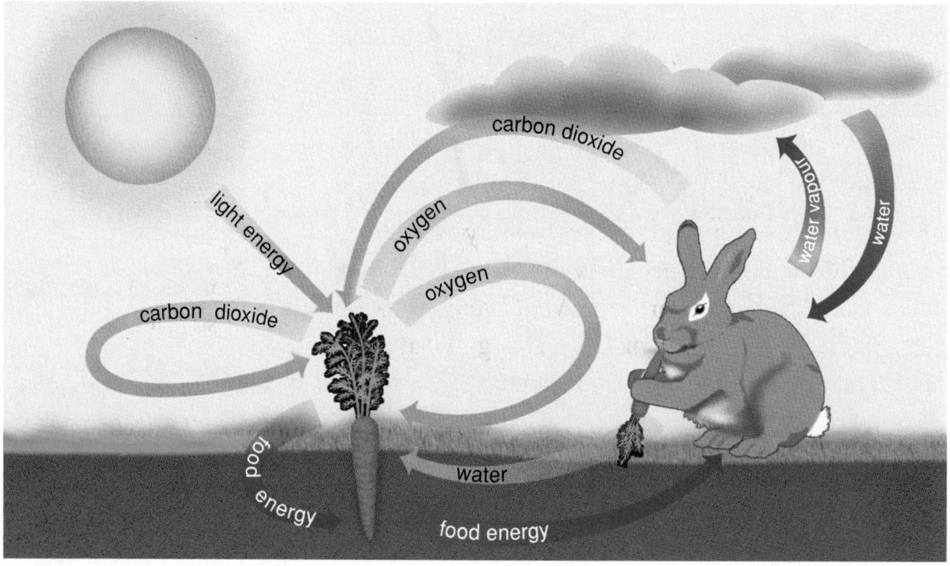

Figure 2.4 Photosynthesis and cellular respiration work together in a cycle.

A Very Special Food Chain

Scientists have discovered that not all producers rely on the Sun for energy. No sunlight reaches the deep ocean yet an ecosystem teeming with life exists 3 km below the surface of the Pacific Ocean off British Columbia's west coast. Water over 370°C bubbles up through deep cracks, called *hydrothermal vents*, in the ocean floor. A certain type of bacteria gains its energy from chemicals released by the vents. This bacterium is the first link in the food chain. Other organisms then eat the bacteria. These organisms include clusters of tubeworms that look like huge lipsticks (see Figure 2.5), giant white crabs, blind shrimp, and yellow mats of bacteria. British Columbian scientists have been leaders in exploring this new frontier.

Figure 2.5 Producers in the deep ocean hydrothermal vents are bacteria that obtain energy from chemicals in a process called *chemosynthesis*. These tubeworms are lined with bacteria, which provide them with nutrients.

38 MHR • Unit 1 Ecosystems

DESIGN YOUR OWN
INVESTIGATION 2-B

SKILLCHECK
- Predicting
- Designing Experiments
- Controlling Variables
- Observing

A Mealworm's Food Chain

Mealworms are not really worms. They are the larvae of a species of beetle. What type of food do you think mealworms prefer? What would a food chain for a mealworm look like? Design your own investigation to find out.

Question
What type of food do mealworms prefer?

Prediction
Make a prediction about the type of food that mealworms prefer.

Safety Precautions

- Never eat any food during science class.
- Wash your hands thoroughly, using soap, when you have completed this investigation.
- Dispose of all the food you used immediately after completing this investigation, according to your teacher's instructions.

Materials
15 mealworms
approximately 50 mL of 5 different foods, such as oats, bran, flour, grass, shredded lettuce, and cornmeal
aluminum pie plate
plastic wrap
masking tape

Procedure

1. Read through the Procedure. **Predict** the types of food that you think mealworms might prefer.

2. Use the mealworms, pie plate, and five different foods to design your investigation.

3. Write your experimental procedure, and review it with your teacher before beginning. In your procedure, you must indicate how you will keep the mealworms from escaping from the pie plate.

4. Conduct your investigation, and **record** your observations.

Skill POWER

For tips on making predictions and designing experiments, turn to SkillPower 6.

Analyze

1. (a) What variables did you control in this investigation?
   (b) What was the independent variable in this investigation?

2. Summarize your results in words or with a clearly labelled diagram.

3. Compare your results with the results of other students in your class. Did everyone have the same results? If not, what could have caused the differences?

Conclude and Apply

4. Based on your observations, which food(s) do mealworms prefer?

5. Draw at least one food chain that includes mealworms.

Chapter 2 Cycles in Ecosystems • MHR 39

Food Webs

In an alpine meadow, high in British Columbia's Rocky Mountains, a hoary marmot (see Figure 2.6) whistles an alarm. Perhaps the marmot sees a golden eagle soaring overhead or a grizzly bear or wolf nearby. These three animals are potential predators of the marmot. As well, the marmot itself eats more than one type of food. Its diet includes grasses, roots, flowers, and berries. To model all the possible food chains that include marmots and other organisms, you can use a food web. A **food web**, such as the one in Figure 2.7, shows the network of interconnected food chains in an ecosystem.

Figure 2.6 The food web of a hoary marmot includes all its predators, as well as all the plants that the marmot eats.

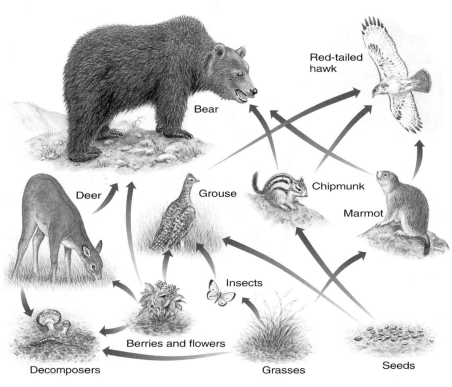

Figure 2.7 A food web provides a more complete model of the feeding relationships in a community than does a food chain.

A food web is more realistic than a food chain for showing the feeding relationships in ecosystems. Producers are usually eaten by many different consumers. Most consumers are eaten by more than one kind of predator.

READING check

Why is a food web a better model than a food chain for describing the relationships among organisms in an ecosystem?

40 MHR • Unit 1 Ecosystems

THINK & LINK

INVESTIGATION 2-C

SKILLCHECK
☼ Observing
☼ Controlling Variables
☼ Interpreting Data
☼ Modelling

Dinner at the Copepod Café

Think About It

Copepods [KOH-puh-podz] are tiny floating animals that are related to shrimp and crabs. They are an important food source for many creatures in the ocean and in estuaries. (An estuary is a place where a river meets the ocean.) Copepods are part of the diet of young salmon, called smolts. In this activity, you will play a game that models a food chain in an estuary.

What You Need

20 strips of cloth, 30 cm long (10 of one colour and 10 of a second colour)

1 large plastic self-sealing bag per student

4–5 L of popped popcorn or foam "peanuts" used for packing

stopwatch

whistle

Safety Precautions

- This activity involves tagging other people as you play a game. Make sure that you tag people gently.
- Play this game in a large area, such as a field or gymnasium.
- Do NOT eat the popcorn you use in this activity.

What to Do *Group Work*

1. Before you begin the game, copy the following data table into your science notebook. Give your table a title.

Number of live organisms	copepods	salmon smolts	herons
Trial 1			
Start of game (0 min)			
After 1 min			
After 2 min			
After 3 min			
Trial 2			
Start of game (0 min)			
After 1 min			
After 2 min			
After 3 min			

2. Your teacher will divide you into three groups of equal size. Copepods, salmon smolts, and great blue herons. Tie strips of cloth of one colour around the arms of all the students who are copepods. Tie strips of cloth of the second colour around the arms of all the students who are salmon smolts. The students who are herons will not wear arm bands.

3. Mark off the boundary of your playing area. Spread the popcorn throughout the playing area.

4. Play the game as follows:
   (a) Spread throughout the playing area.
   (b) When your teacher blows the whistle, the copepods can begin to "eat" decaying plants (popcorn) off the ground by picking up the popcorn and putting it in their "stomachs" (plastic bags). After 30 s, your teacher will blow the whistle again. The copepods stop feeding and stand still.
   (c) When your teacher blows the whistle a third time, everyone can begin to feed.

continued

- Copepods continue to "eat" the popcorn.
- Salmon smolts "eat" copepods by tagging them and putting the contents of the copepods' plastic bags into their own plastic bags. Salmon smolts cannot eat herons.
- Herons "eat" salmon smolts by tagging them and putting the contents of the salmons' plastic bags into their own plastic bags. Herons cannot eat copepods.
- If you are a copepod or salmon smolt that has been eaten, you are out of the game. After you have given the contents of your plastic bag to the predator, wait on the sidelines.

(d) After 1 min, your teacher will blow the whistle. Stop where you are. Count and record the number of copepods, salmon, and herons that are still alive.

(e) When your teacher blows the whistle, continue to "hunt" until the whistle sounds again (after 1 min). Count and record the number of copepods, salmon, and herons that are still alive. Play one or two more rounds.

5 Play the game again, but this time your class will be divided into three groups with the approximate ratio of 9 copepods: 3 salmon: 1 heron. So, for a class of 26 students, there will be 18 copepods, 6 salmon smolts, and 2 herons.

6 Graph the results for each game.

Analyze

1. In Trial 1, the populations of copepods, salmon, and herons were the same size. In Trial 2, the population of copepods was larger than the population of salmon, which was larger than the population of herons. Which trial is closer to what actually happens in nature? Explain your answer.

2. Copepods feed on decaying plants. What might happen to the food chain if only half as much food (popcorn) was available to the copepods?

3. Suppose that there were no salmon in the estuary one year.
   (a) What might happen to the copepod population?
   (b) What might happen to the heron population?

Skill
P O W E R

For tips on drawing graphs, turn to SkillPower 5.

Figure 2.8 Food webs show *how* energy is transferred, but not how many organisms are involved in each step of the energy transfer.

Pyramids of Numbers

Food chains and food webs show how food energy moves through an ecosystem. They do not, however, show the *number* of organisms that are involved in each step. For example, the food web in Figure 2.8 shows that snakes eat grasshoppers. It does not show how many grasshoppers the snakes eat.

To show how many organisms are at each level in a food chain, ecologists use a model called a **pyramid of numbers.** A pyramid of numbers includes the same organisms that are in a food chain, but the size of each level shows the number of organisms involved. There is always a large number of producers at the bottom of a pyramid and fewer organisms at the top. For example, in Figure 2.9, the hawk is the top consumer at the peak of the pyramid. There can be one hawk eating three woodpeckers, but not three hawks eating one woodpecker.

Figure 2.9 A pyramid of numbers is a model of an ecosystem that represents the number of organisms at each level. Producers always form the broad base in a pyramid of numbers.

Energy Flow

Animals high on a food chain must eat many organisms on lower levels because they need the energy. **Energy flow** is the transfer of energy that begins with the Sun and passes from one organism to the next in a food chain. Only a small amount of energy is stored in the body tissues because most of it is used for life processes. Consider the cow in Figure 2.10. About 4 percent of the stored energy in the grass goes to build and repair the cow's body tissues and, thus, stays in the tissues. This is the "stored" energy, which is available to an organism that eats the cow. Most of the energy in the grass that the cow eats is not passed to the next animal in the food chain.

Did You Know?

Sea otters are the only marine mammals without blubber (a thick layer of fat). Instead of blubber, sea otters have a special, thick fur with more than 100 000 hairs per square centimetre! The mass of the food an otter eats every day is equal to about 25 percent of its own body mass.

Figure 2.10 Only 4 percent of the energy stored in the grass that a cow eats is eventually stored in the cow's body.

Reading Check

Why are there always fewer organisms at the top of a pyramid of numbers?

Section 2.1 Summary

Food chains, food webs, and pyramids of numbers are models that show how energy passes from one organism to the next in an ecosystem.

- A food chain shows how energy that is stored in food passes from organism to organism.
- A food web shows a number of interconnected food chains. It gives a more accurate picture of what really happens in an ecosystem.
- A pyramid of numbers is a model that illustrates approximately how many organisms are at each level in a food chain.

Producers are necessary in all food chains. Producers use energy from the Sun to make their own food through a process called photosynthesis. Not all the energy that is stored in food passes directly to the next organism in a food chain. A lot of the energy is used to fuel daily activities, such as breathing and digesting food.

Key Terms

food chain
chlorophyll
photosynthesis
top consumer
cellular respiration
food web
pyramid of numbers
energy flow

Skill POWER

For tips on using Venn diagrams, turn to SkillPower 1.

Check Your Understanding

1. Use a Venn diagram to compare a pyramid of numbers with a food chain.

2. Why is all the energy in one level of a food pyramid *not* available to the organisms at higher levels of the pyramid? Explain.

3. Most humans eat foods from all levels of a food chain, such as the foods in the photograph shown here. Construct two different food chains based on foods you typically eat. Include four or more levels in at least one of your food chains. Use words and diagrams to describe your food chains.

4. **Apply** Suppose that you found hawks, field mice, and corn in the same ecosystem. What roles would each organism have in the food chain?

5. **Thinking Critically** Changes to the population of a species can affect other organisms in the ecosystem. Describe one situation in which changes in a food chain have had an impact on people. Explain your answer.

Section 2.2 Cycles of Matter

In section 2.1, you learned how plants absorb energy from the Sun and convert it into food energy. Animals eat the plants to obtain the stored energy. Eventually, the stored energy is converted into heat and lost to the abiotic environment. Plants must then trap more energy from the Sun to replace the lost energy.

Unlike energy, many types of matter are used over and over again by living systems. In other words, they are cycled through the environment. In this section, you will learn about two important cycles: the carbon cycle and the water cycle.

The Carbon Cycle

Plants use energy from the Sun to convert water and carbon dioxide into foods. These foods contain the carbon from the carbon dioxide. As one organism becomes food for the next organism in the food chain, the carbon-containing materials are passed along. When organisms use the food for energy, the carbon is converted back into carbon dioxide, and is available for plants to use again. Scientists use a model called the **carbon cycle** (Figure 2.11) to show how carbon is used over and over again in ecosystems.

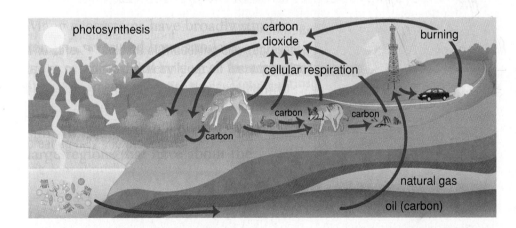

Figure 2.11 The amount of carbon in the environment does not change. It is used over and over again by organisms. Some carbon is stored in the ground for millions of years in the form of coal, oil, and natural gas.

Not all the carbon in plants and animals is converted back into carbon dioxide immediately. When organisms in oceans and lakes die, their tissues often drift to the bottom and form a thick layer of carbon-containing materials. These materials are covered with sand and silt, and buried deeper and deeper. After millions of years, under a lot of pressure, the carbon-containing materials are converted into coal, oil, and natural gas. When people burn the coal, oil, and natural gas for fuel, the energy is released and the carbon is converted into carbon dioxide. These processes also contribute to the carbon cycle.

The Water Cycle

Carbon is not the only type of matter that is cycled through the environment. The **water cycle** is the continuous movement of water through the environment. There are four main processes in the water cycle:

1. **Evaporation** is the process in which liquid water changes into an invisible gas called water vapour. Water is always evaporating from streams, lakes, and ponds, as shown in Figure 2.12. Water is also evaporating from your skin.

Figure 2.12 Water evaporates from streams, lakes, ponds, and other bodies of water. It forms invisible water vapour. Although the water is constantly evaporating, the marsh will probably never be empty because rain will fill it up again.

2. **Transpiration** is the process in which water that is taken in through a plant's roots evaporates from the plant's leaves, stem, and flowers.

3. **Condensation** is the process in which water vapour in the air changes back to tiny droplets of liquid water when the air cools. The droplets of water are so small that they remain suspended in the air as clouds or fog.

4. **Precipitation** is the process in which the tiny droplets inside clouds combine to form large drops. These drops then fall to Earth as rain, sleet, snow, or hail.

The water cycle is shown in Figure 2.13. **Ground water** is water in the soil. The roots of plants can grow down to reach the ground water. People can reach the ground water by digging wells. **Run-off** is water that runs along the surface of the ground into lakes and rivers.

What is the difference between precipitation and evaporation?

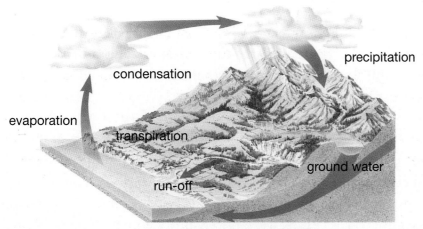

Figure 2.13 The water cycle includes evaporation, transpiration, condensation, and precipitation.

DidYouKnow?

The total amount of water on Earth has not changed for billions of years. A glass of water that you drink today might include some of the same water that cleaned and cooled a dinosaur's skin millions of years ago.

Find Out ACTIVITY 2-D

Modelling the Water Cycle

Water is the only substance that exists naturally on Earth in all three states: solid, liquid, and gas (water vapour). In this activity, you will see water changing from a liquid to a gas and then back again.

Safety Precautions

- Steam can cause severe burns. Take extra care when working with hot water and steam.
- Do this activity only under the supervision of your teacher or another adult.

Note: Your teacher may wish to do this activity as a class demonstration.

What You Need

kettle
cold water
ice
small saucepan
oven mitt or safety glove

What to Do

1. Fill the kettle half full with water. Plug in the kettle.

2. When the kettle has started to boil, place the ice in the saucepan. Put on the oven mitt. Take the saucepan by the handle, and hold it over the steam. **Observe** what happens on the bottom of the saucepan.

What Did You Find Out?

1. How does this activity model the water cycle in nature? Draw a labelled diagram, showing the water cycle you observed in this activity.

2. Which part of your model played a role that is similar to the role of the Sun?

3. Why did you put ice, rather than warm water, in the saucepan?

READING check

How would you define the word "pollution"?

DidYouKnow?

A scientist named Rachel Carson (1907–1954) wrote the now famous book *Silent Spring*. In this book, she warned about the dangers of DDT. Her observations led to studies that showed that DDT caused the deaths of thousands of birds. Partly as a result of her writing, DDT was eventually banned in many countries, including Canada.

Pollution in the Environment

Water, carbon, and other substances constantly cycle through ecosystems. Unfortunately, some unwanted substances can enter these cycles. **Pollution** refers to any substance that is added to the environment so fast that it cannot be broken down, stored, or converted into a form that is not harmful. **Pollutants** are substances that cause pollution.

Many artificial substances that are harmful to organisms have entered ecosystems. In the past, PCBs (polychlorinated biphenyls) were commonly used for a variety of purposes, including paints and packaging materials. PCBs were never meant to enter the environment, but they have leaked into the water and ground. PCBs break down very slowly, so they remain in the ground and in water for years. Once in the ground and in water, they can cause harm to organisms.

DDT (dichloro-diphenyl-trichloroethane) is a pesticide that was commonly used across North America, including Canada. From the 1940s to 1960s, it was sprayed on crops and trees to kill harmful insects. DDT did control insect populations. It killed many insects that could destroy crops. It also killed disease-carrying mosquitoes and, thus, saved millions of human lives. Unfortunately, DDT is harmful to many living organisms — especially birds. DDT is now banned in Canada and many other countries but not everywhere in the world.

Bioaccumulation

Pollutants move from level to level in a food web. Pollutants are not broken down like food, however. If an organism consumes a pollutant, most of it remains in the tissues. Each animal in the food chain receives nearly all the pollutant that was consumed by the organisms below it. The build up of pollutants in organisms is called **bioaccumulation.**

Figure 2.14 Killer whales in waters off southern British Columbia are becoming sick, and their population is getting smaller. One reason is the bioaccumulation of pollutants.

The killer whale in Figure 2.14 lives in the ocean off the coast of southern British Columbia. It is the top consumer in a food chain like the one in Figure 2.15. Pollutants in the water can enter the food chain at any level, but they often enter through organisms that are low on the food chain. For example, a pollutant may enter the food chain when it is absorbed by phytoplankton or zooplankton. The pollutant is passed along and concentrated in the tissues of the next organism in the food chain. Therefore, the amount of pollutant increases at each level in the food chain. As a result, the top consumer ends up with a very large amount of the pollutant in its body.

Many of the killer whales in British Columbia are becoming ill partly due to the bioaccumulation of pollutants, such as PCBs. The population of killer whales that spends most of its time near southern Vancouver Island, Vancouver, and northern Washington State is in the greatest danger. Pollutants in the food chain, as well as declining salmon populations (a source of food), are leading to the decline of the killer whales.

Human activities have caused harm to the environment. Now, however, people who are concerned about the environment are working to improve the habitats of killer whales and other endangered species. For example, populations of peregrine falcons, such as the one in Figure 2.16, were once in danger in British Columbia and throughout North America due to DDT. The birds that the falcons preyed on had eaten insects or seeds poisoned with DDT. The DDT made the falcons less successful at producing young. As a result, their populations began to decline. After DDT was banned in Canada, the populations of peregrine falcons slowly recovered.

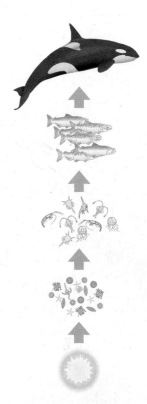

Figure 2.15 Killer whales are the top consumer in a food chain that usually includes phytoplankton, zooplankton, and salmon.

Figure 2.16 Peregrine falcons were once in danger, due to the bioaccumulation of DDT.

In which link of the food chain, producers or consumers, do pollutants accumulate in the largest amounts? Explain your answer.

Section 2.2 Summary

Carbon cycles through the biotic and abiotic parts of an ecosystem. Carbon is found in carbon dioxide in the abiotic part of an ecosystem. In the biotic part of an ecosystem, carbon is in high-energy materials, such as sugar and starch.

Water cycles between the liquid, solid, and gaseous states. The water cycle has four main processes:

- Evaporation is the process in which liquid water changes into a gas.
- Transpiration is the process in which water evaporates from the leaves, stems, and flowers of plants.
- Condensation is the process in which water vapour changes into a liquid.
- Precipitation is the process in which tiny droplets of water inside clouds combine to form large drops that fall to Earth.

Pollution can also cycle through ecosystems. Pollutants are substances that cannot be broken down, stored, or recycled in the environment in a way that is not damaging. Bioaccumulation is the accumulation of pollutants in an organism after eating several polluted small organisms. The top consumers in food chains are affected most by bioaccumulation.

Key Terms

carbon cycle
water cycle
evaporation
transpiration
condensation
precipitation
ground water
run-off
pollution
pollutants
bioaccumulation

Check Your Understanding

1. In what forms do you find carbon in the carbon cycle?

2. Draw a diagram of the water cycle, showing the four main processes. Label and describe each process.

3. What are the similarities and differences between pollution and bioaccumulation?

4. **Apply** Pesticides are chemicals that are commonly used to control agricultural pests, such as insects. How might pesticides affect carnivores, which are higher up in food chains? Explain.

5. **Thinking Critically** What are the advantages of a human, or another omnivore, eating organisms that are lower rather than higher in the food chain?

6. **Thinking Critically** How do pollutants accumulate in food chains? Design a game that shows the process of bioaccumulation.

CHAPTER 2

Section 2.3 Factors in Ecosystems that Limit Populations

Many species, such as rabbits, have an amazing capability to reproduce quickly. In any ecosystem, however, there are limits to the amount of food, water, living space, mates, nesting sites, and other resources that are available. Populations cannot continue growing larger forever. A population eventually reaches the largest number of individuals that an environment can support over a long period of time. This maximum number of individuals is called the **carrying capacity** of the ecosystem. Some common limiting factors are discussed on the following pages.

READING check

How are limiting factors related to carrying capacity?

Limiting Factors

A **limiting factor** is any abiotic or biotic factor that controls the number of individuals in a population. In Figure 2.17, for example, you can see a distinct line across the mountainside. Trees do not grow above this line. The temperature is too cold for the trees to survive above this elevation, called the *tree line*. Temperature is one limiting factor for this population of trees. The strength of the winds and the thin soil may also limit the growth of these trees.

Figure 2.17 The growth of trees can be limited by temperature. The trees in this ecosystem will not grow above a certain elevation.

Chapter 2 Cycles in Ecosystems • MHR 51

Figure 2.18 Predation is one type of limiting factor in a population.

Predator-Prey Cycles

In Chapter 1, you studied the populations of hare and lynx. You learned that predator-prey populations often increase and decrease in cycles. Predation is one way in which one population can limit another. The wolf in Figure 2.18 may catch one of the elk in the herd it is chasing. If so, predation by the wolf has limited the population of elk. Predator–prey cycles limit other things in ecosystems. When populations of deer, elk, and caribou are large, they eat more grasses and limit the population of the grasses.

Diseases and Parasites

Diseases and parasites can limit populations. In most of North America, a disease called canine distemper can infect and kill animals such as wolves, coyotes, foxes, raccoons, and weasels. Diseases and parasites limit populations of plants, as well. For example, a fungus called spruce broom rust has infected the spruce tree shown in Figure 2.19.

Figure 2.19 White spruce can be found throughout most of the interior of British Columbia. Trees that are infected with spruce broom rust have strange yellow growths, which are sometimes called "witches' brooms."

Figure 2.20 Grizzly bears need a large amount of space for their habitat.

Competition for Resources

Only limited amounts of food and living space are available in any ecosystem. The amount of food and the type of den site that a grizzly bear (see Figure 2.20) needs requires a lot of space. In an ecosystem, grizzly bears must compete for food and den sites. Loss of habitat is one of the main reasons that wildlife populations become threatened. Cities, highways, campgrounds, and orchards reduce the available habitat for many animals. For example, bighorn sheep live high in the mountains in the summer. In the winter, however, the sheep must go down into the valleys. Development in the valley bottoms has limited the sheep's winter range in some places.

CONDUCT AN INVESTIGATION 2-E

SKILLCHECK
- Controlling Variables
- Predicting
- Observing
- Interpreting Data

What's the Limit?

Organisms depend on many biotic and abiotic factors in their environment. When these factors are limited or not available, their survival can be threatened. In this investigation, you will experiment with some limiting factors to see how they affect the growth of plants.

Question

How do abiotic factors, such as light, water, and temperature, affect the germination of seeds?

Safety Precautions
- Wash your hands when you have completed each part of this investigation.

Apparatus
waterproof marker
6 small plastic plant pots
refrigerator
small watering can
ruler

Materials
masking tape
potting soil
18 bean seeds
cardboard box

Procedure

1. Make "DO NOT WATER" labels for three of the pots, using the marker and masking tape. Put the labels on the three pots.

2. Fill all six pots with potting soil. Plant three bean seeds in each pot.

3. Place one labelled pot and one unlabelled pot in:
   - the sunshine near a window.
   - a dark closet or under the cardboard box.
   - a refrigerator.

4. **Predict** what will happen to the growth of the seeds in each pot. Why do you think this will happen? Write down your predictions.

5. Leave the plants for several days. Add water to the three unlabelled pots every other day or as needed. Do not overwater.

6. After the seeds have sprouted, **measure** your plants every day. **Record** your measurements in a data table. Continue to **observe** your plants until some of the plants have two fully formed leaves.

Skill POWER

For tips on graphing, turn to SkillPower 5.

Analyze

1. Create a bar graph to show your results.

2. (a) What independent variables did you test in this investigation?
   (b) What was the dependent variable in this investigation?
   (c) What variables did you control in this investigation?

3. Were there variables that you were unable to control? If so, what were they? What effects do you think they had on your results?

4. Compare your results with the results of other groups in your class.

Conclude and Apply

5. What abiotic factors did you test in this investigation?

6. How did the abiotic factors you tested affect the growth of the bean seeds? (Hint: When you make comparisons, be sure that only one variable was different for the plants you are comparing. For example, do not compare a cold, unwatered plant with a warm, watered plant.)

7. Why did you use more than one seed in each pot?

Natural Disturbances

Avalanches, landslides, forest fires, floods, and extreme weather such as tornadoes, are all examples of natural disturbances that can cause changes in ecosystems. The avalanche, for example, can destroy all the plants in its path. As well, it might have carried sheep and mountain goats down the mountain. In Figure 2.21 you can see where an avalanche killed trees and carried debris down the mountain.

Figure 2.21 Each spring, grizzly bears and wolverine travel to places where avalanches have occurred. There they feed on the carcasses of animals that were caught in the avalanches. It will take many years for the trees to grow large again.

Figure 2.22 With the bark removed, you can see how the beetles dug tunnels in the wood and laid eggs.

Figure 2.23 Large parts of British Columbia's lodgepole, limber, and ponderosa pine forests have been attacked by mountain pine beetles.

Several mild winters in a row have contributed to an outbreak of mountain pine beetles in pine forests throughout most of British Columbia. At least two weeks of very cold weather are needed in the winter to kill the beetles. Without this cold weather, these insects thrive. As shown in Figure 2.22, the beetles bore into the trees and carry with them a fungus. The effects of beetles and the fungus eventually kill the trees. The mild winters with no cold spells have allowed the pine beetles to infest large parts of British Columbia's pine forests. As you can see in Figure 2.23, many trees have turned brown and died.

Lightning strikes have always caused forest fires. Despite what you might expect, deer, bears, blue jays, and many other animals are quick enough to escape forest fires. Also, some species need occasional forest fires to keep their populations healthy. For example, lodgepole pines need fires occasionally, in order to reproduce. As you can see in Figure 2.24, the cones of lodgepole pines are tightly sealed. They remain sealed until they are opened by intense heat.

Figure 2.24 The cones of lodgepole pines need the intense heat of a forest fire to open and release the seeds.

Bushes, trees, and other plants are burned in naturally occurring forest fires in many ecosystems. Usually, burned forests quickly begin to grow again. Fires once occurred on a regular basis. However, in the past 100 years or so, people have put out forest fires as quickly as possible. Without these small fires, large amounts of live and dead vegetation have accumulated on forest floors. This vegetation provides excellent fuel for fires. As a result, uncontrollable forest fires have burned large areas in British Columbia and elsewhere in Canada. The fires that occurred in British Columbia in the summer of 2003 were especially destructive. These uncontrollable fires can be much more damaging to ecosystems than the smaller fires that once occurred on a regular basis.

Many scientists feel that **global warming** (the steady increase in the temperature of Earth's atmosphere) has led to reduced moisture in the environment. Less snow in the winter and less rain in summer have caused the forests to be drier and more likely to catch fire.

What are three natural disturbances that can be limiting factors for a population of organisms?

Ecological Succession: Changes Over Time

Ecological succession is the gradual change in the make-up of a biological community over time. In the process of ecological succession, organisms that are present at one stage change the environment in some way. This change makes it possible for other species to move in, because the new conditions are more suited to their needs. Figure 2.2 shows one example of succession.

A Lichens [LIH–kuhns] are plant–like organisms that can grow on rocks. They produce acids that help to break down the rock. The broken-down rock and the decomposing bodies of dead lichens contribute to soil formation.

B The resulting soil is poor and thin. However, mosses and ferns grow and slowly replace the lichens.

C The soil layer thickens, which means it can hold more water. Plants that need more soil and moisture, such as grasses and flowering weeds, take root and grow. They attract insects, such as bees and butterflies.

D Since the soil is now thicker and richer, bushes and trees take root. They provide shelter and food for birds, mammals, and other organisms, which now start moving in.

Figure 2.26 Succession is a long, slow process in which a natural ecosystem gradually develops and changes over time.

All the natural disturbances you read about on the last few pages (such as fires, winds, and avalanches) are followed by succession. In the succession of the burned forest in Figure 2.27 wildflowers and other plants that need strong sunlight are among the first to grow. Blueberry bushes also thrive, since they are adapted to grow better in soil that contains ash. In the next investigation, you will watch succession in action.

Figure 2.27 These wildflowers, called fireweed, are thriving in an area that has been recently burned.

CONDUCT AN INVESTIGATION 2-F

SKILLCHECK
- Modelling
- Observing
- Interpreting Data
- Communicating

Succession in a Bottle

Succession can take place in any area, large or small, in a short period of time or over many years. For instance, weeds quickly grow when a patch of soil is left alone. Trees, however, take many years to grow back in an area that was cleared by a forest fire or logging.

Question
How does succession take place in an ecosystem?

Safety Precautions

Material
- 2 L clear plastic soda bottle (with the top cut off) or large-mouthed jar
- potting soil
- ruler
- water
- small aquatic plant
- 50 mL wild birdseed mix

Skill POWER
For tips on designing experiments, turn to SkillPower 6.

Procedure

1. (a) Put soil in the bottom of the container, to a depth of 5 cm.
   (b) Fill the container with water, to a depth of 7.5 cm, thus covering the soil.
   (c) Place the container, uncovered, close to a window. Allow the contents to settle overnight.

2. The next day, plant an aquatic plant in the container. Although the water will evaporate over time, do *not* add more water to the container.

3. (a) Once a week, add three or four seeds from the wild birdseed mix to the container. **Record** any observations.
   (b) Continue adding seeds weekly, even though the water evaporates. What, if anything, happens to these seeds?
   (c) After a few weeks, gradually start to add water again, as you would when watering a plant. Continue to **record** your observations.

Analyze
1. Describe what is occurring in your container.
2. (a) Describe your observations during step 3.
   (b) What was the significance of not adding water to your ecosystem?
   (c) What happened to the aquatic plant?
3. Compare your ecosystem at the beginning of the investigation with your ecosystem at the end. How was your ecosystem similar at the beginning and the end of the investigation? How was it different?

Conclude and Apply
4. How well does this investigation demonstrate succession?
5. Consider what you have learned about succession. What would you expect to happen after a fire burns through a forest, destroying most of the older trees and vegetation?

Extend Your Skills
6. Design your own investigation to determine the effects of different environmental factors on your ecosystem. Have your teacher approve your procedure before you carry out your investigation.

Section 2.3 Summary

A limiting factor is an abiotic or biotic factor that limits the number of organisms in a population. Three examples of limiting factors are
- predation
- disease and parasites
- competition for resources, such as food and habitat

Weather, climate change, and natural disturbances (such as avalanches, forest fires, floods, and extreme weather) can cause changes in ecosystems.

Ecological succession is the gradual change in the structure of a community over time. Succession occurs after changes in the environment, such as those caused by a forest fire or avalanche.

Key Terms
limiting factor
carrying capacity
global warming
ecological succession

Check Your Understanding

1. How does carrying capacity influence the number of organisms in an ecosystem?

2. Explain the process of succession.

3. How would a shrinking habitat be a limiting factor for an animal population, such as a grizzly bear or wolf population?

4. What are four different limiting factors that might affect a population of organisms living in a lake? Describe each limiting factor.

5. **Apply** A rancher notices that shrubs and small trees are starting to fill in the pasture used for cattle. What could the rancher do to open up the pasture?

6. **Apply** In the early 1900s, many predators were hunted so much that their populations declined. Suppose that most of the cougars in an ecosystem were killed. How might this affect the deer population (the cougars' main prey) and the plants that the deer eat?

7. **Thinking Critically** Why do you think the supply of food and water in an ecosystem usually affects population size more than other limiting factors do?

8. **Thinking Critically** An avalanche removes all the plants and animals from a large area on a mountainside.
   (a) How is this natural disturbance a limiting factor for populations in the ecosystem?
   (b) What changes might take place over the next few years in the area that has been disturbed?

CHAPTER at a glance

Now that you have completed this chapter, try to do the following. If you cannot, go back to the sections indicated in brackets.

(a) Define the terms "food chain" and "food web." (2.1)

(b) Explain how energy flows in a food chain. (2.1)

(c) Explain energy flow. (2.1)

(d) Describe a pyramid of numbers. What information does it provide that a food chain or food web does not? (2.1)

(e) In what states of matter can you find water in the environment? (2.2)

(f) List three sources of carbon dioxide in the carbon cycle. (2.2)

(g) Describe the four main processes in the water cycle. (2.2)

(h) Give an example of bioaccumulation in a food chain. (2.2)

(i) Explain why species of plants and animals that reproduce quickly do not spread and take over Earth. (2.3)

(j) Describe the different kinds of limiting factors that are found in nature. (2.3)

(k) Suppose that a landslide clears all the plants from part of a hillside. Will this part of the hillside still be without plants in 50 years? Explain your answer. (2.3)

Prepare Your Own Summary

Summarize this chapter by doing one of the following. Use a graphic organizer (such as a concept map), create a poster, or write a summary to include the key chapter ideas. Here are a few ideas to use as a guide:

- Draw a diagram that illustrates energy flow. In your diagram, show the source of the energy and include at least five organisms. Indicate what happens to the energy as it is transferred from one organism to the next.
- Why do most humans have more food choices than many animals have? Brainstorm answers to this question, alone or in a group. Summarize your answers in a written report.

- Make a travel log of the journey of a water droplet as it goes through the water cycle.
- Use diagrams to show ways in which energy is transferred and matter is cycled through ecosystems.
- Imagine that you are a real estate agent, advertising a piece of newly burned forest to animals. In your sales pitch to potential "land owners," describe why this burned area might be a great place to live in a few years.

CHAPTER 2 Review

Key Terms

food chain
chlorophyll
photosynthesis
top consumer
cellular respiration
food web
energy flow
pyramid of numbers
carbon cycle
water cycle
evaporation
transpiration
condensation
precipitation
ground water
run-off
pollution
pollutants
bioaccumulation
limiting factor
carrying capacity
global warming
ecological succession

Reviewing Key Terms

If you need to review, the section numbers show you where these terms were introduced.

1. In your notebook, match each description in column A with the correct term in column B. Use each description only once.

   A
   (a) a change from a gas to a liquid
   (b) transfer of energy through ecosystem
   (c) forest renewal is an example
   (d) need other organisms to survive
   (e) water that falls to Earth
   (f) interconnected network of organisms in an ecosystem
   (g) make food using energy from the Sun
   (h) evaporation from a plant

   B
   - condensation (2.2)
   - transpiration (2.2)
   - food web (2.1)
   - consumers (2.1)
   - succession (2.3)
   - food chain (2.1)
   - producers (2.1)
   - precipitation (2.2)

2. For each of the following pairs, how are the meanings of the two terms different? Explain.

   (a) limiting factor, carrying capacity (2.3)
   (b) condensation, evaporation (2.2)
   (c) food web, food chain (2.1)
   (d) energy flow, pyramid of numbers (2.1)

Understanding Key Ideas

Section numbers are provided (if you need to review).

3. Why do *all* consumers depend on producers for food? Explain your answer. (2.1)

4. What are your three favourite foods? Explain how these foods provide you with energy from the Sun. (2.1)

5. Based on your knowledge of cycles, explain the slogan "Have you thanked a plant today?" (2.1)

6. Why is a food web a better model than a food chain for describing the relationships among what an organism eats and what eats the organism? (2.1)

7. Water is not always a liquid. In what other states of matter can it exist? How does water change during the water cycle? (2.2)

8. Label this diagram with explanations of what is happening in each stage of succession. (2.3)

9. What limiting factor(s) might affect a population of fish living in a lake? (2.3)

Developing Skills

10. Use arrows and words to draw five food chains, based on the food web below.

11. Imagine that you are teaching a class of younger students about food webs. How could you explain this idea to them?

12. Design a poster, game, or model to explain bioaccumulation.

Problem Solving

13. Why do some people in the world have limited food choices? Discuss this question in a group or as a class. What might happen if their food supplies were damaged by a natural disturbance, such as a hurricane or a drought?

14. A sailor survived a shipwreck. She managed to save several hens and a bag of grain from the cargo. She is now on an island that is far from the closest mainland, in an area where there are no other people. It may be months before she is rescued. To survive as long as possible, what should she do? Explain why you think your ideas will help her survive.

15. Imagine that you have just eaten a meal consisting of a salmon sandwich (with lettuce and tomato), a cup of mushroom soup, an almond cookie, and a glass of apple juice. What niche in the food chain was occupied by each organism that made up your meal? Draw and label a food chain that shows possible relationships among the organisms.

Critical Thinking

16. Why is there a limit to the number of links in a food chain?

17. Think about a food chain that includes grass → field mice → snakes → owls. Describe what would happen if many mice died as a result of disease. What would happen to the owls? What would happen to the snakes? What would be the probable result in the ecosystem?

18. Think about the food web shown in question 10. What would happen if each of the following changes occurred?
    (a) The number of deer decreases due to disease.
    (b) A subdivision of new homes is built in the habitat, reducing its size by one half.
    (c) An insect pest kills many of the trees.
    (d) A poison is used on the grass to kill weeds.

Pause & Reflect

Go back to the beginning of this chapter on page 34, and check your original answers to the Getting Ready questions. How has your thinking changed? How would you answer those questions now that you have investigated the topics in this chapter?

CHAPTER 3
Ecosystems and

Getting Ready...

- How can counting dragonflies help the dragonflies?
- How does riding your bike help ecosystems?
- Why is it important to not release an unwanted pet into the wild?

Can you see the leg band on this harlequin duck?

This flashy bird not only has some of the brightest feathers in nature, it has its own jewellery, too. Researchers use the leg bands to check, or monitor, the population size and movements of these birds from year to year. When ecologists learn more about species and ecosystems, they can help people make better choices about things that affect the environment.

Harlequin ducks have an aerial view of British Columbia each spring as they fly from their wintering grounds on the coast, to their summer breeding grounds high in the mountains. Researchers are monitoring these birds because their populations may be at risk. Habitat loss, particularly the mountain streams where they nest and raise their young, is one of the main threats to this species. How do people contribute to habitat loss and other changes that can affect organisms and ecosystems? How do people try to help damaged ecosystems? You will explore these questions in this chapter.

People

What You Will Learn

In this chapter, you will learn

- ways in which people learn about ecosystems
- how people can affect ecosystems in negative ways
- how people can help ecosystems

Why It Is Important

- Sometimes your actions and the choices you make can have a negative impact on ecosystems. Understanding the possible impacts can help you make choices that are better for the environment.
- Ecosystems cannot speak for themselves. If you understand how people can harm or help ecosystems, you can help "speak out" for ecosystems.

Skills You Will Use

In this chapter, you will

- conduct an interview to gather information about ecosystems
- draw a map to record and analyze information
- graph data about endangered species in British Columbia
- research the effect of non-native species on British Columbia ecosystems
- create and maintain a model of an ecosystem

Students help monitor the environment.

Starting Point ACTIVITY 3-A

Learning, Harming, Helping

How do people gather information and learn about ecosystems? What are some ways in which people harm ecosystems? What are people doing to help damaged ecosystems recover? Take out your scissors and glue and start clipping to find some answers to these questions.

What to Do

1. Divide a large sheet of paper into three equal-sized columns. At the top of the columns write: LEARNING, HARMING, and HELPING.

2. Look through newspapers and magazines, listen to the radio and television news, and speak with people in your family and community to gather information on the following topics.
   - How people gather information and learn about organisms and ecosystems
   - How the actions of people can harm organisms and ecosystems
   - Ways in which people are helping to conserve and protect organisms and ecosystems

3. Cut out articles and pictures and glue them into the appropriate column on your poster. Add your own notes and sketches, too. At the bottom of each column, summarize the ways in which people are learning about, harming, or helping the environment.

Extension

4. What are some ways in which you could learn about, harm, or help your local environment? Add your ideas to your poster.

5. How could you inform other people in your community about ways in which your local environment is being harmed?

Chapter 3 Ecosystems and People • MHR 63

CHAPTER 3

Section 3.1 Learning About Ecosystems

Imagine that the ecosystem shown in Figure 3.1 is part of the area in which your home is located. You have decided to compile a booklet that describes the natural features of this land. You want to include information about the plants, animals, bodies of water, as well as any changes that have occurred over the years. How would you find this information? Perhaps you could count the birds you see, record weather patterns, or even do a quadrat study of plants. You could also talk to people who have lived in the area for a long time to learn what they know about the land. These are some of the ways in which people learn more about ecosystems.

INTERNET CONNECT

www.mcgrawhill.ca/links/BCscience7

To learn more about the Garry oak ecosystem, go to the web site above. Click on **Web Links** to find out where to go next.

Figure 3.1 Garry oak ecosystems have a dry, warm climate. This Garry oak forest is filled with Scotch broom, an introduced species that has endangered this ecosystem because it takes habitat away from native species.

Figure 3.2 Blue camas was an important food and trade plant for Coast Salish people.

The Garry oak ecosystem, shown in Figure 3.1, is one of the most endangered ecosystems in Canada. The oak trees of this beautiful ecosystem are rapidly being removed for the expanding cities on south-east Vancouver Island. These meadows were important to the Coast Salish people, who harvested the edible bulbs of the camas flower (Figure 3.2) and other species from this ecosystem. The local Songhees people called the place where Victoria was originally built, Camosun, or "place to gather camas."

Today, the Garry oak ecosystem includes a very large number of rare plant and animal species compared to the rest of British Columbia. Plant studies, bird counts, amphibian monitoring, and ecosystem mapping are just some of the ways in which people are learning more about the Garry oak ecosystem.

Traditional Ecological Knowledge

Traditional ecological knowledge is the knowledge that Aboriginal peoples have gathered about their home environments. Traditional ecological knowledge is based upon the beliefs, cultures, and values of Aboriginal peoples and their thousands of years of experience in their environments.

Aboriginal peoples consider themselves as part of their environments. According to Aboriginal teachings, all parts of the natural world are made by a Creator. All of these parts have life, are related, and belong together in an interconnected whole. For the whole to remain healthy, balance and harmony between all life forms needs to be maintained.

READING check

What is traditional ecological knowledge?

Figure 3.3 This Haida Gwaii Watchman is discussing the history, culture and beliefs of the Haida people.

Pause & Reflect

When people understand more about their environment, they can use this information to make better decisions about how they treat their environment. Traditional ecological knowledge and scientific knowledge are both ways in which people learn about the environment. What could you learn from traditional ecological knowledge?

By combining traditional ecological knowledge with scientific knowledge, we can expand our understanding of the environment. Today, some scientific studies and government plans for land use include traditional ecological knowledge. Researchers work with elders and other knowledgeable people in Aboriginal communities to learn more about the environment. For example, they learn about traditional ways of using and caring for lands and water without causing long-term harm to the environment. They learn about wildlife populations in the past and present. Researchers learn the locations of important habitats for particular species.

Find Out ACTIVITY 3-B

Ecosystem Stories

Your community has probably gone through many changes over the last century. These changes might have occurred for a variety of reasons. These reasons might include advances in technology, increasing numbers of people, and other activities such as forestry, farming, mining, and the building of cities and roads. Older people in your community can often tell you a lot about changes they have seen in their environment.

Materials
notepaper
pencil
video camera (optional)
tape or mini-disc recorder (optional)
camera (optional)

What to Do

1. Individually, or in a small group, prepare to interview a person in your community to learn more about some of the changes he or she has seen in the environment.

2. Prepare a list of questions to ask in your interview. Questions might include: What was the environment in our community like when you were younger? What kind of wildlife did you see? What changes to the natural environment have you seen as the community has grown?

3. Before you conduct your interview, review your questions with your teacher. Your teacher will discuss guidelines for interviewing before you begin.

4. Conduct your interview. You could use a tape or mini-disc recorder, or a video camera if you choose. You may also want to photograph the person you interview.

5. Prepare a three-part poster or make a presentation using slideshow software that describes the natural world of your community's past, present, and your predictions for the future. Include information you gathered from your interview, your own observations, and any other items such as photocopies of old photographs or pictures from magazines or newspapers.

6. Present your project to your class.

What Did You Find Out?

1. How has the environment of your community changed over the years?

2. Do you think interviewing people is a good way to gather information about ecosystems? Suggest several reasons that support your answer.

3. How could an interview with a person who has lived in a place for many years be an important part of a scientist's study of an organism or ecosystem?

Traditional Ecological Knowledge and Science

Clayoquot Sound on Vancouver Island (see page xxiii) has been the subject of many conflicts about the environment over the past several years. People have disagreed about whether this land should be protected or available for logging. The government established a special panel to make recommendations on forest practices for Clayoquot Sound. This panel included Aboriginal experts so they could contribute their knowledge of the region's ecosystems.

Ecosystem Monitoring

When you are feeling sick, you might take your temperature frequently to monitor your condition. **Ecosystem monitoring** (also called environmental monitoring) is a way to check the condition of an ecosystem. By comparing the results of investigations done at different times, ecologists can track changes in ecosystems. The information helps them reverse these changes or reduce the damage they cause. For example, ecologists can record temperatures over a long period of time. Someone might count the number of birds or butterflies in an area once a month for several years. Figure 3.4 shows some of the different types of ecosystem monitoring that ecologists can do. You will learn about one type of monitoring in the next investigation.

READING check

What are two ways in which you could monitor plants or animals in an ecosystem?

(A) **Physical monitoring** uses satellites to track the changes in the landscape over time. For example, satellite maps can show the changes to the land that occur due to the construction of cities or logging.

(B) **Environmental monitoring** tracks changes in climate, temperature, and weather patterns.

(C) **Chemical monitoring** assesses the quality of air, soil, and water.

(D) **Biological monitoring** tracks the changes in organisms or populations of organisms.

Figure 3.4 Ecosystem monitoring can be done in many ways. Four common types of monitoring are physical, environmental, chemical, and biological as described above.

CONDUCT AN INVESTIGATION 3-C

SKILLCHECK
- Interpreting Data
- Predicting
- Communicating
- Modelling

Counting Caribou

Woodland caribou are found in scattered populations in northern and eastern British Columbia. As the seasons change, caribou move from their winter range in the forested valleys to their summer range in high, alpine meadows. Imagine that a proposed new highway will cut through their travel routes. Many people question how the highway will affect the caribou. Your task is to gather some data and find out how many caribou live in the area.

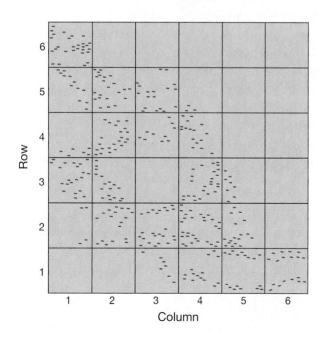

Question
How can you estimate the number of caribou in a particular area?

Apparatus
2 dice

Materials
notebook

Procedure

1. In your notebook, make a table like the one shown below. Give your table a title.

Sample	Row	Column	Number of caribou
A			
B			
C			
D			

2. Graphs such as the one in the lower left corner are made from aerial photographs of herds of caribou. This graph shows the caribou in a 1 km² area. Each mark represents one caribou. Instead of counting all of the caribou, you will take random samples. You will then estimate the total number of caribou.

3. Throw dice to determine which square to count. Roll the first die to select a row. Roll the second die to select a column. **Record** the row and column numbers in your table. For example, if you roll a 2 and a 4, you would find the square to sample by going up to the second row in the grid and then over to the fourth column.

4. Count and **record** the number of caribou in the square you are sampling.

5. Repeat steps 3 and 4 at least three more times.

6. Find the average number of caribou per sample and then multiply this average by the number of squares in the grid.

68 MHR • Unit 1 Ecosystems

Analyze

1. What is the average number of caribou in your study area?

Conclude and Apply

2. You found the average for a 1 km² area. Suppose that the area of the caribou habitat was 500 km². How many caribou would you estimate are in the whole area? Show your calculations.

3. Suppose that the highway will be built through the middle of the area. The caribou will not be able to cross over the highway because it will be fenced.
   (a) What might happen to the caribou? Explain your answer.
   (b) What might happen to the caribou if the highway did not have a fence?

Extend Your Knowledge

4. (a) How could you find out more about the caribou in your study area?
   (b) What information might local Aboriginal peoples have about caribou populations and habitats? Explain your answer.
   (c) How could you use this information in the research study?

5. Research to find out more about caribou in British Columbia. Find out where they live, what they eat, and any threats to their habitat. Prepare a short report summarizing your findings.

Long-Term Monitoring

Imagine you are a member of a research team. You are counting mountain bluebirds such as the one in Figure 3.5. These bluebirds are known to nest in the British Columbia interior. One summer you count 82 bluebirds. The next summer, you count 57 bluebirds. Would this be enough information to tell you about the health of this population? It would probably not be enough. The populations of many species change naturally from year to year. Thus it is important to monitor populations over longer periods of time. Studying organisms for many years is called **long-term monitoring**.

Figure 3.5 Mountain bluebirds are found throughout the southern half of British Columbia.

Why is it important to monitor ecosystems over a long period of time?

Helping with Monitoring Projects

Often volunteers work closely with researchers doing long-term monitoring. Many Aboriginal groups in British Columbia are involved with monitoring projects in their traditional territories. Projects include counting fish returning to local rivers, measuring clams on a stretch of beach, or observing changes in populations of beaver or deer. Across Canada, thousands of people, including the students in Figure 3.6, are helping scientists monitor organisms and ecosystems. For example, one program, called Project Feeder Watch, has volunteers record the numbers and species of birds that visit their bird feeders throughout the year. This helps scientists monitor trends in population sizes, as well any changes in the birds' habitat.

The goal of all these projects is to gather information that is needed in order to keep ecosystems healthy and to restore damaged ecosystems to health.

Populations change naturally depending on many different factors. Think back to the limiting factors you learned about in Chapter 2. Changes in food supply, predators, habitat size, and weather patterns can all cause populations to change from year to year. Long-term monitoring of an ecosystem gives a clearer overall picture of what is really happening.

Figure 3.6 These students are participating in a River Watch program. They monitor water quality and observe wildlife and human activities on a river.

Methods of Monitoring

In the summer of 2003, many large wildfires, like the one in Figure 3.7, burned in British Columbia. Now that the fires are out, ecosystems will begin to grow again. Many changes happen in an ecosystem after a natural event such as a flood or fire. To determine the types of changes, you need to know what the habitat was like before the event occurred. To do this, scientists gather baseline data.

INTERNET CONNECT

www.mcgrawhill.ca/links/BCscience7

To find out how you can participate in a monitoring program, go to the web site above. Click on **Web Links** to find out where to go next.

Figure 3.7 Scientists are monitoring changes in the environment that are taking place after the fires that burned throughout British Columbia in the summer of 2003.

Baseline data give scientists information about ecosystems before any events occur. After an event such as a flood or fire, scientists can compare the ecosystem with its conditions before the changes took place. The data that you analyzed in Investigation 3-C was baseline data. There are many other ways to gather baseline data such as monitoring permanent plots and doing yearly surveys.

Permanent plots are study sites that scientists monitor year after year. For example, scientists might count the number and types of plants in the plot every year to monitor any changes. Yearly counts, or **annual surveys,** of species such as birds and other animals are conducted at the same time and at the same place each year. Over time, scientists can see if there are any changes in populations. If necessary, they can take action to help the organisms.

Sometimes scientists capture animals and attach a radio collar or other device that can transmit a radio signal. Then the researchers use special listening devices, such as those shown in Figure 3.8, to track the movement of the animals. Biologists use this method to track animals as large as whales and as small as butterflies. Researchers must follow strict guidelines to ensure that they do not harm any animals when using all of these sampling methods.

Some researchers use maps to record the data that they collect each year. Maps are good tools to help researchers see any changes that are happening to populations. They use computers to generate these maps. Create your own map in the next At Home Activity.

How is a permanent plot one way of gathering baseline data?

Figure 3.8 Scientists use special antennae to track the movements of animals with radio collars or other types of transmitters.

At Home ACTIVITY 3-D

Mapping Home

Mapping is one tool that people use to record data about ecosystems. The information recorded on maps varies, but could include natural features such as rivers and mountains or important areas for wildlife. What would you include if you drew a map of your ecosystem?

Materials
large sheets of paper
coloured markers

What to Do

1. On a large sheet of paper, use the coloured markers to make a map of some of the parts of the ecosystem in an area that includes your home. Follow the steps below. Use words and pictures on your map. Complete as many of the following steps as you can.

   Draw and label:
   (a) your home in the centre of the page
   (b) the nearest body of fresh water
   (c) the source of your drinking water
   (d) where your household waste water and sewage drain
   (e) any landforms, such as mountains, fields, or valleys
   (f) two plants that are natural to the area
   (g) one land animal and one bird or aquatic animal that are natural to the area
   (h) the direction in which the Sun rises

2. If possible, add the following information to the map:
   (a) Find the name of the Aboriginal peoples who live in the area. Name two plants or animals that are important to these Aboriginal peoples. Describe how they use these plants or animals.
   (b) Draw and label one positive thing that might be happening to the environment in your neighbourhood.
   (c) Draw and label one negative thing that might be happening to the environment in your neighbourhood.

3. Share your map with your classmates. Discuss any questions that you had difficulty answering. Add to your map any new information that you found. If you still have questions that you cannot answer, do research in a library or on the Internet. You can also ask your teacher or another adult to help you find the answer.

What Did You Find Out?

1. How well do you think you understand the environment in which you live? Explain your answer.

2. What are some ways in which you could increase your understanding of your environment?

3. What long-term study could ecologists do that would be beneficial to your environment?

4. How could you use your map as baseline data for a long-term study?

5. What are two ways that you could monitor plants or animals in your ecosystem?

Predicting Changes to Ecosystems

Ecologists can use the information collected from monitoring studies to predict changes in the environment. For example, if someone wants to drain a wetland to build a shopping mall, scientists can examine studies from that area, or from a similar area, to predict the effects. Developers could use the scientists' results to prepare a plan that would minimize the damage to the ecosystem. If the predicted changes are determined to be too serious, local citizens could try to have the development cancelled. A report that outlines how an activity will affect the environment is called an **environmental impact assessment**.

In the Career Connect feature below, you can read about someone who has made a career of doing research and developing environmental impact assessments.

Career CONNECT

What's the Count?

Linda Söber is an environmental biologist who helps governments and developers use their land in a way that preserves the existing wildlife. "I count each type of plant and animal I see. Once I know what's there, I can suggest ways to protect the natural environment." When doing a survey of an area, Linda does not look for just the animals themselves. She looks for tracks, droppings, nests or bedding sites, and fish eggs on plants along the water's edge or in a swamp. She also listens for identifying sounds of certain bird calls.

Linda Söber

Developers often want to fill in wetland areas on their land, to make the areas solid ground on which they can build. Unfortunately this destroys the wetlands and almost everything that lives there. According to Linda, wetlands have more wildlife than either fields or forests.

Do some research at the library or on the Internet, talk to somebody at a wetland reserve if there is one in your area, or contact a wildlife organization, such as Ducks Unlimited. Identify ten animals and plants that live in wetlands (also called swamps, bogs, or marshes). For each animal or plant, write a sentence about how filling in the wetland will affect it. Will filling in the wetland remove its food supply or breeding ground? What are some other ways that wetland animals could be affected?

Section 3.1 Summary

There are many different ways in which people can gather information about ecosystems. These include:

- learning from the traditional ecological knowledge of Aboriginal communities
- ecosystem monitoring, including physical, environmental, chemical, and biological monitoring
- long-term monitoring projects (for example, annual bird counts), and
- gathering of baseline data, such as the number of animals in a particular ecosystem.

Scientists can collect data in a variety of ways, including aerial surveys and the use of bird bands or radio collars. As well, thousands of volunteers participate in programs to monitor particular species.

People can use traditional ecological knowledge and scientific knowledge together to learn more about ecosystems and to make wise decisions about land use.

Key Terms

traditional ecological knowledge
ecosystem monitoring
long-term monitoring
baseline data
permanent plots
annual surveys
environmental impact assessment

Check Your Understanding

1. How can people use long-term monitoring to help protect natural ecosystems? Explain your answer.

2. What are the four types of ecosystem monitoring? Give an example of something that could be measured or monitored for each type of monitoring.

3. What is baseline data?

4. **Apply** Choose a species of plant or animal in your community. How would you monitor this species over a long period of time?

5. **Thinking Critically** The harlequin duck shown here was caught on the west coast of British Columbia. A band describing when and where it was caught was put on its leg. Months later, the bird was sighted resting on a rock near a stream in the Rocky Mountains. A researcher could tell where the duck had been originally caught by the colour and number-letter code on its band. Why do you think this information is important in helping ecologists understand species of duck and the ecosystem(s) in which it lives?

CHAPTER 3

Section 3.2 Human Impacts on Ecosystems

Figure 3.9 How do you think the changes to Vancouver might have affected the plants and animals that lived in this ecosystem?

When you look out the window of your home or classroom, are you looking at the same scene that you might have seen 100 years ago? You probably are not. As British Columbia's population began to increase, people cleared land for homes and farms. Eventually some of these settlements grew into the cities and towns we know today. Trees were cut for fuel and buildings. Roads were built and eventually paved. In some places, trees and grasslands were ploughed under to create farmland or orchards. Human societies affect the environment around them in an effort to meet their needs. What types of changes to the environment can you see in the two pictures in Figure 3.9?

Making Choices

Many people seem so far removed from nature that they forget how dependent they are on the environment. Most people buy food from the grocery store and spend most of their time in houses or other buildings. People put trash out at the curb for collection and never think about where it is going. However, everything people use for food and shelter comes from natural resources that Earth provides. You and everyone you know not only *depend* on nature but are *part* of nature.

Natural resources are objects found in nature, such as trees, water, oil, fish, and minerals, which people use to meet their basic needs. Resources that can be replaced are called **renewable resources**. For example, new trees grow, fish have offspring, and food crops grow from seed every year. Resources that cannot be replaced are called **non-renewable resources**. These include fossil fuels such as coal and oil.

DidYouKnow?

Did you know that leaving the lights on in highrise buildings can result in the deaths of thousands of songbirds? At night, the birds are attracted to the lights of buildings and crash into the glass. Now some building owners voluntarily turn off the lights when there are large numbers of birds in the area.

Chapter 3 Ecosystems and People • MHR 75

INTERNET CONNECT

www.mcgrawhill.ca/links/BCscience7

To find out more about ecological footprints, visit the web site above. Click on **Web Links** to find out where to go next.

People in North American and a few other countries use far more than their share of Earth's natural resources. Most of these societies do not live in a sustainable manner. **Sustainability** means that the resources of nature are being renewed or replaced at least as quickly as they are used. In addition, it means that all wastes can be absorbed or recycled without harming the environment. Today, people have many concerns over loss of Earth's resources. One way to determine how much of an impact you have on the environment is to determine your ecological footprint. An ecological footprint is the total area of land and water needed to supply all of the materials and energy that you use. As well, it must absorb all of the waste that you produce.

Habitat Loss

Figure 3.10 The western rattlesnake is losing habitat due to expanding towns and other developments.

When people drain water from a marsh or clear trees from a forest, the environment loses habitats. When habitats are gone, organisms that depend on these habitats must find another habitat to meet their basic needs. When too much habitat disappears, species have nowhere else to go and often disappear. Loss of habitat is the biggest threat facing living organisms today. The populations of western rattlesnakes (see Figure 3.10), for example, are threatened in British Columbia due to loss of habitat. Snakes live in the dry interior of southern British Columbia and have lost much of their natural habitat to housing subdivisions, farms, highways, and other developments. Many rattlesnakes are also killed on the highway or by people who are worried the snakes will harm people or animals.

Sometimes people affect habitats when roads or other large projects expand into them. This separates one part of an animal's habitat from another part or breaks it up into smaller pieces. In some cases, the habitat may be reduced to an "island" of land surrounded by development. When one part of a habitat is separated from another, it is called **habitat fragmentation.**

Introduced Species

How does an introduced species differ from a native species?

Sometimes people, either on purpose or by accident, bring a new species into an ecosystem. These new species are called **introduced species** (also called exotic or alien species). Introduced species can cause problems for **native species,** the organisms that naturally occur in that ecosystem. Introduced species occur naturally in another part of the world where their populations are controlled, or limited, by predators and other natural factors there. If introduced species are able to survive and reproduce better than naturally occurring species, they can crowd out one or more native species.

The plant in Figure 3.11 is an introduced species to North America. Only three seeds of Scotch broom were planted on Vancouver Island in 1850! Today this plant is widespread. It has taken over the habitats of many native plants in southwest British Columbia. In the next Find Out Activity you will discover more about introduced species in British Columbia.

Figure 3.11 The introduced species, Scotch broom, now grows over most of southern Vancouver Island and parts of southwest British Columbia. It started from just three seeds that sprouted on Vancouver Island, near Sooke.

Find Out ACTIVITY 3-E

Alien Invaders

Many animals and plants have been introduced to British Columbia. Some quickly die out and make no impact. Others, however, are very destructive to native species and ecosystems. How did some of the invaders come to British Columbia? What impact are they having here?

Materials
art materials of your choice

What to Do

1. Choose a British Columbia introduced species to investigate. You may choose from the following list or select another species: common dandelion, ring-necked pheasant, brown trout, black slug, European starling, cowbird, knapweed, bullfrog, purple loosestrife, Scotch broom, European earwig, tent caterpillar, sowbug, gray squirrel, Asian long-horn beetle, English ivy, fallow deer, thistle, gorse, Norway rat, crested mynah, Asian or European gypsy moth.

2. Research your species using the library or the Internet. Try to find answers to the following questions.

    (a) The species is native to what country or region?

    (b) When did the species come to British Columbia (or Canada or North America)?

    (c) Where in North America did the species first arrive?

    (d) What impact, if any, did the species have on natural ecosystems in British Columbia?

    (e) What are some of your ideas for preventing the spread of this introduced species?

    (f) What are people doing to stop the spread of this introduced species?

What Did You Find Out?

Create a cartoon strip, story, short play, or other type of presentation to summarize your answers to these questions.

Air Pollution

Ecosystems can be damaged by air, water, and land pollution. The burning of **fossil fuels** — coal, oil, and natural gas — contributes to air pollution. When fossil fuels burn, they release large amounts of carbon dioxide gas as well as gases containing sulfur and nitrogen compounds. While plants require carbon dioxide gas, the increased amounts have resulted in more carbon dioxide gas than the plants on Earth can use. As a result, carbon dioxide gas accumulates in the atmosphere. Carbon dioxide gas as well as a few other gases trap heat from Earth, much like the glass in a greenhouse traps heat as shown in Figure 3.12. This condition is called the **greenhouse effect**. Many scientists believe that the greenhouse effect is causing a warming of Earth's atmosphere.

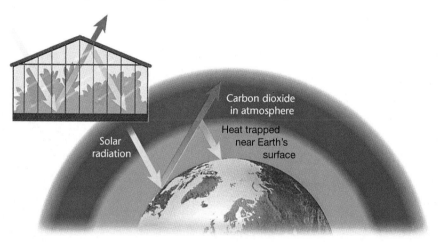

Figure 3.12 This model shows how many scientists believe the greenhouse effect works.

What are the similarities and differences between the terms "global warming" and "greenhouse effect"?

The warming of Earth's atmosphere is called **global warming**. Normal amounts of carbon dioxide help to keep Earth warm enough to support life. Many scientists believe, however, that increased amounts of carbon dioxide and other greenhouse gases are warming Earth at a rapid rate.

During the twentieth century, the average global air temperature increased by 0.6°C. This may not sound like too much, but it is a faster increase in temperature than at any other time during the last 1000 years. Even a small rise in temperature of a few degrees can change the climate and the ability of plants to grow in certain places. This, in turn, could impact food webs and other parts of ecosystems.

Many scientists feel that global warming is one of the most serious environmental issues of our times. Computer models suggest that the average temperature in British Columbia will increase by another 1°C to 4°C during the next 100 years.

Water Pollution

Acid rain occurs when pollutants containing sulfur and nitrogen are found in large amounts in the air. When fossil fuels are burned, gases containing sulfur and nitrogen compounds are released as waste. These pollutants mix with water vapour, making it acidic. When it falls from the atmosphere as precipitation, it damages ecosystems. For example, entire lakes can "die" because the water is too acidic for fish or plants to survive.

Other types of water pollution can harm **aquatic** (water) habitats, such as streams, rivers, lakes, and oceans. A variety of pollutants, including fertilizers, pesticides, sewage, detergents, and oil, can run off the land into water. These pollutants often have a harmful effect on aquatic life. In turn, this would affect food webs in aquatic habitats.

Land Pollution

Garbage, or **solid waste,** is largely made up of plastic, paper, cans, bottles, metals, food, and other items that people discard every day. Each Canadian throws away about 1.5 kg of solid waste every day. That is about 547 kg of waste per person per year! Most of this solid waste is buried in landfills such as the one in Figure 3.13. Landfills take up wildlife habitat and some can contribute to land and water pollution. If we reduce our solid waste by reducing, reusing, and recycling, we can help minimize the impact of our waste on ecosystems.

Figure 3.13
Solid waste is buried in landfills such as this one.

Endangered Species

Have you ever seen a spotted owl or white pelican? If not, the photographs in Figure 3.14 might be your only chance. Both of these birds are endangered species in British Columbia. An **endangered species** is one that is nearly extinct. **Extinct** means that it will no longer exist. The population of an endangered species is so small that unless immediate steps are taken to increase the population, all individuals will die. A **threatened species** is one that could become endangered if the factors limiting its population are not reversed. Habitat loss is the main reason why so many species are at risk, but it's not the only reason. Some biologists think that the problem of introduced species is so serious that it will one day be the main cause of extinction. Pollution, changes in climate, and overharvesting or hunting are other factors that can also result in loss of species.

What is the difference between an endangered species and a threatened species?

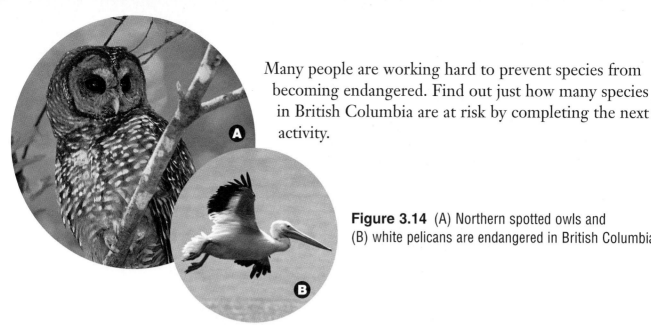

Many people are working hard to prevent species from becoming endangered. Find out just how many species in British Columbia are at risk by completing the next activity.

Figure 3.14 (A) Northern spotted owls and (B) white pelicans are endangered in British Columbia.

Find Out ACTIVITY 3-F

Risky Business

British Columbia designates threatened and endangered species as being "red-listed." What percentage of animals and plants in British Columbia are on the red list? Find out in this activity.

Materials

graph paper

Threatened or Endangered Species (as of 2001)		
Type of organism	Red-listed species	Total number of species
Freshwater fish	24	80
Amphibians	5	19
Reptiles	6	16
Birds	34	465
Terrestrial mammals	11	104
Marine mammals	3	29
Plants	257	2333
Butterflies	12	187
Dragonflies	9	87

(Source: BC Ministry of Sustainable Resource Management, Conservation Data Centre)

Skill POWER

For tips on making bar graphs, turn to SkillPower 5.

What to Do

1. Calculate the percentage of species from each type of organism that are red-listed. Sample Calculation:

$$\frac{24 \text{ red-listed freshwater fish in BC}}{80 \text{ species of freshwater fish in BC}} \times 100\% = 30\%$$

2. Construct a bar graph of the percentages of each type of organism that are red-listed in British Columbia.

What Did You Find Out?

1. Which type of organism has the highest percentage of species that are threatened or endangered?

2. What do you think might have led to some of these species becoming threatened or endangered?

3. What percentage of vertebrates in British Columbia are threatened or endangered? (Vertebrates are animals that have a backbone. They include fish, amphibians, reptiles, birds, and mammals.)

Section 3.2 Summary

There are only a limited number of natural resources on Earth. These must be used sustainably to maintain the health of ecosystems. In North America, natural resources are being used at a rate that may not be sustainable. This has impacts on the environment, such as pollution and global warming. One way to reduce our impact is to use fewer resources and less energy. We can achieve this by making better choices that have less environmental impact.

Other human-caused changes to ecosystems include the following:
- habitat loss
- introduction of alien or exotic species
- overharvesting of resources such as trees or fish
- air, water, and land pollution

All of these impacts can lead to species becoming endangered or threatened. In section 3.3, you will learn about ways to help protect ecosystems and the species that live in them.

Check Your Understanding

1. What are four threats that can lead to the extinction of a species?

2. Describe the relationship between each pair of terms in the list below.
   (a) introduced species and extinction
   (b) habitat loss and endangered species
   (c) greenhouse effect and global warming
   (d) renewable resources and non-renewable resources

3. **Apply** A species of wildflower grows only by the shores of a certain marsh in the British Columbia interior. Recent observations indicate the population is about half the size it was last year. Parts of the marsh are being drained and filled with dirt to make room for a new housing development.
   (a) What could be happening to the wildflower?
   (b) What could be done to stop or slow down the decline in wildflower population in this area?

4. **Thinking Critically** Why should people not release exotic pets, such as snakes, hedgehogs, or tarantulas, into the wild if they decide they do not want to keep them any longer?

5. **Thinking Critically** Sometimes our "needs" conflict with our "wants." What is the difference between a need and a want? How does satisfying all of our "wants" make an impact on the environment?

Key Terms
natural resources
renewable resources
non-renewable resources
sustainability
habitat fragmentation
introduced species
native species
fossil fuels
greenhouse effect
global warming
acid rain
aquatic
solid waste
endangered species
extinct
threatened species

CHAPTER 3

Section 3.3

Conserving and Protecting Ecosystems

Pollution, introduced species, habitat loss, and endangered species are problems that can seem overwhelming. Nevertheless, some people are making positive changes to help protect ecosystems and the organisms that live in them. Recognizing problems and monitoring changes are the first steps. In this section you will learn how some people are putting this information into action.

Preserving Habitats

Creating parks and other protected areas is one way to preserve ecosystems. The government of British Columbia creates provincial parks. The Canadian government creates national parks. In British Columbia there are seven national parks and over 600 provincial parks. Wells Gray Provincial Park, in the eastern part of the province, is shown in Figure 3.15. One goal of the Canadian government is to have national parks in all of the major ecosystems across Canada. Park managers try to balance the protection of habitat while still allowing people to visit and enjoy parts of these wild places.

British Columbia also has 134 **ecological reserves.** These areas are set aside to protect examples of the different habitats in the province as well as rare and endangered plants and animals. Access to most ecological reserves is restricted and people go to them for research or educational purposes only. Most ecological reserves have volunteer wardens that monitor activity in the reserves. High school students from Lester Pearson College near Victoria are the wardens for Race Rocks Ecological Reserve (Figure 3.16). The students work closely with the local Coast Salish people to manage the site and also contribute to research and education projects.

Figure 3.15 Wells Gray Provincial Park is one of the largest protected areas in British Columbia.

What is the main difference between a park and an ecological reserve?

Figure 3.16 Race Rocks Ecological Reserve is near Victoria.

Habitat Restoration and Enhancement

Habitat restoration projects try to restore or improve habitats that have been damaged. For example, groups sometimes "adopt" a stream. They might remove garbage or plant trees along the banks.

Habitat enhancement projects improve existing habitat. One project might be to provide artificial structures such as nest boxes. Nest boxes are platforms that birds use for their nests. The people in Figure 3.17 are putting up bird boxes for purple martins, a bird that nests in British Columbia.

Another project to enhance habitat is to provide safe routes for animals to cross busy highways. In some places in British Columbia, for example, special culverts or fencing has been used to direct frogs, snakes, and other amphibians and reptiles safely across the road. The structures are placed along the routes that animals regularly use to travel between feeding, resting, and breeding sites.

Figure 3.17 Nest boxes provide nesting habitat for many birds in British Columbia.

Figure 3.18 Several groups are working together to restore the health of the Salmon River.

Often, several groups work together to improve damaged habitats. This is the case on the Salmon River (shown in Figure 3.18), which runs into Shuswap Lake near Salmon Arm. Over the years, the river's habitat has been damaged. Several groups have joined together to help restore the health of the Salmon River. People from the Neskonlith band raised 2000 seedlings of the shrub red osier dogwood to plant along the riverbanks. Volunteers from the Salmon Arm Fish and Game Club put up fences and rebuilt stream banks. The Environmental Youth Corp surveyed landowners to get their suggestions for solutions to pollution and erosion problems. It may take a long time to restore the Salmon River, but the people of the Salmon Valley are determined to try.

How would you care for a habitat for which you have the responsibility? Try your ideas in the next investigation.

CONDUCT AN INVESTIGATION 3-G

SKILLCHECK
- Predicting
- Modelling
- Observing
- Controlling Variables

Model Ecosystem in a Bottle

If you could design an ecosystem, what would you include? How would you care for the ecosystem to ensure that it stayed healthy? Try your ideas by making a terrarium, a small garden in a bottle.

Question
What abiotic factors do you need to grow healthy plants?

Safety Precautions

Materials
potting soil
seeds or small plants (**Note:** Do not remove plants from parks or other protected areas.)
masking tape

Apparatus
scissors
a clean 2-L plastic pop bottle

Procedure

1. Predict the abiotic factors that you will need in order to grow healthy plants in a plastic bottle. Write your prediction in your notebook.

2. Use the scissors to carefully cut off the neck of the plastic pop bottle.

3. Add potting soil to the bottom half of the bottle until it measures about 10 cm deep.

4. Plant your seeds or small plants. Place the top of the bottle back onto the bottom half and secure the two-halves together with masking tape.

5. Write a plan that you will follow to care for your terrarium.

6. Choose a good location for your terrarium and monitor it over the next two to three weeks. **Record** your observations of your terrarium during this time.

7. Adjust the plan you developed in step 5 if you find that the requirements for your terrarium change.

Analyze

1. How successful were the plants in your terrarium?

2. What did not work well in your terrarium?

Conclude and Apply

3. How would you adjust the growing conditions for your next terrarium?

4. What changes to the abiotic factors would cause your plants to become "extinct"?

5. If there were other organisms in your terrarium, what biotic factors would cause your plants to become "extinct" in your model excosystem?

6. How is your model like an actual ecosystem? How is it different?

Endangered Species Protection

In British Columbia, over 350 species of plants and animals are threatened or endangered. Therefore, special consideration is needed to protect their habitat. When a species is endangered, biologists might start a **captive breeding program**. In these programs, the animals are mated in captive situations, such as zoos. Biologists then release the offspring back into the wild. By raising the babies in captivity, more young will survive than would in the wild.

The Vancouver Island marmot, shown in Figure 3.19, is endangered. Scientists estimate that there are less than 50 marmots left in the wild. Habitat loss has been the major factor that has led to the marmot's decline. The Vancouver Island Marmot Recovery Team is working to protect marmot habitat. They have also started a captive breeding program to try and increase the population size of this rare animal.

> **INTERNET CONNECT**
>
> www.mcgrawhill.ca/links/BCscience7
>
> To learn more about Vancouver Island marmots and the attempts being made to prevent them from becoming extinct, go to the web site above. Click on **Web Links** to find out where to go next. At the site you can meet some of the researchers, learn about marmot biology and habitat needs, and learn how you can help this endangered species.

Figure 3.19 The Vancouver Island marmot is one of the most endangered species in North America.

Environmental Stewardship

Stewardship is the careful and responsible management of something for which you are responsible. Aboriginal peoples in British Columbia practised stewardship by developing methods of harvesting resources that allowed them to take only specific numbers or amounts of resources. For example, they used fish traps, fish weirs (see Figure 3.20) or fences, and nets to harvest fish from the rivers and shorelines of British Columbia. The extra fish that they caught when they used these methods could be released alive when enough fish had been caught. As well, if any fish were too small or of the wrong species, the people would release them.

Figure 3.20 Wooden weirs like this one used by Aboriginal peoples trap fish in a sustainable way.

Chapter 3 Ecosystems and People • MHR **85**

Did You Know?

The Nisga'a people used fish wheels on the Nass River. The river current spins the wheel's curved panels, which scoop up fish and drop them, unharmed, into pens. Today, the Nisga'a operate fish wheels on both the lower and upper Nass River, allowing biologists to tag fish at the lower wheel and monitor how many of them are caught upstream. This is a form of environmental monitoring.

Figure 3.21 This woman is removing a strip of bark from a cedar tree to be used for weaving. This practice does not kill the tree.

Aboriginal peoples also practised stewardship when harvesting trees. Coastal Aboriginal peoples, for example, used cedar for a wide variety of purposes such as houses and clothing. Planks of wood were wedged from living trees. Bark was stripped in a way that did not harm the trees (see Figure 3.21). Many trees with strips of bark or planks removed can be seen today, alive and healthy, growing in the forest.

Aboriginal peoples of the British Columbia interior harvested white birch bark for baskets, roofs, canoes, and many other items (see Figure 3.22). When harvesting the bark, people made careful cuts so that the inner bark, which protected the birch's living tissues, was left on the tree. This method ensured that the tree would continue to grow.

READING check

What is an example of good environmental stewardship?

Figure 3.22 Aboriginal peoples learned how to harvest birch bark without killing the tree.

Today, responsible stewardship of natural resources is more important than ever. The total human population is so large that just meeting basic needs requires more resources than were needed in the past. One method that some people are using is **selective harvesting** (taking only some of the resources). For example, loggers harvest only certain trees or smaller stands of trees at any one time. There are specific hunting and fishing seasons. By acting as good stewards of the environment, we can ensure that resources will be available for future generations.

Educating People About Wildlife

Fortunately for everyone, a large part of British Columbia is still in its natural state. Sometimes, though, the needs of wildlife conflict with the needs of people. For example, when settlers first came to North America, badgers (Figure 3.23) were seen as a nuisance. People shot, trapped, and poisoned them. People did not understand how badgers benefit the environment. The burrows they dig help air and water penetrate into the soil. As well, badgers prey on rodents, which eat grain crops. Badgers are now endangered in British Columbia. One of the important parts of the plan to help their populations recover is to educate people about them.

> **Pause & Reflect**
>
> Sometimes people value things only if they are useful to them. For example, some people wonder, "What good is an animal such as a slug, or a badger, or a wolf?" Should you value only those organisms that have an obvious value to you? Write your thoughts in your science notebook.

Figure 3.23 Badgers are an endangered species in British Columbia. Some people are working to educate others about the importance of this species.

Natural history clubs, nature centres, zoos, and aquariums all educate people about the species and wild places in British Columbia. A "Living with Wildlife" program in Revelstoke is educating people about how to live safely around bears. They encourage people to keep all food, garbage, compost, and extra fruit on fruit trees, away from the bears. If the bears have no reason to come into town, they will probably stay in the wild. Once a bear becomes accustomed to human food, it will often lose its fear of people. Sadly, the bear usually ends up being shot.

One of the best ways for people to learn about ecosystems and how to care for them is to spend time outdoors. Hiking, camping, canoeing, and biking are ways to enjoy and learn about the natural world. The student in Figure 3.24 is taking part in a culture and science camp on the Koeye River near Bella Bella. Near an ancient village, children learn more about their traditional lands, waters, culture, and values.

Even if you cannot go out to wilderness areas, you can still learn about ecosystems by observing your neighbourhood. Watch the birds, take a walk in a park, or look at the organisms wriggling in a ditch! The more you understand about the natural world, the better you can take care of all living things.

Section 3.3 Summary

There are many ways in which people are working to conserve and protect ecosystems. Some of these ways include:

- protecting habitats in national and provincial parks as well as in ecological reserves
- becoming involved in habitat restoration or enhancement projects
- working to help threatened and endangered species with habitat protection and captive breeding projects
- practising environmental stewardship
- educating the public about the needs of species and ecosystems

Figure 3.24 This Heiltsuk youth is at an outdoor camp in the Koeye River Valley on B.C.'s central coast.

Key Terms
ecological reserves
habitat restoration projects
habitat enhancement projects
captive breeding program
stewardship
selective harvesting

Check Your Understanding

1. What are three ways that people can restore or enhance habitat?

2. (a) Give an example of selective harvesting.
   (b) Give an example of stewardship.

3. What is one way that you could be involved in a project that helps a species or habitat?

4. **Apply** Design a brochure to educate people about a plant or animal of your choice.

5. **Thinking Critically** Should humans simply leave all natural ecosystems exactly as they are, or is it acceptable to make changes? What kinds of changes are acceptable? Use these questions as the basis for an article or letter to a local or regional newspaper.

CHAPTER at a glance

Now that you have completed this chapter, try to do the following. If you cannot, go back to the sections indicated in brackets after each part.

(a) Describe the meaning of traditional ecological knowledge. (3.1)

(b) Give two examples of ways in which scientists can monitor ecosystems. (3.1)

(c) Describe three ways in which ecologists can survey species. (3.1)

(d) Explain ways in which human activities can impact ecosystems. (3.2)

(e) Give an example of how an introduced species can impact an ecosystem. (3.2)

(f) Give an example of a renewable resource and of a non-renewable resource. (3.2)

(g) Use a greenhouse as a model to explain the process of global warming. (3.2)

(h) Describe human actions that can help conserve and protect ecosystems. (3.3)

(i) Explain what is meant by responsible stewardship. (3.3)

(j) Give an example of a habitat enhancement or restoration project. (3.3)

(k) Explain the role of education in conserving and protecting ecosystems. (3.3)

Prepare Your Own Summary

Summarize this chapter by doing one of the following. Use a graphic organizer (such as a concept map), create a poster, or write a summary to include the key chapter ideas. Here are a few ideas to use as a guide:

- Make a three-part presentation showing how humans learn about ecosystems, how humans can harm ecosystems, and how humans can help ecosystems.
- Write a plan with your ideas for helping an endangered species. Include background information explaining why the species is endangered, how you would monitor this species, and how you could help it.
- Create a talk, skit, game, or other way of teaching a group of young people about how the choices we make can affect the environment.

CHAPTER 3 Review

Key Terms

traditional ecological knowledge
ecosystem monitoring
long-term monitoring
baseline data
permanent plots
annual surveys
environmental impact assessment
natural resources
renewable resources
non-renewable resources
sustainability
habitat fragmentation
introduced species
native species
fossil fuels
greenhouse effect
global warming
aquatic
acid rain
solid waste
endangered species
extinct
threatened species
ecological reserves
habitat restoration projects
habitat enhancement projects
captive breeding program
stewardship
selective harvesting

Reviewing Key Terms

If you need to review, the section numbers show you where these terms were introduced. Use the clues below to complete this crossword puzzle. Do not write in your textbook.

Across

1. When two parts of an animal's habitat are separated, it is called habitat _____. (3.2)

4. The information gathered at the beginning of a monitoring project is _____ data. (3.1)

5. When a species is almost extinct, it is _____. (3.2)

8. Responsible use of resources and care for ecosystems is called _____. (3.2)

9. Coal, oil, and natural gas are called _____ fuels. (3.2)

Down

2. A species that is _____ can become endangered if the causes for it being at risk are not reversed. (3.2)

3. We use _____ to supply our basic needs. (3.2)

6. A species that no longer exists is _____. (3.2)

7. A species that is not native to an ecosystem is called a(n) _____ species. (3.2)

8. When people use resources in a way that they are not used up, they are practising _____. (3.3)

Understanding Key Ideas

Section numbers are provided if you need to review.

11. What are two reasons that might cause a species to become threatened or endangered? (3.2)

12. How can introduced species affect populations of native species? (3.2)

13. How would a road through a wilderness area affect the organisms? (3.2)

14. How can traditional ecological knowledge be used to help make decisions over land and resource use? (3.1)

15. What are the differences between extinct, endangered, and threatened species? (3.2)

16. List one way you could help to conserve each of the following: (a) animals, (b) energy, (c) trees, (d) water. (3.3)

17. Why is it important to monitor changes in ecosystems over a long period of time? (3.1)

Developing Skills

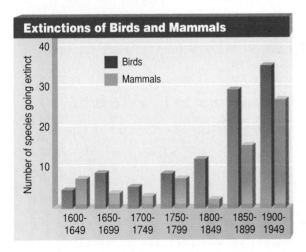

The graph above shows extinction rates for species of birds and mammals since 1600. Use the graph to answer questions 18-21.

18. In what interval did the most extinctions of mammalian species occur?
    (a) 1600–1649   (c) 1850–1899
    (b) 1650–1699   (d) 1900–1949

19. Approximately how many species of birds became extinct in the interval 1650–1699?
    (a) 7   (c) 15
    (b) 10   (d) 20

20. Approximately how many species of birds became extinct during the interval from 1900–1949?
    (a) 37   (c) 115
    (b) 30   (d) 300

21. How many species of birds and how many species of mammals do you predict became extinct in the years 1959–1999?

Problem Solving

22. Create a monitoring plan for a bird that lives in your area.

23. Create a plan for reducing your impact on the environment.

Critical Thinking

24. Why should British Columbians who lives in cities be concerned about the loss of habitats in a forest in a remote part of the province that is uninhabited by people? Explain your answer.

25. Some people think that parks are a waste of land and money because few people visit them. How would you respond to this statement?

Pause & Reflect

Go back to the beginning of this chapter on page 62, and check your original answers to the Getting Ready questions. How has your thinking changed? How would you answer those questions now that you have investigated the topics in this chapter?

UNIT 1
Ask an Elder

Trudy Frank

In many Aboriginal cultures, certain Elders teach young people about the natural world. Elders also pass on important knowledge about plants, animals and weather patterns. Some Elders, for example, have specialized knowledge about how plants can be used for food, medicine, and cultural traditions. Trudy Frank is an Elder from the Ahousaht First Nation, one of the Nuu-chah-nulth First Nations. She was born at Kelthsmaht, an island near Tofino, and has eight children and many grandchildren and great-grandchildren.

Q. What is your Nuu-chah-nulth name?

A. My grandfather named me Hai-u-pitl'aht. It means ten of something, like a ten-stringed instrument. I like to think of it that way. We are all instruments of something.

Q. Which of your grandparents or Elders were particularly important to you?

A. When I was young, I spent a lot of time with my father's parents. My grandfather's name was Keesta, and my grandmother's name was Mary. We used to call her Queen Mary because she was the head chief of her tribe. That was unusual for a woman. My grandparents taught me about fishing and trapping, and about preserving food. You have to know which fish you can smoke or dry to put away for the winter, because some fish keep better than others. We didn't eat anything from the store in those days. My grandmother was also a very skilled basket maker.

Q. Are there different kinds of Elders in your community?

A. Yes. Some, like my grandparents, know about gathering and preserving food. Others are knowledgeable about our culture. We depend on them when we are preparing special events. They hold the memory of songs and dances. We also have medicine women and other kinds of teachers.

Q. What does *hishuk ish ts'awalk* mean?

A. It means we are all one. The people, the animals, the plants, even the rocks and water are all one. Everything is connected. This is how we survive. The name also refers to the connections among our communities. There are 14 Nuu-chah-nulth tribes, but we are all one people.

These connections are an important part of our culture. For example, many of our traditional medicines are parts of plants that you first have to chew. You mix the plant with your own saliva, so the medicine comes both from the plant and from you. Then you also have to believe in the medicine. The medicine might not be any good without this element of respect.

Q. What did you and your community do in the past to live in harmony with plants and animals?

A. Our Hereditary Chiefs, our *Ha-huulthi*, manage our territories. There are three Hereditary Chiefs in Ahousaht. However, we have amalgamated with other tribes so we now have many Chiefs. Each has a responsibility, along with his tribe, to look after a specific area.

We have always relied on our spirituality in caring for our territories. This spirituality says that everything comes from the Creator. I was taught that before you went out to gather anything you must always take the time to prepare. For instance, if you are going to gather plants for medicine, you should take the time to meditate first. During this time you ask for guidance about how to find what you need. You ask how to gather it, how to care for the plant, and how to leave the place in its original healthy state. You must also always thank the Creator for providing for the community.

Traditionally, people used natural pools in the rainforest or the ocean as places for this preparation. The bark of a young yew tree lathers, just like soap, and it has a very nice odour. So people would bathe and meditate, sometimes for days.

Q. Are the young people in your community doing these things?

A. Yes, the young people are just starting to get back to looking after their own areas. Traditional activities like spirit quests are coming back again.

Q. How do you pass on your knowledge to young people in your community?

A. I work as an Elder at a family development centre. That is where I pass on a lot of my knowledge. I have also been involved in some programs to share traditional knowledge. We have a project about medicinal plants that describes what they are, where to find them, and how to prepare them. We also have a project about traditional foods.

It used to be that young people were expected to go to the Elders and ask questions. For a long time young people were not doing that. Now they are starting to ask questions again. It is nice to see. It is very promising.

EXPLORING Further

Using Natural Resources

People from different regions develop skills for using available resources.

Select two different regions of the province. Research the main biotic and abiotic elements of each region. Then imagine how these features might affect people. For example, how would the type of landscape influence how people travel? What would be the sources of food and building materials? For each region, research some examples of Aboriginal uses of the natural resources. Then write a brief report discussing the skills of each Aboriginal community.

The Secwepemc peoples of south-central British Columbia developed techniques for making baskets from bark and roots that can carry water.

UNIT 1
Ask a Biologist

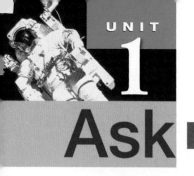

Hilda Ching

Have you ever heard of parasites? They are small organisms that live outside or inside other animals and obtain their food from these animal hosts. Hilda Ching can tell you all about parasites. She is a parasitologist. She studies parasitic worms and the way they use food chains to get to their hosts.

Q. How did you become interested in parasites?

A. Growing up in Hawaii, I was surrounded by coral reefs, tropical fish, and colourful plants. They inspired my interest in nature. Then when I was in Grade 11, we learned about a parasite called the cattle liver fluke. I was fascinated that what looked like a tiny speck on a watercress leaf could develop into a huge worm inside the liver of a cow or a person. I knew, then, that I wanted to study parasites.

Q. A tiny speck can grow into a huge worm? How does this happen?

A. Parasites can take dramatically different forms at different stages of their lives. It's a bit like a caterpillar turning into a butterfly. In order to develop, however, some parasites have to "hitchhike" their way up the food chain. Unless they find their way into the right host animal, they can't develop to the next stage of life.

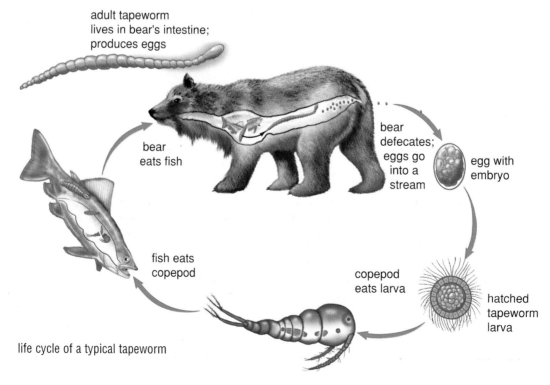

life cycle of a typical tapeworm

94 MHR • Unit 1 Ecosystems

Q. How can something so tiny find its way into the right animal? Are they intelligent?

A. They are, in a way, by being at the right place at the right time. It's a natural process. Recently I've been studying a type of tapeworm, called *Diphyllobothrium dendriticum* [die-FI-lo-BAW-three-um den-DRI-ti-COME], here in British Columbia. Its life cycle begins in fresh water when its eggs hatch into larvae and tiny crustaceans called copepods [KOPE-uh-pods] eat them.

The copepods, with the tapeworm larvae inside them, are eaten by salmon or trout. Inside a fish, the larvae grow a little more, then remain in the fish for the rest of its life, up to about four years. When the infected fish gets eaten by a larger animal, say a bear or a gull, the larvae develop into mature tapeworms in their host's intestines.

Q. That's quite a journey!

A. The journey's not over yet! The tapeworms grow as much as 30 cm a week inside their final host. As they grow they release millions of eggs. The bear's or the gull's feces, with the eggs inside them, get washed into the river. In the water, the eggs hatch into larvae which get eaten by copepods, and the cycle begins again.

Q. People eat salmon and trout, too. Can these tapeworms affect us?

A. Yes. This particular tapeworm can infect people, but the salmon and trout we buy are pretty clean of tapeworms. When you catch a salmon or trout yourself, make sure that you clean it promptly and cook it thoroughly to kill any parasites that might be lurking. If you want to eat raw salmon, such as in sushi or sashimi, make sure that the fish has been previously frozen.

Q. Who benefits from the work that you do?

A. Doctors often use the information that parasitologists provide to find out what parasite might be causing a health problem in a patient. I am often asked to identify parasites from fish or other animals. Also, companies and the government ask me to assess changes in the environment that may cause diseases in fish. One recent concern that we have is the introduction of new, possibly harmful, parasites into our marine and fresh-water environment.

The work I did, studying *Diphyllobothrium dendriticum* with colleagues in other parts of the world, showed that it is a common tapeworm in North America, Europe, and Japan. Like many other kinds of zoologists, parasitologists try to learn more about the creatures with which we share the planet Earth.

EXPLORING Further

Parasites and You

Although most people in Canada are fairly safe from parasites, there are still some, such as tapeworms, that pose a risk. There are others, such as heartworms and roundworms, that are a serious threat to our pets. Contact a family doctor or a veterinarian in your area, and get some information about these parasites.

Consider asking questions like the ones below:
- How does a person or the pet get a parasite?
- How can people avoid contracting parasites?
- How do you treat patients to get rid of a parasite?

UNIT 1

Project

Hands-on Habitat

SKILLCHECK
- Measuring
- Observing
- Interpreting Data
- Communicating

You can enhance or improve habitats in your community in many ways. You could plant native plant gardens or plants that birds and butterflies prefer for food or nests, monitor bird or butterfly populations, or clean up damaged habitats. These are some of the ways that you could contribute to your community. People are helping habitats in many ways, large and small. In this project, you will design and carry out your own plan for enhancing your community.

Challenge

In a group, create a plan for monitoring and improving some natural habitat in your schoolyard or elsewhere in your community.

Materials
2 sheets of poster board
art supplies
other supplies depending on student projects

Design Criteria

A. Choose a project that you could carry out that would enhance a habitat in your schoolyard or community. For example, you could build and install nesting boxes for birds that nest in your area. You might design and plant a garden of native plants in your schoolyard. You could also help clean up a local pond or stream. Ensure that you choose a project that is realistic for your group to carry out.

B. Decide on a way to gather baseline (starting) data as well as a way to monitor your habitat after you have carried out your plan.

C. At the top of one piece of poster board, put the title "BEFORE." Create a presentation that shows how on your habitat looked before you enhance it. Include labelled drawings, photographs, and/or baseline data on the BEFORE board.

D. At the top of a second piece of poster board, put the title "AFTER." Create a presentation that shows your habitat after you enhance it. Include labelled drawings, photographs, and/or monitoring data on this AFTER board.

Plan and Construct *Group Work*

1. Work in a group of 4 to 6 students. Decide on the sort of project you would like to plan. You could use ideas that you learned about in this unit. You might choose to do research on the Internet. A local agency might provide ideas and help you with your plans.

2. To create a plan for your project, list:
   - the goals of your project
   - how you will collect baseline data
   - how you will monitor your project
   - the materials you need

3. Have your teacher approve your plan. Adjust your plan if necessary.

4. Create a list of jobs you need to do to complete the project. Decide which group member will be responsible for each job.

5. Create your BEFORE display of the habitat. You can use photographs, labelled diagrams, words, graphs, charts and any other methods of your choice. You may also want to conduct interviews with knowledgeable people to include as part of your project. For example, learn about what this place might have looked like 20 years ago from a person who knows the area well. You could interview an avid birdwatcher to learn about the types of birds that lived in this habitat.

6. Gather your baseline data.

7. Carry out your project.

8. Carry out your monitoring.

9. Create your AFTER display of the habitat. You can use photographs, labelled diagrams, words, graphs, charts and any other methods of your choice.

Evaluate

1. Explain how your project improved a habitat.

2. What would you change about your project if you were to repeat it?

UNIT 2

Chemistry

Rafting is fun—as long as the raft does not flip over! If these rafters fall into the river, their life jackets will keep them floating. What if one of the rafters has coins in a pocket, and the coins drop into the water? The coins will sink. Why do some objects float while others do not? The answer depends partly on the type of matter that the objects are made of.

People, rafts, coins, and life jackets are made up of matter. Mountains, rivers, and air are made up of matter, too. In fact, there are millions of different types of matter. Some matter is solid, like rocks and rafts. Some matter is liquid, like water. Other matter is usually gas, like air. Why are there so many types of matter? What happens when one type of matter changes to become another type of matter? For example, liquid water changes to solid ice when the temperature falls below freezing. A log of wood changes to charcoal, ashes, and gases when it is put on a campfire. Are the changes permanent? Why do the changes occur? Chemisty is the science that tries to answer these questions

In this unit, you will explore these and other questions by observing and investigating matter. By the end of this unit, you will have a better understanding of the matter that is all around you.

Unit Contents

Chapter 4
Characteristics and Properties of Matter   **100**

Chapter 5
Classifying Matter   **130**

Chapter 6
Exploring Solutions   **158**

Getting Ready...

- Are water, ice, and water vapour the same type of matter?
- How is a mixture of sugar and water different from a mixture of sand and water?
- How does pure water differ from tap water?

CHAPTER 4 Characteristics and

Getting Ready...

- What words can you use to describe an object?
- In what ways can a material change? What does it mean when you say that a material is changing?
- How can you explain what happens when solid ice melts to form liquid water?

What materials are used to make a bicycle? Why are these materials good choices? What properties do these materials have?

Early bicycles were made entirely of wood—including the wheels. Later models had wheels made of iron. Not surprisingly, the popular name for an early bicycle was "the boneshaker." It must have been very uncomfortable for the cyclist. Wood and iron are hard, rigid materials. Today's cyclists ride on rubber tires that are filled with air. Rubber is soft and flexible. It bends and bounces back into shape easily. The air in the tires allows the tires to change shape with bumps in the road, so the ride is smoother.

Hardness is a property (a feature) of some kinds of metal. Flexibility (the ability to bend easily) is a property of rubber. You rely on properties of matter, such as hardness and flexibility. Properties determine how you use different materials. As well, they help you to describe and identify unknown materials. When you say that water is "wet" and gold is "shiny," you are describing their properties. Other liquids, such as milk and apple juice, are also wet. Other materials, such as silver and glass, are also shiny. You may need a list of properties, or perhaps one very specific property, to identify a material.

In this chapter, you will make careful observations of matter and its properties. As well, you will think about how matter changes, and how these changes and the properties of matter can be explained.

Properties of Matter

What You Will Learn

In this chapter, you will learn

- what properties you can use to describe matter
- how physical changes of matter are different from chemical changes of matter
- how to measure mass and volume, and calculate density
- how the particle model of matter explains the states and changes of matter

Why It Is Important

- Properties of matter determine how different materials can be used.
- Describing and measuring matter enables you to communicate your observations to others.
- You and all the materials and living things in the world are made of matter.

Skills You Will Use

In this chapter, you will

- observe and describe different samples of solids, liquids, and gases
- measure mass and volume
- investigate properties and changes of matter
- research the way materials are used, based on their properties

Why is wood or iron not a good choice for making bicycle wheels?

Starting Point ACTIVITY 4-A

Mystery Materials

How well can you use your experiences to infer the identity of different materials?

What You Need

5 film containers, each filled with a "mystery material"

What to Do

1. Make a table of observations like the one below. Leave space to record your observations of five film containers. Give your table a title.

Container	Ranking from lightest (1) to heaviest (5)	Identity of material in container	
		Your inference	Actual identity

2. Do not open the containers. Work with a partner. Pick up the film containers, one by one. **Observe** how heavy or light each container feels. Rank the containers from lightest (1) to heaviest (5). **Record** your ranking in your table.

3. Use your experiences of everyday materials to **infer** what might be inside each container. **Record** each of your inferences.

4. Check with your teacher to find out the identity of each material. **Record** this in your table.

What Did You Find Out?

1. Which materials did you infer correctly or closely? Did any of the materials surprise you? Why or why not?

CHAPTER 4

Section 4.1 Describing Matter

READING check

Name ten examples of matter.

Here are some things you probably see around you almost every day: people, clothes, desks, chairs, walls, and books. What would be different if you stood outside? If you were outside, you might see trees, buildings, pavement, cars, snow, rain, mountains, the Sun, and clouds. You could name dozens of different things that you might see. Perhaps you could name hundreds.

Everything that you can see, as well as touch, smell, and taste, is made of matter. **Matter** makes up every living thing and every material object. **Chemistry** is the study of matter and its changes. Use Figure 4.1 to help you build a clearer picture for yourself about what matter is. Later in this chapter, you will see how scientists clarify their ideas about what matter is.

Figure 4.1 Which words on the sign describe matter? Which words do *not* describe matter? What is the difference?

Figure 4.2 Is one of these liquids water? Maybe both are. Perhaps neither is. How can you find out?

Using Your Senses to Observe Matter

Study Figure 4.2, and think about these observations of the two liquids in the beakers:

- Both liquids are colourless.
- Both liquids are transparent. (This means that they both let light travel through them.)
- Both liquids are odourless.

Based on these observations of the two liquids, do you have enough information to say that they are the same? No. The next paragraph explains why.

Your senses let you observe matter in the world around you. In science class, you may find that you rely mostly on your sense of sight. Often, you may use your sense of touch. In some cases, you may use your sense of smell. Figure 4.3 shows one way that you may use your sense of hearing. You will *never* use your sense of taste, however, to observe matter in the classroom, unless your teacher tells you it is safe to do so.

Why should you never taste matter in your classroom? In science class, you may work with samples of matter that seem safe to taste but are not. For example, one of the liquids in Figure 4.2 is water. It is safe to drink. The other liquid contains a chemical called hydrogen peroxide. People use hydrogen peroxide [HY-druh-jen purr-OX-eyed] to clean and disinfect cuts and wounds. It can be poisonous if you drink it.

Figure 4.3 How can your sense of hearing help you observe what is happening in this photograph?

You can use your senses to observe matter safely in the science classroom. Figures 4.4A and 4.4B show several methods that you may have used already. Remember, though, that if you need to observe an unknown material—in class or outside school—always use your common sense, too.

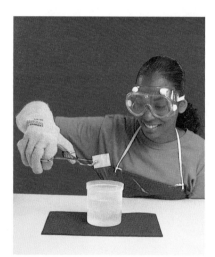

Figure 4.4A How is this student protecting her senses of sight and touch?

Figure 4.4B This student is using the safe method to smell materials. What is this method? How else is he protecting his senses?

Describing What You Observe

Think about the following observations:

- You observe that your friend's shirt is inside out.
- You smell dinner burning in the oven.
- You slip on a patch of ice that is covered by snow.
- You eat the best homemade pie that you have ever tasted.
- You hear the cupboard door squeaking each time it opens.

Why is it important to use your common sense when you observe an unknown material?

To share these observations, or any other observations you make with your senses, you must communicate them. Table 4.1 lists some words that you can use to describe the things you observe with your senses of sight, touch, and smell. In the next activity, you will use sense-related words to communicate your observations of common objects.

Table 4.1 Some Words to Describe Matter

Sense used to observe matter	Examples of words that describe matter	
sight	orange	transparent
	shiny	fast-moving
	bubbly	grainy
	furry	oval
touch	warm	cottony
	smooth	greasy
	rough	feathery
	jagged	slippery
smell	sweet	putrid
	odourless	acrid
	overpowering	cheesy
	floral	cedar-scented

Find Out ACTIVITY 4-B

Describing Matter

In this activity, you will observe objects closely to describe them as fully as you can.

What You Need

selection of objects, such as those in the photograph

What to Do Group Work

1. Decide how you want to record your descriptions of the objects you will be observing. For example, you could design a table, make index cards, or set up a computer database.

2. Choose one object. Use your senses of sight, touch, and smell to **observe** the object closely.

3. Describe the object. Use as many words as you can to describe it. **Record** your description in the format you chose in step 1.

4. Repeat steps 2 and 3 for at least five other objects.

5. Meet with the members of another group. Read each of your descriptions, and have the other group **infer** the identity of the object you are describing. Then listen as the other group reads its descriptions to your group.

What Did You Find Out?

1. How successful was the other group at inferring the identities of your objects?

2. How successful was your group?

3. Did you find that some sense-related words led to more successful inferences than other sense-related words? Explain what you found out.

Extension

4. Could you describe one rock with enough detail to locate it after it was mixed up with other rocks? Collect some rocks and find out.

Did you notice the word "putrid" in Table 4.1? There's a chemical that makes things smell putrid: putrescine! The smell of rotting meat is caused by putrescine. Strangely, your cells need putrescine to reproduce properly. Fruit and cheese are good sources of putrescine.

Chapter 4 Characteristics and Properties of Matter • MHR 105

Using Properties to Describe Matter

If you say that window glass is solid, brittle, and transparent, you are describing some of its characteristics or features. Scientists use the word **property** when they want to talk about any characteristic or feature that helps to describe matter. When you use your senses to observe a material, you are observing its properties.

Each kind of matter has its own set of properties. This means that you can use properties to help you identify a specific type of matter. For example, Figure 4.5 shows two materials that share *some* of the same properties. One is gold. The other is pyrite, which is also known as "fool's gold," because it looks so much like gold. Both materials are solid, heavy, and yellow-coloured. Gold, however, melts at a temperature of 1063°C, and pyrite does not.

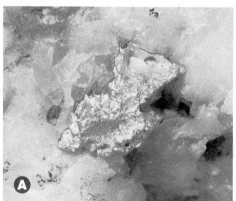 

Figure 4.5 The property of melting temperature can help you tell gold (A) from pyrite (B). What other properties could help you decide which of the two solids is gold?

Pause & Reflect

Sometimes, qualitative properties of matter are called *observable* properties. Explain why the term "observable property" applies to colour, odour, and shape. Could the term apply to any other properties? Give reasons to support your answer.

You can measure the temperature at which a material melts. You can measure other properties of materials, too. Any property of matter that you can measure or describe with a numerical value is called a **quantitative property**. You will learn some measurable, quantitative properties for describing matter later in this section, and in section 4.2.

Any property that you can observe directly with your senses is called a **qualitative property**. Usually, you describe qualitative properties with words. Colour, odour, and shape are examples of qualitative properties, because you describe them using words. State of matter is an important qualitative property of matter. You will explore this property next.

State of Matter: A Qualitative Property

All matter on Earth normally exists in one of three forms. These forms are called states of matter. The three **states of matter** are solid, liquid, and gas.

A **solid** is a form of matter that has a fixed shape. This means that its shape stays the same (unless the solid is acted on forcefully with a tool such as a hammer). Gold is a solid at room temperature. So are rubber and wood. Water is a solid (ice) at temperatures below 0°C.

A **liquid** is a fluid (flowing) form of matter. A liquid has no shape of its own. It takes the shape of its container and forms a surface within its container. Engine oil is a liquid at room temperature. So are rain (liquid water), milk, and juice.

A **gas** is also a fluid form of matter. A gas has no shape of its own. It takes the shape of its container and fills its container completely. A gas, therefore, does not form a surface in its container. Air is a gas at room temperature. So are water vapour (gaseous water), carbon dioxide, and natural gas. Figure 4.6 summarizes these key ideas about solids, liquids, and gases.

Did You Know?

Many scientists agree that there is a fourth state of matter, called *plasma*. It is like a gas, but it can carry an electric current and respond to magnetic forces. The Sun and other stars are mainly plasma. On Earth, plasma forms when you turn on fluorescent lights and when lightning electrifies the sky.

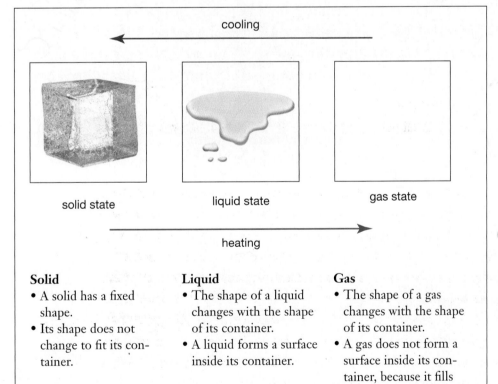

Solid
- A solid has a fixed shape.
- Its shape does not change to fit its container.

Liquid
- The shape of a liquid changes with the shape of its container.
- A liquid forms a surface inside its container.

Gas
- The shape of a gas changes with the shape of its container.
- A gas does not form a surface inside its container, because it fills its container completely.

Figure 4.6 These drawings represent the three common states of matter. The state of a material is a qualitative property.

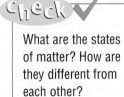

What are the states of matter? How are they different from each other?

Did You Know?

The lowest temperature of air ever recorded in Canada is −62.8°C, at Snag in the Yukon Territory. The highest temperature of air ever recorded in Canada is 45°C, at Midale and Yellow Grass, Saskatchewan.

Quantitative Properties of Matter

There are many quantitative properties of matter. You use some of these properties each day. Others are used mainly by people in specialized fields such as metalworking or materials science. In the remainder of this section, you will learn about a variety of quantitative properties. Afterwards, in section 4.2, you will measure and calculate three of the most common quantitative properties.

Temperature

Temperature measures how hot or cold matter is. Temperature is an example of a quantitative property of matter. It is described with a number that is expressed in units of degrees Celsius (°C). Healthy human body temperature, for example, is 37°C. The temperature of a freezer for food storage is −17°C.

Melting Point and Boiling Point

Solid water (ice) melts to become liquid water at a temperature of 0°C. Liquid water becomes solid water at this same temperature. All materials have a characteristic temperature at which they melt to become a liquid or freeze to become a solid. Scientists call this temperature a material's **melting point**. (You could also call it the material's freezing point.) Similarly, all materials have a characteristic temperature at which they boil to become a gas. Scientists call this temperature a material's **boiling point**. The boiling point of water, for example, is 100°C. Table 4.2 shows the melting points and boiling points of some common materials.

Table 4.2 Melting Points and Boiling Points of Common Materials

Material	Melting point (°C)	Boiling point (°C)
oxygen	−218	−183
mercury	−39	357
water	0	100
lead	328	2602
aluminum	660	2519
table salt	801	1413
silver	962	2162
gold	1064	2856
iron	1535	2861

READING Check

What do the terms "melting point" and "boiling point" mean? Which of these terms is the same as "freezing point"?

Other Quantitative Properties of Matter

People depend on properties to identify matter and to describe it. People also think about properties because they determine how matter can be used. For example, people who make jewelry or weave blankets use the *colour* of their materials to create certain effects. Figures 4.7 to 4.12 show a variety of other quantitative properties of matter.

Figure 4.7 *Viscosity* describes how thick or thin a liquid is. A thicker liquid, such as maple syrup, is more viscous than a thinner liquid, such as water.

Figure 4.8 The *strength* of a material is its ability to resist forces that squeeze or press on it. Strength is a property that is important when designing vehicles, buildings and structures such as bridges.

Figure 4.9 *Elasticity* is the ability of a material to go back to its original shape after it has been stretched or compressed. This property is important for making golf balls, bedsprings, bicycle tires, and floorboards.

Figure 4.10 The property of *hardness* describes how well a material resists being scratched or dented permanently. Diamond is the hardest natural material on Earth. This makes diamond-tipped drills ideal for cutting other hard materials, such as rock and steel.

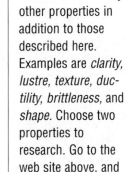

INTERNET CONNECT

www.mcgrawhill.ca/links/BCscience7

Materials have many other properties in addition to those described here. Examples are *clarity, lustre, texture, ductility, brittleness,* and *shape.* Choose two properties to research. Go to the web site above, and click on **Web Links** to find out where to go next. Share what you learn with the class.

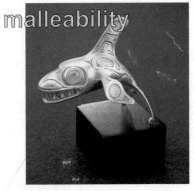

Figure 4.11 *Malleability* is the ability of a material to be hammered, stretched, or rolled in all directions, without breaking or cracking. People have used gold, copper, and silver for centuries to make coins and sculptures because these metals are so malleable.

Figure 4.12 A material that can be stretched or bent into another shape, and can keep this shape without breaking or cracking, has the property of *plasticity.* Basket makers depend on plasticity.

Chapter 4 Characteristics and Properties of Matter • MHR

Section 4.1 Summary

Matter is all around you. You eat, drink, and wear matter. You *are* matter, too. In this section, you learned about some of the properties of matter.

- Matter exists in three states: solid, liquid, and gas.
- Any material can be described by its qualitative properties, such as its state and colour, and by its quantitative properties, such as melting point and hardness.
- Properties determine the ways in which different materials can be used.
- Temperature is a quantitative expression of the hotness or coldness of a material.
- All materials have a characteristic temperature at which they melt and boil.

Key Terms
chemistry
matter
property
quantitative property
qualitative property
states of matter
solid
liquid
gas
temperature
melting point
boiling point

Check Your Understanding

1. How would you define matter in your own words? What are three examples of matter?

2. Use your imagination to do one of the following tasks:
   - Write a story about what you could do if you were a solid or a gas.
   - Write a riddle about a solid, a liquid, or a gas. In your riddle, describe the properties of the material. Ask a friend to guess the material you have described.

3. In what state would each type of matter be at the given temperature? (Refer to Table 4.2. on page 108)
   - (a) oxygen at –50°C
   - (b) aluminum at 800°C
   - (c) gold at 3000°C
   - (d) iron at 1534°C

4. What is the difference between a qualitative property and a quantitative property? Give two examples of each to explain your answer.

5. **Apply** Aluminum is a lightweight but strong material. How do these properties make aluminum a good choice for building racing bicycles?

6. **Apply** Very few materials exist in all three states at everyday temperatures on Earth. Table 4.2 on page 108 lists one of these materials. Which is it? Why do you encounter most other materials in their solid state?

7. **Thinking Critically** What property do gases and liquids share? How do gases and liquids differ from solids? How do gases differ from liquids?

CHAPTER 4

Section 4.2 Measuring Matter

As you can see in Figure 4.13, qualitative properties are not very exact. They depend too much on your opinions about what words such as "small" or "green" mean. Scientists prefer to measure matter so they can use quantitative properties as much as possible to describe matter. Recall that quantitative properties are properties that you can measure and describe using numerical values.

All **matter**—large or small, heavy or light, solid, liquid, or gas—has two things in common: mass and volume. The **mass** of an object is the quantity of matter that makes up the object. The **volume** of an object is the quantity of space that the object takes up. In this section, you will learn about quantitative properties. You will focus especially on mass and volume, and develop a better idea of what these terms mean. You also will explore a property of matter that is related to mass and volume. This property is density.

Figure 4.13 There are limits to how useful qualitative properties can be.

Measuring Mass

A lion is made up of a lot of matter. It has a large mass. A mouse is made up of much less matter than a lion is. A mouse has a smaller mass than a lion does. How can you describe mass using numbers?

You usually describe large masses with kilograms (kg). You usually describe small masses with grams (g). The *kilo-* part of the word "kilogram" means "thousand." So one kilogram is just another way to say one thousand grams (1 kg = 1000 g).

For much smaller masses, you use milligrams (mg). The *milli-* part of the word "milligram" means "thousandth." So one milligram is one thousandth of one gram (1 mg = 0.001 g). Figure 4.14 shows the masses of some common objects.

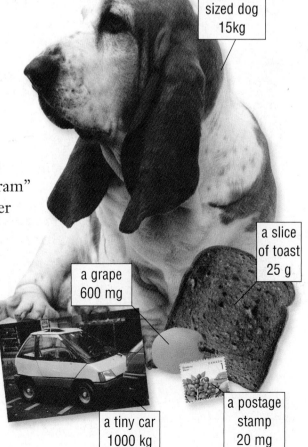

Figure 4.14 What is the total mass of all the matter that is shown here?

Chapter 4 Characteristics and Properties of Matter • MHR **111**

Find Out ACTIVITY 4-C

Practise Measuring Mass

People usually measure mass using a device called a balance. Your teacher will explain how to use the type of balance available in your classroom. It may be one of the types of balances shown in the photographs.

What You Need

balance
samples of regular and irregular solids
beaker or measuring cup
salt
water

equal-arm balance

triple-beam balance (low form)

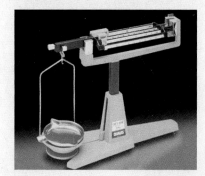
triple-beam balance (high form)

What to Do

1. Examine the solid samples, one at a time. **Estimate** and **record** the mass of each sample.

2. Follow your teacher's directions to **measure** the mass of each sample. **Record** your measurement.

3. **(a)** Place the beaker or measuring cup on the balance. **Measure** and **record** its mass.
   (b) Pour salt into the beaker. **Measure** and **record** the mass of the container and salt together.
   (c) To find the mass of the salt alone, subtract the mass of the container from the combined mass of the container and salt. Record the mass of the salt.

mass of container and salt
− mass of container
= mass of salt

What Did You Find Out?

1. How accurate were your estimates?

2. Which sample had the largest mass? Which sample had the smallest mass? Rank your samples in order from least mass to most mass.

3. How would you measure the mass of a sample of liquid? Describe the method you would use.

Extension

4. How could you measure the mass of a sample of air? Describe the method you would use.

Skill POWER

Refer to SkillPower 3 for tips on measuring mass.

The Volume of Matter

Recall that volume is a measure of the amount of space that a sample of matter takes up. All matter—regardless of its state—has volume. You measure the volume of a solid in cubic units, such as cubic metres (m^3) or cubic centimetres (cm^3). For example, a refrigerator has a volume of about 1 m^3. A sugar cube has a volume of about 1 cm^3.

You usually measure the volume of a liquid or a gas in litres (L) or millilitres (mL). For example, you can buy milk in a 1 L carton. A common medicine dropper holds about 1 mL of liquid.

The cubic units for measuring the volumes of solids are related to the units for measuring the volumes of liquids and gases. One cubic centimetre is the same volume as one millilitre (1 cm^3 = 1 mL).

Measuring Volume

What is the volume of the rectangular solid in Figure 4.15? You can see that the solid is made up of 12 small cubes. Each cube has a volume of one cubic centimetre (1 cm^3). Thus, the solid has a volume of 12 cm^3. What if you did not know that the solid was made up of 12 cubes? You can calculate the volume of any cube or rectangular solid by using this formula:

V = $l \times w \times h$

In this formula, the symbol *V* stands for volume. The symbol *l* stands for length. The symbol *w* stands for width. The symbol *h* stands for height.

To determine the volume of the solid in Figure 4.15, you can multiply its length (2 cm) by its width (3 cm) by its height (2 cm):

V = $l \times w \times h$
= 2 cm × 3 cm × 2 cm
= 12 cm^3

You can take measurements and use a formula to calculate the volume of any regular solid, such as a sugar cube or a cardboard box. How do you measure the volume of a solid that has an irregular shape, such as a stapler or a rock? You use a method called *displacement*. It involves measuring how much water a solid object displaces (pushes aside) when it is submerged in water. You will use this method in the next investigation.

In ancient Greece, a man named Archimedes discovered the method of displacement when he sat down in a full bathtub, as shown in Figure 4.16.

READING check

What units would you use to measure the volume of water in a small glass? What units would you use to measure the volume of a giant boulder? How are cm^3 and mL related?

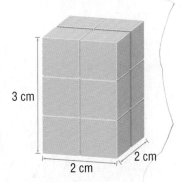

Figure 4.15 You can calculate the volume of a rectangular solid by multiplying its length by its width by its height.

Figure 4.16 When the water spilled out of the tub, Archimedes realized that the volume of spilled water was equal to the volume of his body that was underwater.

CONDUCT AN INVESTIGATION 4-D

SKILLCHECK
- Observing
- Measuring
- Predicting
- Interpreting Data

Practise Measuring Volume

The method you use to measure volume depends on the sample of matter that you are measuring. You must match the method you use to the properties of the sample.

Question
How do you measure the volumes of solids, liquids, and gases?

Safety Precautions

Apparatus
ruler
large tray
graduated cylinder or large measuring cup
cup
shallow dish
bicycle pump

Materials
cardboard box
small metal toy
water
balloon

Procedure

Part 1: Measuring the Volume of a Regular Solid

1 **Measure** and **record** the length, width, and height of a cardboard box.

2 Calculate the volume of the box. To do this, multiply the measurements together: length × width × height.
(Length is usually considered to be the longest side.) **Record** the volume that you calculate.

Part 2: Measuring the Volume of an Irregular Solid

1 When an object is placed in water, it displaces (pushes aside) a volume of water that is equal to the volume of the object. Study the method shown in the photographs. It shows how to find the volume of an irregular object.

2 Find the volume of the toy. **Record** your measurements and calculations.

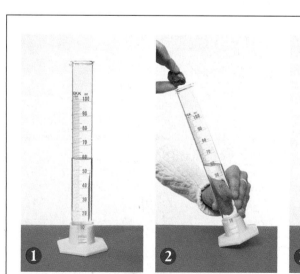

1 Record the volume of the liquid.

2 Carefully lower the object into the cylinder containing the liquid. Record the volume again.

3 The volume of the object is equal to the difference between the two volumes, e.g.:

Volume of object = 63 mL − 60 mL
= 3 mL

Measuring the volume of an irregularly shaped solid

Part 3: Estimating the Volume of a Liquid

1 **Predict** whether the cup or the dish holds a greater volume of water. **Record** your prediction.

2 Fill the cup to the brim with water. Without spilling any water, carefully transfer the water to the graduated cylinder. **Measure** and **record** the volume of the water.

3 Empty the graduated cylinder.

4 Repeat steps 1 and 2, using the dish instead of the cup.

Skill POWER

Refer to SkillPower 3 for tips on measuring volume.

Part 4: Studying the Volume of a Sample of Air

1 Put your hand near the opening of the bicycle pump as you pump it. What do you feel? **Record** your observations. What is coming out of the opening of the bicycle pump?

2 Pump the bicycle pump four or five times into a balloon, and then tie the balloon. What do you **observe**? **Record** your observations.

3 Repeat step 2, using a balloon that is a different shape. Make sure that you pump the bicycle pump the same number of times. **Record** your observations. **Predict** whether you will be able to squeeze the balloon and change its shape. Test your prediction.

4 Design a method for measuring the volume of air in a balloon. If your teacher approves, carry out your method.

Analyze

1. Think about Parts 1 and 2.
   (a) Why did you use two different methods to measure the volumes of solids?
   (b) Which method would you use to estimate the volume of a pencil? Explain.

2. Does it matter if you use different units for the height, width, and thickness of a solid when you calculate the volume of the solid? For example, could you calculate the volume of a solid using a height measured in metres and a width and a depth measured in centimetres? Explain your answer.

3. What is the volume of air inside the cardboard box you measured in Part 1? Explain.

4. Explain how you might use the steps in Part 4 to estimate the volume of air in one pump of the bicycle pump. What errors might be involved in this method?

5. Suppose that you have two identical balloons. You fill one balloon as full as possible with air. You fill the other balloon as full as possible with water. Do you think the volume of air will be the same as the volume of water, or will it be different? Give reasons for your answer.

Conclude and Apply

6. Summarize the different methods you used to observe and measure the volumes of matter in this investigation.

7. Suppose that there is a cube-shaped hole in the ground. The hole measures 20 cm by 20 cm by 20 cm. What volume of soil do you need to fill the hole? Justify your answer.

8. Design a method you could use to measure the volume of a carrot.

Density: A Property That Links Mass and Volume

Have you ever seen an adventure film in which a character's life is threatened by a huge boulder rolling down a hill? Filmmakers do not use real boulders, of course. They use boulders that are made of a lightweight material called Styrofoam™. The Styrofoam™ boulders are shaped and painted to look like the real thing. A fake boulder is much less dangerous than a real boulder. A fake boulder is also much easier to move. It may have the same size, shape, and volume as a real boulder, but it has much less mass.

In the activity at the start of this chapter, you tried to infer the identities of five samples based on how heavy or light they felt. Your teacher prepared the samples by measuring equal volumes of five materials. In other words, each sample had the same volume. The masses of the samples differed, though. As a result, some felt heavier than others. The quantity of mass in a certain volume of a material is a property called **density**. To understand what this means, try the following "thought experiment":

Picture yourself cutting a piece of lead into a cube that measures 1 cm by 1 cm by 1 cm. The volume of this piece of lead is 1 cm^3. Picture yourself doing the same thing with a piece of aluminum. Now imagine measuring the mass of each cube. What do you think you would find? The cube of lead would have a much greater mass than the cube of aluminum. That is because the density of lead is much greater than the density of aluminum. Use these results to help you answer the questions in Figure 4.17.

READING check

Which material in Figure 4.17 has the largest density? Which material has the smallest density?

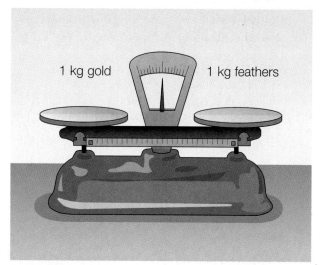

Figure 4.17A Which side of the balance should have materials with a larger *volume?* Should the balance be tipped to this side? Why or why not?

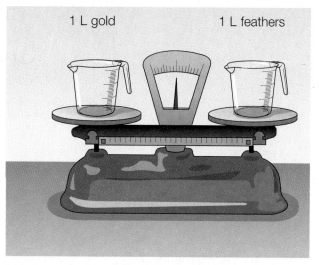

Figure 4.17B Which side of the balance should have materials with a larger *mass?* Should the balance be tipped to this side? Why or why not?

CONDUCT AN INVESTIGATION 4-E

SKILLCHECK
- Predicting
- Observing
- Inferring
- Communicating

Building a Density Tower

In this investigation, you will find out how the densities of different materials vary.

Question
How can you compare the densities of various liquids and solids?

Safety Precautions

Apparatus
tall plastic jar with lid or transparent container with lid
rubber gloves

Materials
cork
toothpick or wood chip
paper clips
water, with food colouring added
vegetable oil

Procedure

1. Study the list of materials. **Predict** the order of their densities. **Record** your prediction by ranking the materials from the most dense (sinks to the bottom) to the least dense (floats on top of everything else).

2. Place the oil, cork, wood chip, and paper clips in the container. Fill the container with the coloured water, and put on the lid.

3. Allow the materials to settle (stop moving). Sketch and label the tower and its contents.

4. Shake the tower. Allow the materials to settle again. If the tower looks different than it looked in step 3, sketch and label a new drawing of the tower.

Analyze

1. Make a data table. Rank the materials in the density tower from least dense (1) to most dense (5).

2. Which of the materials you used are less dense than water? Which are more dense than water?

Conclude and Apply

3. How did the densities that you observed compare with the rankings you predicted? If some densities surprised you, explain why.

4. Are solids always more dense than liquids? What evidence supports your answer?

5. Does the volume of an object determine its density? Explain your answer.

Extend Your Skills

6. Choose more items to add to the density tower. For example, you could add a rubber stopper, a small plastic toy, a candle stub, baby oil, and a piece of soap. Predict where you think these items will settle in the tower. Then test your prediction.

Did You Know?

Hydrogen sulfide is colourless, but it certainly has a characteristic smell. This gas smells like rotten eggs.

READING Check

When a bucket is "empty," it is really filled with air. Is a bucket of air heavier or lighter than a bucket of water? What can you infer about the densities of air and water?

Comparing Densities

In Conduct an Investigation 4-E, you learned that less dense liquids and solids float in liquids that are more dense. You also learned that some liquids are more dense than some solids. For example, mercury is a very dense liquid. A chunk of solid iron will float in mercury. In Figures 4.18 and 4.19, you can see that if a solid is less dense than a liquid, the solid will float in the liquid.

You can compare the densities of gases in a similar way. If gas A is less dense than gas B, then gas A will rise in gas B. For example, helium gas is less dense than air. A balloon that is filled with helium gas will rise when it is released.

Are any gases more dense than air? Yes, hydrogen sulfide is one example. Sources of this toxic gas include pulp and paper mills, crude oil refineries, and sewage treatment plants. If a balloon was filled with this very toxic gas, it would drop to the ground when it was released.

How else can you compare the densities of materials? Another way is to observe if something feels heavy or light for its size. For example, a bucket full of water is quite heavy. If you fill the same bucket with rocks, it is even heavier. If you fill it with feathers instead of water or rocks, it is much lighter. The volume of the bucket and its contents stays the same in each case. Only the mass varies. The mass depends on the density of the contents. Based on how heavy the bucket feels, you can infer that rocks are denser than water. You can also infer that a feather is less dense than water.

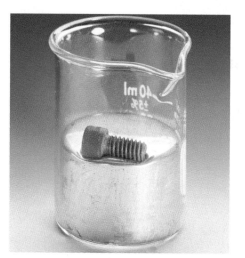

Figure 4.18 Liquid mercury is denser than solid iron. Therefore, an iron bolt will float in liquid mercury. In this photograph, a layer of oil has been placed on top of the mercury to prevent toxic mercury vapour from polluting the air. How does the density of the oil compare with the density of the mercury?

Figure 4.19 This solid block of wood will float in liquid water. The wood is less dense than the water.

A Formula for Density

Density is represented by the symbol D. Its units are units of mass divided by units of volume. You use grams per millilitre (g/mL) to express the density of a liquid or gas. You use grams per cubic centimetre (g/cm³) to express the density of a solid. Notice the use of these units in Table 4.3.

You can express the relationship among density, mass, and volume with the following formula:

$$D = \frac{m}{V} \quad \frac{g}{cm^3}$$

In this formula, D stands for density, m stands for mass, and V stands for volume. Work through the examples below to see how to use the formula for density. Then try the practice questions that follow.

Table 4.3
Densities of Some Common Materials

Material	Density
Gases	
hydrogen	0.00009 g/mL
air	0.0013 g/mL
carbon dioxide	0.0020 g/mL
Liquids	
water	1.00 g/mL
seawater	1.03 g/mL
mercury	13.55 g/mL
Solids	
cork	0.24 g/cm³
sugar	1.59 cm³
red cedar	0.35 g/cm³
table salt (solid, not grains)	2.16 g/cm³
copper	8.92 g/cm³
lead	11.34 g/cm³

Example 1

Aluminum is useful for making many objects, from cans to foil to bicycles. It is a strong material, but it is less dense than other metals, such as iron. A sample of aluminum has a mass of 5.40 g and a volume of 2.00 cm³. What is its density?

What Information Are You Given?

You are given the mass and the volume of aluminum.

$m = 5.40$ g
$V = 2.00$ cm³

How Can You Solve the Problem?

You know the formula for density. Insert the values that you are given in the formula. Then solve for density by dividing the mass value by the volume value.

$$D = \frac{m}{V}$$
$$= \frac{5.40 \text{ g}}{2.00 \text{ g/cm}^3}$$
$$= 2.70 \text{ g/cm}^3$$

The density of aluminum is 2.70 g/cm³.

Which has a higher density, wood or mercury? Does carbon dioxide or hydrogen have a higher density? Use Table 4.3 to decide.

Example 2

Glycerol is a material that is used to make many products, such as cosmetics. For example, if you look at the ingredients in hand cream, you will probably find glycerol. (It may be listed by its commercial name, glycerin.) A sample of glycerol has a volume of 3.00 mL and a mass of 3.78 g. Is glycerol denser than water?

Glycerol is an ingredient of most shampoos—including pet shampoos.

What Information Are You Given?

You are given the mass and the volume of glycerol.

$m = 3.78$ g

$V = 3.00$ mL

How Can You Solve the Problem?

You know the formula for density. As in Example 1, insert the given values in the formula. Then solve the formula by dividing the mass value by the volume value. Once you know the density of glycerol, compare it with the density of water, which is given in Table 4.3.

$$D = \frac{m}{V}$$

$$= \frac{3.78 \text{ g}}{3.00 \text{ mL}}$$

$$= 1.26 \text{ g/mL}$$

The density of glycerol is 1.26 g/mL. From Table 4.3, the density of water is 1.0 g/mL. Therefore, glycerol is denser than water. (It has more mass for each millilitre of volume.)

Practice

In your answers to the questions below, show how you calculated density. Include the units for density in your answer.

1. A 25 mL sample of vegetable oil has a mass of 23 g. What is the density of the vegetable oil?

2. Chloroform is a liquid that was once used as an anesthetic during surgery. (Researchers later found out that it causes cancer, so it is no longer used for this purpose.) A 250 mL sample of chloroform has a mass of 373 g.
   (a) What is the density of chloroform?
   (b) Do you think chloroform floats or sinks in water? Explain your reasoning.

3. A piece of metal has a mass of 147 g. Suppose that you place it in a graduated cylinder containing 20 mL of water. The water level rises from 20 mL to 41 mL. What is the density of the metal?

4. A rectangular solid is made of metal. It measures 2 cm by 2 cm by 5 cm. It has a mass of 227 g.
   (a) What is the density of the metal?
   (b) Look back at Table 4.3 on the previous page. What kind of metal do you think the solid is made of?

INTERNET CONNECT

www.mcgrawhill.ca/links/BCscience7

Try some quizzes and puzzles about mass, volume, and density to get more practice and test your knowledge. Go to the web site above. Click on **Web Links** to find out where to go next.

Section 4.2 Summary

Quantitative properties enable you to describe matter more precisely than you could if you used only qualitative properties.

- All matter has mass and volume.
- Mass is the amount of matter in a material. Mass is measured using a balance.
- Volume is the amount of space that is occupied by a material.
- Different methods are used to measure and calculate the volumes of solids, liquids, and gases.
- Density is a mathematical calculation of the mass of a material divided by its volume.
- Different samples of the same material may have different masses. Their density does not change, however.

Key Terms
mass
volume
density

Check Your Understanding

1. What is volume? Define it in your own words.

2. (a) What units would you use to measure the volume of a solid?
   (b) What units would you use to measure the volume of a gas?

3. (a) What is mass?
   (b) What instrument would you use to measure mass?
   (c) What units would you use to measure mass?

4. What is the volume of each object described below? Show your work.
   (a) A book has a length of 55 cm, a width of 30 cm, and a thickness of 8 cm. What is its volume in cm^3?
   (b) What volume of air is contained in an "empty" 250 mL aluminum can?
   (c) A cherry tomato displaces 6 mL of water. What is its volume in cm^3?

5. How can you calculate density?

6. **Apply** A bar of iron has a volume of 12 cm^3 and a mass of 95 g. A bar of lead has a volume of 7 cm^3 and a mass of 79.38 g.
   (a) Which bar has a greater mass?
   (b) Which bar is more dense? (Show your work.)

7. **Thinking Critically** When you shake oil-and-vinegar salad dressing, the oil and vinegar seem to mix. After a while, however, the oil and vinegar separate. The oil rises to the top. Which has a lower density, oil or vinegar?

How would you determine the volume of a book?

Section 4.3 Changes in Matter

Figure 4.20A If you heat a metal, it becomes more malleable, so it can be bent into a different shape. The metal keeps its identity. It is still the same type of matter. The change is physical.

Picture a smooth, flat sheet of writing paper. Now picture crumpling it into a ball. You have changed the shape of the paper. You have changed one of paper's properties. What if you tear the single sheet of paper into pieces? You have changed the size of the paper. You have changed another of paper's properties. Have you changed the type of matter that paper is? No. Crumpled paper and smooth paper are still the same material: paper. Torn paper and whole paper are still the same material: paper. The identity of the matter has not changed. Scientists use a special term to describe changes in matter in which identity remains the same: physical change. During a **physical change,** a material may change its shape or its state, but it keeps its identity.

What other physical changes of matter are there? Figure 4.20 shows three examples.

 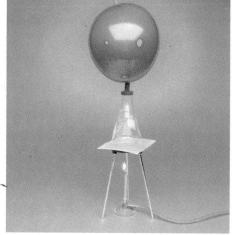

Figure 4.20B There is air inside the flask on the left. When you heat the air, it expands and fills the balloon. (Its volume increases.) If you cool the air, the air contracts and the balloon grows smaller. (Its volume decreases.) In both cases, the air keeps its identity. The changes in volume are physical.

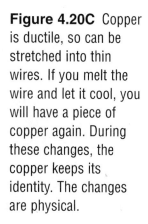

Figure 4.20C Copper is ductile, so can be stretched into thin wires. If you melt the wire and let it cool, you will have a piece of copper again. During these changes, the copper keeps its identity. The changes are physical.

THINK & LINK
INVESTIGATION 4-F

SKILLCHECK
- Communicating
- Interpreting Data
- Inferring
- Classifying

Name the Change

Think About It

You know that matter commonly exists in three states on Earth: solid, liquid, and gas. The arrows in the diagram below show that there are six possible ways that the state of matter can change. You know most of these changes from experience. This investigation will help you think about what you may take for granted.

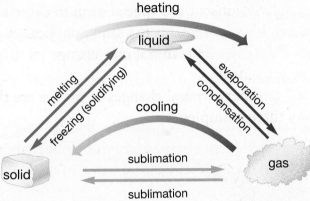

What to Do

1. Sketch the diagram on a clean sheet of paper, so you have your own copy.

2. Now copy the table below into your science notebook. Give the table a title.

3. Fill in the blank boxes in the table. Use the completed row as a guide to help you. (Notice that there are two changes to fill in for the last row.)

Name of change of state	Change that occurs
freezing (solidifying)	• from liquid to solid
condensation	•
evaporation	•
melting	•
sublimation	•
	•

4. Read statements (a) to (d) in the next column. Then describe each statement using a change-of-state term. Here are two examples:

Statement: Wet clothes dry in sunshine.
Description: evaporation of water

Statement: Solid carbon dioxide (dry ice) becomes gaseous carbon dioxide at room temperature.

Description: sublimation of carbon dioxide

(a) Melted candle wax hardens when you blow out the candle.
(b) A warm wind makes snow on the ground seem to disappear, but no puddles of water form.
(c) In the winter, invisible moisture in the air forms frost on glass windows.
(d) On a cold day, you can "see your breath" when you exhale.
(e) Ice cubes in the refrigerator shrink over time as some of the ice becomes water vapour.

Analyze

1. Classify the six changes of state according to whether they involve heating matter or cooling matter.

2. Changes of state are *reversible* physical changes. What do you think this means?

How Can You Explain Physical Changes of Matter?

What makes matter in the solid state different from matter in the liquid state or the gaseous state? Why does the volume of a material get larger (expand) when it is heated? What happens when a material changes state? Why do materials keep their identity when they undergo a physical change?

To answer these questions, scientists use the idea that all matter is made up of extremely tiny particles. Usually, this idea is presented as a model, called the *particle model of matter*. In science, a **model** is anything that helps you understand, communicate, or test an idea. A model may be a picture, a mathematical formula, words, a three-dimensional object or a combination of some or all of these.

The Particle Model of Matter

The **particle model of matter** is a scientific description of many different features of matter. Here are the key points of this model.

- All matter is made up of particles that are much too small to be seen.
- The particles are always in motion. They vibrate, rotate, and (in liquids and gases) move from place to place.

continued

Using the Particle Model

The particle model of matter helps to explain the properties and changes of solids, liquids, and gases. Find out how the model matches your observations.

What to Do

1. The diagrams show how particles are arranged in the three states of matter. Study these diagrams, and review the key points of the particle model listed on this page and the next page.

2. Identify the state in which matter
   (a) has a definite shape
   (b) takes the shape of its container
   (c) always fills the container it is in

3. Identify the state in which the particles
   (a) are far apart from each other
   (b) are fairly close together
   (c) are closely packed in fixed positions
   (d) are free to move around

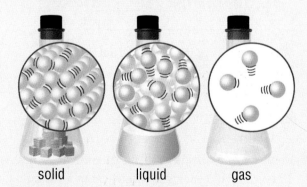

solid   liquid   gas

What Did You Find Out?

1. Use the particle model to explain your answers for steps 2 and 3.

- The particles have empty spaces between them. There is a lot of empty space between the particles of a gas. There is very little empty space between the particles in a solid and between the particles of a liquid.
- Particles at a higher temperature are moving faster than particles at a lower temperature.
- Each pure substance has its own kind of particle, which is different from the kinds of particles that make up any other pure substances.

You will explore the last key point in Chapter 5. Use the activity on page 124 to gain a better understanding of the first four key points.

Figure 4.21 How do the properties of paper change when it burns?

Other Ways That Matter Can Change

Imagine burning a piece of paper, as shown in Figure 4.21. While the paper burns, it gets smaller. Thus, its size changes. Its colour also changes, and smoke rises from the paper. When the paper stops burning, you observe a small quantity of ashes that are less dense than the original paper. Burning changes the identity of the paper. The ashes and the gaseous smoke are not the same types of matter as the original paper. The properties of the ashes and the smoke are different from the properties of the paper. You cannot collect the ashes and the smoke and make them into paper again. This change is *non-reversible*.

Burning paper is an example of a chemical change. During a **chemical change**, one type of matter changes to produce one or more different types of matter. The matter that is produced has a different identity and different properties from the original matter.

The ability to burn is a chemical property of matter. The term **chemical property** refers to the ability of a material to take part in a chemical change. Chemical properties describe how one type of matter interacts with other types of matter, or with energy, during a chemical change. Table 4.4 lists several other chemical properties, as well as evidence that indicates a chemical change has occurred.

Table 4.4 Examples of Chemical Properties

Chemical property	Example	Change that occurs
reacts with acid	baking soda	Bubbles form when you mix baking soda and vinegar.
corrodes when exposed to water and air	iron	Rust is a crumbly, red coating that forms as iron corrodes.
reacts with water	antacid tablet	Bubbles form, and fizzing sounds are heard. Soon, the tablet seems to disappear.

Section 4.3 Summary

Matter can change physically and chemically. The particle model of matter helps you visualize what happens when matter changes.

- Physical changes do not change the identity of a material.
- Changes of state occur when a material is heated or cooled. Changes of state are reversible changes.
- Chemical changes result in types of matter with properties that are different from the properties of the original matter. Chemical changes of matter are non-reversible.
- The particle model of matter pictures matter as being made up of particles that are too small to see.

Key Terms
physical change
model
particle model of matter
chemical change
chemical property

Check Your Understanding

1. In what ways are physical changes and chemical changes similar? In what ways are they different?

2. Read each statement below. Does it describe a physical change or a chemical change? Explain your reasoning.
   (a) A steak is well cooked.
   (b) A piece of chalk is crushed.
   (c) A green coating forms on a copper statue when it is exposed to air.
   (d) The filament (thin wire) inside a light bulb gets hot and glows when an electric current moves through it.
   (e) Sweat evaporates from your skin.
   (f) Frost forms on the inside of a freezer.

3. What is a scientific model?

4. In your own words, describe the particle model of matter. See if you can summarize the model in just two or three sentences.

5. **Apply** Think about each type of diagram below. Is it an example of a model? Give reasons to justify your answer.
   (a) map of Canada
   (b) pie chart
   (c) food chain
   (d) diagram showing a hiking path

6. **Thinking Critically** The ability to burn is a chemical property. Another word for this property is "flammability." Matter that has the property of flammability is said to be flammable.
   (a) What are four different types of flammable matter (e.g., paper)?
   (b) In a chemistry laboratory, flammable materials are stored in a metal cabinet. Suggest a reason why.
   (c) Materials that are not flammable are said to be "non-flammable." What uses might these materials have?

CHAPTER at a glance

Now that you have completed this chapter, try to do the following. If you cannot, go back to the sections indicated in brackets.

(a) Define matter, and give examples of matter. (4.1)

(b) Name the senses that people use to observe matter. What precautions should you take when using your senses? (4.1)

(c) List the common states of matter on Earth. Give two examples of each state. (4.1)

(d) Explain the difference between qualitative properties and quantitative properties. (4.1)

(e) Name four properties of matter.

(f) Define volume. What units would you use to express volume? (4.2)

(g) Define mass. What units would you use to express mass? (4.2)

(h) Define density. Explain how density is related to mass and volume. (4.2)

(i) Give two ways to compare the densities of materials, without doing any calculations. (4.2)

(j) Explain how melting point and boiling point can help you identify matter. (4.2)

(k) Describe a reversible change of matter and a non-reversible change of matter. (4.3)

(l) Explain the difference between a physical change of matter and a chemical change of matter. (4.3)

(m) Explain what the particle model of matter says about matter. (4.3)

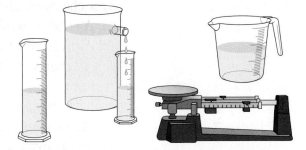

Prepare Your Own Summary

Summarize this chapter by doing one of the following. Use a graphic organizer (such as a concept map), create a poster, or write a summary to include the key chapter ideas. Here are a few ideas to use as a guide:

- Sketch a series of pictures to illustrate different properties. Write two or three sentences to explain each picture.
- Write an instruction manual explaining how to measure volume and mass and how to calculate density. Include labelled diagrams.
- Illustrate the three states of matter, based on the particle theory of matter.
- Use words, pictures, or both to show how the particle model of matter explains what happens when a solid changes state to become a liquid.
- Design a series of index cards or a computer database that lists the properties of a variety of different materials.

CHAPTER 4 Review

Key Terms

chemistry
matter
property
quantitative property
qualitative property
states of matter
solid
liquid
gas
temperature
melting point
boiling point
mass
volume
density
physical change
model
particle model of matter
chemical change
chemical property

Reviewing Key Terms

If you need to review, the section numbers show you where these terms were introduced.

1. What are three properties of water? (4.1)

2. Divide a sheet of paper into three columns. Label the columns "Solid," "Liquid," and "Gas." In each column, write one statement that always applies to the shape of matter in this state. Then write one statement that always applies to the volume of matter in this state. (4.1)

3. For each of the following, what is the relationship between the two terms?
   (a) solid and liquid (4.1, 4.3)
   (b) liquid and gas (4.1, 4.3)
   (c) quantitative property and qualitative property (4.1)
   (d) melting point and boiling point (4.1)
   (e) mass and volume (4.2)
   (f) mass and density (4.2)
   (g) physical change and chemical change (4.3)

Understanding Key Ideas

Section numbers are provided if you need to review.

4. For each of the following, what properties do the two materials have in common? What properties would you use to tell one from the other? (4.1, 4.2, 4.3)
   (a) water and vinegar
   (b) soil and instant coffee
   (c) glass and ice
   (d) air and carbon monoxide

5. What properties would you use to describe a baseball? (4.1)

6. What formula shows how mass, volume, and density are related to each other? What does each symbol in this formula represent? (4.2)

7. What units would you use to describe each of the following? (4.1, 4.2)
   (a) the mass of a sugar cube
   (b) the volume of liquid in a tablespoon
   (c) the volume of lemonade in a pitcher
   (d) the mass of a bicycle
   (e) the melting point of copper
   (f) the density of ocean water
   (g) the density of air
   (h) the density of an iron nail

8. How are mass, volume, and density related? (4.2)

9. How would you use the particle model of matter to describe each of the following? (4.3)
   (a) the spacing between the particles in a solid
   (b) the movement of the particles in a gas
   (c) the movement of the particles in a solid
   (d) the spacing and movement of the particles in a liquid

Developing Skills

10. You need to fill a bucket with water using a 50 mL cup. The volume of the bucket is 5 L. How many cupfuls will you need in order to fill the bucket?

11. In your notebook, draw a concept map using the key terms listed in this chapter.

12. Draw a diagram to show how water changes state from solid to liquid to gas, and then back from gas to liquid to solid. Add labels, naming each process that produces a change of state. Show where energy is absorbed or released.

13. The volume of solid matter is slightly different at different temperatures. Examine the data in the table below. Then answer the questions that follow. (Here, a change in length reflects a change in volume.)

The Effect of Different Temperatures on the Lengths of Solids

Solid material	Length at −100°C	Length at 0°C	Length at 100°C
lead	99.71	100.00	100.29
aluminum	99.77	100.00	100.23
copper	99.83	100.00	100.17
glass	99.91	100.00	100.09
*Pyrex™	99.97	100.00	100.03
steel	99.89	100.00	100.11

*(a glass-like material that is stronger than glass)

(a) Which material has the largest increase in volume as it gets hotter?
(b) Which material has the smallest increase in volume as it gets hotter?
(c) Which material has the largest decrease in volume as it gets colder?
(d) Which material has the smallest decrease in volume as it gets colder?
(e) Which material, glass or Pyrex™, would you choose for making a baking dish? Explain why.

14. Imagine that the students in your class represent particles of matter and the school gym represents a container. Use labelled sketches to show how you would organize the students in the gym to demonstrate changes of state. How would the students move? Where would they stand in relation to one another?

Problem Solving

15. Suppose that you want to measure the volume of an irregularly shaped solid. Using a plastic cup and a drinking straw, design a piece of apparatus that you could use. Sketch your apparatus. Explain how you would use it.

16. Water and gasoline are both clear liquids. What properties could you use to distinguish between them?

Critical Thinking

17. Gwen says that solids are always denser than liquids. Wafik disagrees. Which student do you agree with? What arguments would you use to support your opinion?

18. Which do you think is more dense, cold water or warm water? Use the particle model of matter to make a prediction. Then explain a method you could use to test your prediction.

Pause & Reflect

Go back to the beginning of this chapter on page 100, and check your original answers to the Getting Ready questions. How has your thinking changed? How would you answer those questions now that you have investigated the topics in this chapter?

CHAPTER

5 Classifying Matter

Getting Ready...

- What is a mixture?
- Is any matter not a mixture?
- How can you get useful materials from mixtures?
- What do scientists think the smallest particles of matter are like?

A dipnet is a kind of filter that separates fish from water. What other kinds of filters can you think of?

What do these two photographs have in common? The person in the large photograph is fishing with a dipnet. A dipnet is a traditional Aboriginal tool for catching fish. The dipnet catches large fish but allows smaller fish and gravel to go through the holes. Water flows through the holes, too.

The scientist in the photograph on the next page is separating different kinds of matter with a gas chromatograph (GC). Separating different types of matter is the first step in identifying matter. Forensic scientists use the GC to analyze evidence such as charred wood. The outcome of an arson case may depend on knowing whether or not a sample of charred wood has gasoline on it.

Both the dipnet and the GC use properties to separate mixtures into their parts. The dipnet uses the properties of size and state of matter. The GC uses other properties. In this chapter, you will learn about different methods for separating materials into their parts. As well, you will find out how you can use the properties of materials to classify them.

What You Will Learn

In this chapter, you will learn
- how to identify the mixtures all around you
- how to distinguish between different kinds of mixtures
- what theory will help you visualize particles of matter

Why It Is Important

- You eat, wear, breathe, and use mixtures every day.
- Many mixtures, such as gold ore, contain important substances that people can extract using their knowledge of physical changes.

Skills You Will Use

In this chapter, you will
- classify materials as mixtures or pure substances
- communicate observations about mixtures you use every day
- investigate different ways to separate mixtures

This scientist is using a GC to help identify the types of matter in a sample. Why might this be important evidence in a court case?

Starting Point ACTIVITY 5-A

Mixture or Pure Substance?

Do you read the labels on packages? Some labels claim that the product in the package is "pure." Are these claims accurate? Does "pure" mean the same to a consumer as it does to a scientist? In this activity, you will classify some common materials as either mixtures or pure substances.

What to Do Group Work

1. In groups, decide how you would define the terms "mixture" and "pure substance." **Record** your definitions.
2. Prepare a table of observations like the one below. Give your table a title.

Product	Mixture or pure substance?	Reasons

3. Brainstorm a list of common products that you might find in a bathroom or kitchen.
4. Choose ten products that you want to **classify** as a mixture or pure substance. **Record** the names of the products in your table.
5. As a group, decide whether each product is a pure substance or a mixture. Use the definitions you created in step 1. **Record** your decisions, and the reasons for each decision, in your table.

What Did You Find Out?

1. In a large table, compile a class list of the products examined. How many products were classified as mixtures? How many products were classified as pure substances?
2. Which products were difficult to classify? Which products were easy to classify? Suggest reasons.

Chapter 5 Classifying Matter • MHR **131**

Section 5.1 Pure Substances and Mixtures

Figure 5.1 Is this fabric a mixture? How do you know?

Figure 5.2 Suppose that you examined several different samples of concrete. Do you think all the samples would have the same density? Would they all have the same hardness?

Most solids, liquids, or gases that you see and use every day are mixtures. A **mixture** contains two or more different types of matter. How can you tell that a sample of matter is a mixture? Many products in your home, such as hand lotion and shampoo, have long lists of ingredients. An ingredient list tells you all the ingredients that are contained in a product. If an ingredient list has more than one ingredient, the product is clearly a mixture! For example, examine the clothing tag in Figure 5.1. What is the fabric made of?

Not every object that you encounter has an ingredient list. Is a sandy beach a mixture? Is an ocean a mixture? How can you tell?

Heterogeneous Mixtures

Whenever you see a sample of matter that has more than one set of properties, you know for sure that it is a mixture. For example, a sandy beach may contain white grains of sand, multicoloured seashells, and brownish-green seaweed.

The concrete wall shown in Figure 5.2 is also a mixture. You can see that it contains a variety of stones with different shapes, sizes, and colours. If you have watched someone make concrete, you may know that concrete also contains sand, cement powder, and water.

Concrete is a good example of a heterogeneous mixture. A beach is another example. A **heterogeneous** [het-uhr-oh-JEEN-ee-uhs] mixture is made up of different parts that you can detect quite easily. Often, you can see the different parts just by looking at the mixture. Sometimes you need a microscope to see the different parts.

Pure Substances

Is there anything that is *not* a mixture? What about a bar of pure gold? Pure gold contains nothing but gold. Every sample of pure gold has the same properties as every other sample of pure gold. For example, a nugget of pure gold from Barkerville, British Columbia, has the same melting point, hardness, and density as the nugget of pure gold from Yanacocha, Peru.

Pure gold is an example of a pure substance. A **pure substance** is the same throughout. Every sample of a pure substance always has the same properties. Other examples of pure substances are helium, pure water, and white sugar. Pure substances are homogeneous materials. **Homogeneous** [hoh-moh-JEEN-ee-uhs] means that every part of the material is the same as every other part.

READING check
What is a pure substance? Give three examples of pure substances.

Homogeneous or Heterogeneous?

How do you decide whether a sample of matter is a pure substance or a mixture? One way is to observe whether the sample is homogeneous or heterogeneous. Heterogeneous materials are always mixtures. Sometimes, though, you cannot tell whether a sample of matter is homogeneous or heterogeneous just by looking at it. For example, is milk homogeneous or heterogeneous? Milk appears homogeneous, but under a microscope, you can see that it contains "blobs." These "blobs" are called globules. Figures 5.3 shows a sample of milk under a microscope. One single drop of milk contains about 100 million fat globules. The globules are so tiny that milk seems to be homogeneous even though it is not. Do you think it is possible to have a homogeneous mixture? Find out in the next activity.

Did You Know?
Many years ago, it was easy to see that milk is a heterogeneous mixture. Instead of staying mixed with the milk, the fat globules floated to the top. There, they clustered together, forming a layer of cream. Today, milk is *homogenized*. Homogenized milk is specially prepared so that the fat globules remain mixed with the rest of the liquid.

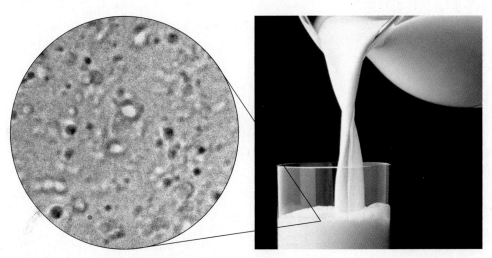

Figure 5.3 The round photograph shows how milk looks under a microscope. The milk is magnified 400 times. How can you tell, from the microscopic image, that milk is heterogeneous?

At Home ACTIVITY 5-B

Making Sugar "Disappear"

Can you make a mixture in which you cannot detect the different parts? The two substances you will mix are sugar and water.

Safety Precautions

Do not taste the samples if you are doing this activity in class. You may taste the samples if you are doing this activity at home.

What You Need

drinking glass
tap water
white sugar
teaspoon

What to Do

1. Make a table of observations like the one below. Give your table a title.

Observable properties	Sugar	Water	Sugar-water mixture
colour			
state			
taste			
transparency			

2. **Observe** the four properties of sugar and water that are listed in the table. **Record** your observations.

3. Fill a glass with cold water. Let it sit for a few seconds, until the water is still. Gently pour one level teaspoon of sugar into the water. **Observe** the appearance of the water and the sugar. Place the glass where it will not be moved for 24 h.

4. **Observe** the contents of the glass the next day. **Record** the properties of the mixture in your table.

What Did You Find Out?

1. Can you detect either of the two substances in your mixture as different parts?

2. Is the sugar present in the mixture or not? How do you know?

3. How do the properties of the sugar-water mixture compare with the properties of the sugar and the properties of the water? Choose one of the following statements. Record it, and give reasons for your choice.
   - The mixture has all the properties of water and *only* these properties.
   - The mixture has all the properties of sugar and *only* these properties.
   - The mixture has a blend of sugar's properties and water's properties.

What kind of mixture is milk? What kind of mixture is sugar and water? Explain the difference.

Homogeneous Mixtures

When you mix sugar with water, the sugar crystals disappear from view. *You cannot see them, even with a microscope.* A sugar-water mixture is an example of a homogeneous mixture. Homogeneous mixtures contain two or more pure substances. The substances are mixed so that their properties are blended and every part of the mixture is the same.

What other homogeneous mixtures can you think of? In the next investigation, you will practise classifying mixtures as either homogeneous or heterogeneous.

CONDUCT AN INVESTIGATION 5-C

SKILLCHECK
- Hypothesizing
- Observing
- Classifying
- Communicating

Inspector's Corner

How can you determine whether a mixture is homogeneous or heterogeneous? Your challenge in this investigation is to inspect several products in your home and classify them.

Question
How can you classify common household products as homogeneous or heterogeneous mixtures?

Hypothesis
Write a hypothesis about the properties of homogeneous and heterogeneous mixtures that will help you classify them. Your hypothesis should take the form "If… then… because…."

Safety Precautions

Materials
variety of common household mixtures, such as orange juice, apple juice, jam, cereal, salsa, toothpaste, skin cream, and soap

Procedure

1. Prepare a table of observations like the one below, to record your observations. Give your table a title.

Product	Observations	Heterogeneous or homogeneous?	Reasons for classification

2. **Observe** each product carefully. **Record** your observations in your table.

3. **Classify** each product as a heterogeneous or homogeneous mixture. **Record** your classification in your table.

Skill POWER
For tips on writing a hypothesis, turn to SkillPower 6.

4. Under "Reasons for classification," identify and describe the properties of each product that helped you classify it.

5. Put away all the products. Clean up any spills, as instructed by your teacher.

Analyze
1. Propose a test that you could use to classify a product as homogeneous or heterogeneous.

Conclude and Apply
2. Re-examine your hypothesis. How well did your results support your hypothesis? Modify your hypothesis to reflect your results.

3. Compare your conclusions with the conclusions of other students. Which products were easier to classify? Which products were more difficult to classify? Why?

Chapter 5 Classifying Matter • MHR **135**

Section 5.1 Summary

In this section, you learned that samples of matter can be either pure substances or mixtures. You also learned how to classify matter as homogeneous or heterogeneous.

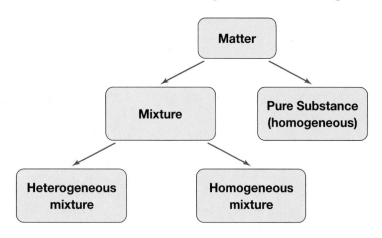

- Homogeneous materials have the same properties throughout.
- Heterogeneous materials have different parts with different properties.
- Pure substances contain only one type of matter. They are homogeneous.
- Mixtures contain two or more types of matter. Mixtures can be heterogeneous or homogeneous.

The chart above summarizes what you have learned so far about matter. You will add to this chart in the sections that follow. In section 5.2, you will learn more about homogeneous and heterogeneous mixtures.

Key Terms
mixture
heterogeneous
pure substance
homogeneous

Check Your Understanding

1. (a) What are two examples of mixtures?
   (b) What are two examples of pure substances?

2. Classify each mixture as homogeneous or heterogeneous. Explain your classification.
   (a) milk
   (b) pulpy orange juice
   (c) apple juice
   (d) window cleaner
   (e) pizza
   (f) sports drink crystals

3. Heterogeneous materials are always mixtures. Are homogeneous materials always pure substances? Explain your answer.

4. **Apply** When you first open a bottle of soda, the liquid is filled with tiny bubbles.
   (a) Is the bubbly soda homogeneous or heterogeneous? Explain your answer.
   (b) If you let the soda sit for a day, what happens? Is the mixture heterogeneous or homogeneous now?

5. **Thinking Critically** The photograph on the left shows polluted air—smog. Smog is air that contains solid and gaseous pollutants.
   (a) Is smog a heterogeneous or homogeneous mixture?
   (b) Why do you think breathing smog can be harmful?

Small bits of solids hang in polluted air. These bits cause the air to appear cloudy from a distance.

Section 5.2 Classifying Mixtures

In Chapter 4, you classified matter as a solid, a liquid, or a gas. In section 5.1 of this chapter, you classified matter as a pure substance or a mixture. What are some different kinds of mixtures?

Mechanical Mixtures

Mixtures that are clearly heterogeneous are called **mechanical mixtures**. You can see the different types of matter in mechanical mixtures. Many foods, such as sandwiches, salads, salsa, and stir-fried vegetables, are mechanical mixtures. Another example of a mechanical mixture is pizza, shown in Figure 5.4.

Pause & Reflect

In your science notebook, list three foods, drinks, or other household mixtures (not already mentioned in this textbook) that are mechanical mixtures. Explain why they are mechanical mixtures.

Figure 5.4 You can see the crust, sauce, cheese, and toppings on a pizza. A pizza is a mechanical mixture.

Suspensions

A heterogeneous mixture in which the particles settle slowly after mixing is called a **suspension**. An example of a suspension is oil and vinegar salad dressing. Before shaking the salad dressing, you can see the oil and vinegar parts of the mixture. Often the dressing contains herbs and spices, as well. When you shake the mixture, the herbs, spices, and vinegar become suspended in the oil. They soon separate, however, when you put down the bottle. How can you make the bits of matter in a suspension remain suspended longer? You will find out in the next activity.

Did You Know?

The sky on Mars is a pinkish-yellow colour. What causes this colour? The Mars atmosphere is actually a suspension of very tiny pieces of red dust.

Find Out ACTIVITY 5-D

Keep It Together!

Oil and vinegar salad dressing is an example of a suspension. Manufacturers of salad dressings often add substances that slow down the separation of the parts of the mixture. These substances are called *emulsifying agents*. In this activity, you will observe the effect of one emulsifying agent.

What happens when you add dishwashing liquid to a suspension?

What You Need

cooking oil
vinegar
jar with tight-fitting lid
dishwashing liquid (not dishwasher detergent)
watch or clock with second hand or counter

What to Do

1. Place equal amounts of cooking oil and vinegar in the jar. Put on the lid to seal the jar tightly.

2. Shake the jar thoroughly, and then put it down. **Measure** how long the two liquids take to separate. **Record** the time in seconds.

3. Add two drops of dishwashing liquid to the mixture in the jar. Again, shake the jar thoroughly. How long do the two liquids take to separate now? **Record** the time.

4. Leave the mixture undisturbed overnight. **Observe** it in the morning. **Record** your observations.

What Did You Discover?

1. Describe the effect of the dishwashing liquid on the mixture.

2. Did adding the dishwashing liquid cause a permanent suspension to form?

READING Check

What is the difference between a suspension and an emulsion? Give one example of each kind of mixture.

Emulsions

In the last activity, you added dishwashing liquid to a suspension of oil and vinegar. You probably observed that the dishwashing liquid enabled the suspended particles to stay suspended longer. As you learned, the dishwashing liquid acted as an *emulsifying agent*. An emulsifying agent allows the parts of a suspension to remain distributed through the mixture for a longer period of time. A suspension that has been treated with an emulsifying agent is called an **emulsion**. Mayonnaise is one example of an emulsion. Paint is another example.

An emulsion may appear to be a homogeneous mixture, but it is not. You can see the different parts of an emulsion under a microscope. Both emulsions and suspensions are heterogeneous mixtures.

Solutions

Homogeneous mixtures are called **solutions.** You made a solution in At Home Activity 5–B. Apple juice, vinegar, and tap water are all solutions. So is the window cleaner shown in Figure 5.5. A solution is made when two or more substances combine to form a mixture that looks the same throughout, even under a microscope.

Solutions are everywhere. Seawater is a solution of salts and water. Air is a solution of nitrogen, oxygen, carbon dioxide, and other gases. There are even solutions of solids, such as the objects shown in Figure 5.6. Solutions that are made from two or more metals are known as **alloys.**

Why make an alloy? You can change the properties of a metal by adding small amounts of other substances. For instance, jewellery made out of pure gold would become scratched very easily, because gold is so soft. Adding silver or copper to gold makes a gold alloy that is harder than pure gold. As an added benefit, silver and copper are cheaper than gold!

Figure 5.5 Window cleaner is a solution of ammonia and other substances in water.

Figure 5.6 The circle graphs show the percentages of gold and other metals found in the different "gold" objects. Which of these objects are pure substances? Which are mixtures?

READING check

What is another way of saying "Water and salt form a solution when mixed"? Use the word "dissolve."

Parts of a Solution

When you mix two substances and they form a solution, you say that one substance **dissolves** in the other substance. As you can see in Figure 5.7,

- the **solute** is the substance that dissolves
- the **solvent** is the substance in which the solute dissolves

When you mix salt and water, for example, the salt is the solute, and the water is the solvent. The salt dissolves in the water.

Figure 5.7 The solute dissolves in the solvent to form a solution.

Table 5.1 Examples of Solutions

Solution	Solute and solvent	State of solute	State of solvent
air	oxygen and other gases in nitrogen	gas	gas
soda water	carbon dioxide in water	gas	liquid
vinegar	acetic acid in water	liquid	liquid
filtered seawater	sodium chloride (salt) and various minerals in water	solid	liquid
brass	zinc in copper	solid	solid

Table 5.1 provides some examples of solutions, with their solutes and solvents. There is usually less solute than solvent in a solution. If you read the labels on liquid products around your home, you will notice that water is the solvent for many different solutes.

Solution or Mechanical Mixture?

How can you tell if a mixture is a solution or a mechanical mixture? Often you can tell by just looking. What if you cannot?

- Use a microscope. If the mixture is a solution, you will be able to see only one type of matter, even under a microscope.
- If the mixture is a liquid, pour it through a filter. If anything is caught in the filter, then the mixture is definitely mechanical.
- Shine a light through the mixture. Solutions contain no undissolved particles and do not scatter light. Therefore, you will *not* see a beam running through a solution. A heterogeneous mixture, however, *does* contain undissolved particles that can scatter light. Which mixture in Figure 5.8 is a solution?

Figure 5.8 The tiny, undissolved particles in the mixture on the right reflect and scatter the light. You see the light as a beam running through the mixture.

CONDUCT AN INVESTIGATION 5-E

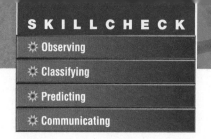

SKILLCHECK
- Observing
- Classifying
- Predicting
- Communicating

What Kind of Mixture?

One way to tell whether a mixture is a solution or a heterogeneous mixture is to pour the mixture through a filter. In this investigation, you will practise using a filter.

Question

How can a filter help you decide whether a mixture is a solution or a heterogeneous mixture?

Safety Precautions

Apparatus
funnel
4 beakers or jars

Materials
4 mixtures
4 pieces of filter paper
cardboard with hole for funnel

Procedure

1. Copy the following table into your science notebook. Give your table a title.

Mixture	Prediction: heterogeneous mixture or solution?	Observations before filtering	Observations after filtering	
			In filter	In beaker

2. Your teacher will provide you with four household mixtures.

3. **Observe** each mixture. **Predict** whether it is a solution or a heterogeneous mixture. Write your prediction in your table.

4. Pour each mixture through a filter.

5. For each mixture, **observe** the substance that went through the filter. Was anything left in the filter? **Record** your observations in your table.

6. Wipe up any spills, and wash your hands thoroughly.

Analyze and Conclude

1. Which of your observations matched your predictions? Did any observations surprise you? Explain your answer.

2. If you observe matter on the filter, can you state that the mixture is definitely a heterogeneous mixture? Explain your answer.

3. If you do not observe any matter on the filter, can you state that the mixture is definitely a solution? Explain your answer. If you answered "no," what are some ways you can tell that a mixture is definitely a solution?

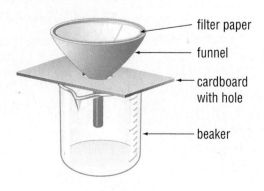

Chapter 5 Classifying Matter • MHR 141

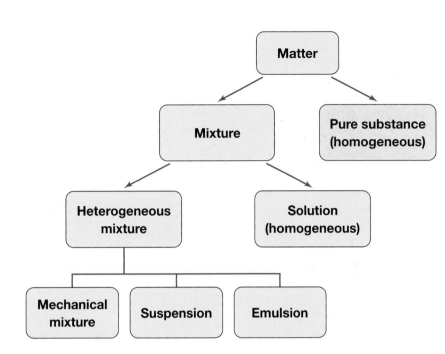

Section 5.2 Summary

In this section, you learned how to classify mixtures as heterogeneous mixtures or solutions. You learned that mechanical mixtures may be ordinary mechanical mixtures, suspensions, or emulsions. You can now add to your classification of matter chart, as shown on the left.

Key Terms
mechanical mixtures
suspension
emulsion
solutions
alloys
dissolves
solute
solvent

Check Your Understanding

1. How would you classify each mixture: as a solution, a mechanical mixture, a suspension, or an emulsion?
   (a) bran cereal with raisins
   (b) paint with an ingredient that prevents separation
   (c) oil and vinegar salad dressing
   (d) 18 karat gold ring
   (e) filtered air

2. (a) What kind of mixture is an alloy?
   (b) What are two different kinds of alloys?
   (c) Why are alloys sometimes more useful than pure metals?

3. How are a solution, a suspension, and an emulsion different? How are they the same? Give one example of each. (Do not choose examples from question 1.)

4. **Apply** Suppose that your teacher gives you a liquid mixture. You cannot see any small pieces of different matter in the mixture. When you filter the mixture, nothing is left on the filter paper.
   (a) Is the mixture a solution or a heterogeneous mixture? Explain your inference.
   (b) How could you be more certain about your classification? What tests would you carry out? What equipment would you need?

CHAPTER 5

Section 5.3 Pure Substances

In Chapter 4, you learned that Archimedes discovered the method of displacement for finding volume. He used the method of displacement to solve a problem for Hiero II, the ruler of Syracuse.

Hiero II suspected that the royal goldsmith had not used pure gold to make his crown. The king asked Archimedes, shown on the right, to determine whether or not the crown was made entirely of gold. Archimedes knew that he could determine whether the density of the crown matched the density of pure gold. He measured the mass of the crown on a balance. Then he determined its volume by displacement. Archimedes found that the crown was less dense than pure gold. The goldsmith had tried to cheat the king by mixing pure gold with silver, which is less dense *and* less valuable!

The crown looked like pure gold, but was it?

Identifying Pure Substances

Archimedes used the property of density to determine whether the crown was made of a pure substance or a mixture. You can also use boiling point or melting point to determine whether a material is a pure substance or a mixture. Pure substances boil and freeze at constant, known temperatures. Mixtures boil and freeze at different temperatures than the pure substances that make them up. This property of mixtures is used to keep roads clear of ice in the winter, as shown in Figure 5.9.

What properties can help you identify a pure substance?

Figure 5.9 Pure water freezes at 0°C. By adding salt, you create a mixture that stays liquid at much lower temperatures.

Chapter 5 Classifying Matter • MHR 143

Did You Know?

Just how tiny are the particles that make up matter? A 250 mL glass of water contains about 8 billion billion billion water particles. That's 8 000 000 000 000 000 000 000 000 000 water particles in a glass of water. No wonder you cannot see them!

Pure Substances and the Particle Model of Matter

Why do pure substances always have the same properties? In Chapter 4, you learned about the first four points in the particle model of matter. Scientists use the fifth point in this model to explain how pure substances are different from mixtures.

- Each pure substance has its own kind of particle, which is different from the kinds of particles that make up all other pure substances.

For example, pure gold contains only gold particles. Pure water contains only water particles. Because a pure substance is always made up of exactly the same kind of particles, it always has the same properties.

The particles that make up matter are so tiny that you cannot see them, even with a microscope. How many sugar particles do you think are in a sugar cube? See Figure 5.10 to find out.

Figure 5.10 A single grain of sugar contains about 1 800 000 000 000 000 000 sugar particles. A sugar cube contains about 2 700 000 000 000 000 000 000 sugar particles.

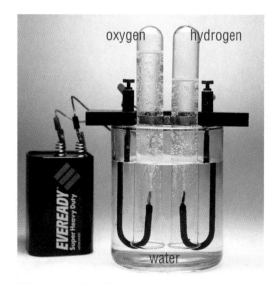

Figure 5.11 When electricity is run through water, hydrogen and oxygen are produced.

Classifying Pure Substances

In section 5.2, you learned that mixtures can be classified as mechanical mixtures, suspensions, emulsions, and solutions. Pure substances can also be classified into different groups. These groups are based on whether the pure substance can be broken down by a chemical change.

Examine Figure 5.11. It shows what happens when electricity is passed through water. The electricity causes a chemical change. As a result of this change, the water is broken down into two different gases, hydrogen and oxygen.

Elements and Compounds

Can you also break down hydrogen and oxygen by a chemical change? No. Hydrogen and oxygen are examples of a group of pure substances called elements. An **element** is a pure substance that cannot be broken down or separated into other pure substances. Other examples of elements are sodium, chlorine, carbon, iron, potassium, aluminum, and helium.

Water is an example of a different group of pure substances called compounds. A **compound** is a substance that is made from two or more elements that have combined in a chemical change. For example, water is made by combining the elements hydrogen and oxygen.

There are millions of compounds in the world. Each compound is a different combination of two or more elements. Salt and sugar are two common examples of compounds. Salt is made from the elements sodium and chlorine. Sugar is made from the elements carbon, hydrogen, and oxygen. Figure 5.12 shows two different compounds that are made from carbon and hydrogen.

The properties of compounds are generally very different from the properties of the elements that formed them. For example, sodium is a dangerously reactive metal. Chlorine is a poisonous gas. Yet these two elements combine to produce a substance that you and most other living things need to survive: salt!

Figure 5.12 Propane is a gas that is used in barbecue tanks (A). Paraffin wax is a solid that is used to make candles (B). Both of these compounds contain only hydrogen and carbon.

Find Out ACTIVITY 5-F

An Element and a Compound

Gold and copper are found as pure elements in Earth's crust. Most other metals, such as iron and aluminum, are not found as pure elements in nature. Instead, they are found combined with other elements in compounds. You are going to investigate how one element, iron, forms a compound.

Safety Precautions

Wash your hands thoroughly after this activity. Dispose of the materials as instructed by your teacher.

What You Need

1 iron nail (plus one more for scraping)
magnet
plastic wrap
jar
tap water
piece of white cardboard

What to Do

1. **Observe** one of the iron nails. **Record** your observations.

2. Obtain a magnet. Wrap your magnet in plastic wrap to keep it clean.

3. Through the cardboard, test the nail with the magnet. **Record** your observations.

4. Place the nail in a jar with a little salty water. Allow the nail to sit for a few days.

5. Remove the nail from the water. **Record** your observations.

6. Use the second iron nail to rub the rusted nail gently, over a piece of white cardboard. **Record** your observations.

7. Hold up the cardboard horizontally. Move the magnet back and forth under the cardboard. **Record** your observations.

What Did You Find Out?

1. How did the magnet affect the new iron nail? Is iron magnetic?

2. How did the nail change after it had been sitting in the water for several days?

3. What did you observe when you moved the magnet under the powder on the cardboard?

4. What evidence do you have that a chemical change occurred?

Is rusting a chemical change or a physical change?

Elements and the Periodic Table

There are 92 elements that exist naturally and 24 elements that can be made in a laboratory. Scientists have listed all these elements in a table called the **periodic table**. For simplicity, the periodic table in Figure 5.13 includes only some of the elements. The full periodic table looks very similar to this table, as you will see in later science courses. Each square in the periodic table represents one element, as in Figure 5.14.

Periodic Patterns

The elements in the periodic table are arranged in columns and rows. The columns are called *groups*. The rows are called *periods*. Elements that have similar properties are close to each other.

Take a look at the simplified periodic table in Figure 5.13.

- Place your finger on the first row. Name the elements in this period.
- Run your finger down the eleventh column. List the elements in this group.
- Find the spot where the fourth row (period) and the seventh column (group) meet. What element is there?

> **INTERNET CONNECT**
>
> www.mcgrawhill.ca/links/BCscience7
>
> To learn more about elements, go to the web site above. Click on **Web Links** to find out where to go next.

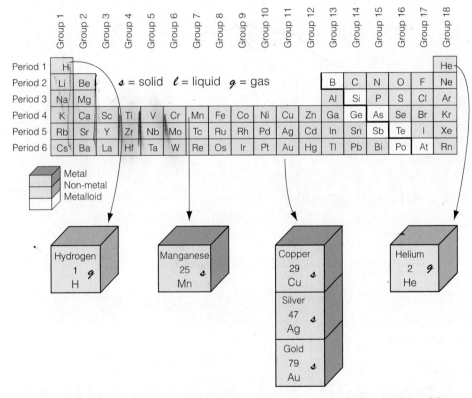

Figure 5.13 The elements are arranged in order across periods, starting with hydrogen. Thus, hydrogen (H) is element number 1. Lithium (Li) is element number 3. What number is beryllium (Be)?

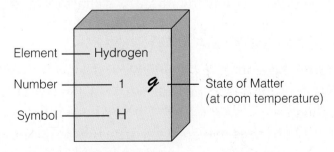

Figure 5.14 Each element has a name, symbol, and number. Oxygen has the symbol O. It is element number 8. Detailed periodic tables are a valuable source of information about elements and their properties.

Define "element" using Dalton's atomic theory.

Explaining Elements and Compounds

How can you explain the differences between elements and compounds using the particle model? Go back to page 144. The particle model of matter states that the particles of a pure substance are the same. It does not tell you what the particles are like. Finding out what the particles are like is key to explaining why elements and compounds are different. There is one problem, however. The particles are far too small to see!

Beyond the Particle Model of Matter

John Dalton (1766–1844), shown in Figure 5.15, was an English teacher and a scholar. In 1808, he introduced a new model of matter. He used his model to develop a theory. A **theory** is an explanation that scientists develop after completing many experiments. Scientists develop theories to explain their observations. Dalton developed his theory as a way to explain the observations about matter that he and other scientists had made. He needed to theorize because he could not actually see the structure of matter.

Figure 5.15 John Dalton developed a theory about matter that is still useful today.

Dalton's Atomic Theory
- All matter is made of small particles, called **atoms.**
- Atoms cannot be created, destroyed, or divided into smaller particles.
- All atoms of the same element are identical. They have the same mass and size. They are different in mass and size from the atoms of other elements.
- Compounds contain atoms of different elements in definite proportions or ratios.

oxygen atom

carbon atom

hydrogen atom

Figure 5.16 These models show that a carbon atom is bigger than a hydrogen atom but smaller than an oxygen atom.

Go back to page 145 to review the definition of an element. Dalton's theory provides a different way to define an element. According to Dalton's theory, an element is a pure substance that is made up of only one type of atom. As well, the atoms in an element cannot be changed, destroyed, or divided. So, Dalton's theory helps to explain why you cannot use a chemical change to get simpler substances from an element. The models of atoms in Figure 5.16 help to illustrate Dalton's theory. Note that the colours are used to help you identify the elements. An oxygen atom is not actually red, and a carbon atom is not actually black. Remember, atoms are millions and millions and millions of times smaller than these models.

Molecules

Dalton theorized that compounds are composed of fixed ratios of elements. For example, water contains fixed ratios of hydrogen and oxygen. If you run electricity through water, you get hydrogen and water. You *always* get *twice* as much hydrogen by volume as oxygen. Hydrogen peroxide also contains hydrogen and oxygen, but in a different ratio. If you decompose hydrogen peroxide, you *always* get *equal* volumes of hydrogen and oxygen.

How is the idea of atoms useful for explaining why compounds contain elements in fixed ratios? Scientists theorize that atoms can link together. When two or more atoms link together, a **molecule** is formed. A molecule of water has two atoms of hydrogen ($2 \times H$) and one atom of oxygen ($1 \times O$). This is why water is sometimes written this way: H_2O. Similarly, a molecule of hydrogen peroxide contains two atoms of hydrogen and two atoms of oxygen (H_2O_2). A chemical change can break apart a water or hydrogen peroxide molecule, forming oxygen and hydrogen.

Four different compounds, and models of the molecules they are made from, are represented in Figures 5.17, 5.18, 5.19, and 5.20.

Dalton's theory and the theory that atoms can form molecules are sometimes grouped together and called the atomic-molecular theory.

DidYouKnow?

Theories are constantly changing. For example, scientists no longer believe that atoms are indestructible. In the late nineteenth century and early twentieth century, scientists found evidence that atoms are made up of still smaller particles. Scientists called these particles *electrons*, *protons*, and *neutrons*. Later in the twentieth century, scientists found evidence that even neutrons and protons are made up of smaller particles. Scientists believe that they still have a lot to learn about atoms.

Figure 5.17 Every water molecule consists of one oxygen atom and two hydrogen atoms.

Figure 5.18 The bubbles that you see in a carbonated drink are carbon dioxide gas. A carbon dioxide molecule consists of one carbon atom and two oxygen atoms.

Figure 5.19 Some stoves burn natural gas to provide heat. Natural gas is mostly methane. Every methane molecule consists of one carbon atom and four hydrogen atoms.

Figure 5.20 Sucrose (white sugar) can be obtained from sugar cane. A sucrose molecule consists of 12 atoms of carbon, 22 atoms of hydrogen, and 11 atoms of oxygen.

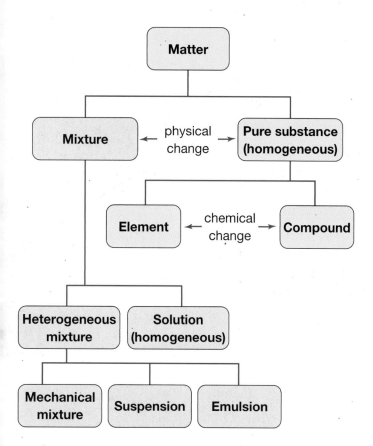

Section 5.3 Summary

In this section, you classified pure substances as elements or compounds.

- An element cannot be broken down into simpler substances by a chemical change.
- Each element is made of a different kind of atom.
- The 116 known elements are listed in the periodic table.
- A molecule is made of more than one atom linked together.
- A compound is made of chemically combined elements.

The chart on the left shows how compounds and elements fit into the classification of matter. In section 5.4, you will learn how you can obtain useful materials, such as pure substances, from mixtures.

Key Terms
element
periodic table
compound
theory
atoms
molecule

Check Your Understanding

1. How can a chemical change help you tell the difference between an element and a compound?

2. Could you use a chemical change to obtain simpler substances from sodium? Explain your answer.

3. Both compounds and mixtures contain two or more elements. How are compounds different from mixtures?

4. **Apply** Atomic-molecular theory is a way to explain what the particles of the particle model of matter are like. For example, according to atomic-molecular theory, a particle of water is a water molecule made of two hydrogen atoms and one oxygen atom.
   (a) According to atomic-molecular theory, what is a particle of carbon?
   (b) According to atomic-molecular theory, what is a particle of white sugar (sucrose)?

5. **Thinking Critically** Both water and hydrogen peroxide contain hydrogen and oxygen. You can drink water, but hydrogen peroxide is poisonous. Why do these compounds have different properties?

CHAPTER 5

Section 5.4 Pure Substances from Mixtures

You probably separate mixtures without even thinking about it. For example, you are separating a mixture when you sort coins based on their colours and sizes. You are separating a mixture when you drain pasta in a colander. Figure 5.21 shows how a washing machine separates a mixture. In the next activity, you will explore methods of separation.

Figure 5.21 The spin cycle on a washing machine separates the clothes from the water.

Find Out ACTIVITY 5-G

Separating Mixtures

What methods can you use to separate different mixtures? Decide how you would recover all the materials from each mixture, in their original form.

What You Need

10 containers of different mixtures, provided by your teacher
dictionary

Safety Precautions

Do not open the containers.

What to Do *Group Work*

1. Make a table that is similar to the one below. Give your table a title.

Mixture	Separation method	Why the method works

2. **Observe** the mixtures. With your group, brainstorm possible methods that you could use to separate each mixture.

For example, you could use:
- filtration
- sifting (using sieves)
- magnetism
- evaporation
- flotation
- dissolving

You may think of other methods you could use. Look up any terms you do not know in a dictionary, and record their meanings.

3. Complete your table. Some mixtures may require more than one method. If a mixture contained marbles, sand, and small pieces of iron, for example, you could use sifting and then a magnet to separate the parts.

What Did You Find Out?

1. Which properties of pure substances did most of the methods involve?

2. Which methods of separation did you list most frequently?

3. (a) Which mixtures would be easier to separate? Why?

   (b) Which mixtures would be more difficult to separate? Why?

Career Connect

Checking Mixtures

Have you ever wondered what is in your milkshake or soft drink, or in the cleaner you use to wash the sink? You can be sure that a chemical research analyst knows. Chemical research analysts examine mixtures. They find out what pure substances are in a mixture to ensure that the mixture contains no harmful substances, such as pesticides.

John Persaud is a chemical research analyst. He works for a petrochemical company that makes products from crude oil, a natural mixture. John's job is to check the contents of these products before they are sold to consumers. He makes sure that the crude oil is good quality and contains no unwanted substances. In the laboratory, he tests samples of the products that are made from the crude oil, such as natural gas, motor oil, gasoline, and diesel fuel. If he finds unwanted substances, he develops ways to separate them from the batch.

John cannot possibly tell what a petrochemical product contains, just by looking at it. Specialized equipment and computers in the laboratory help him separate out all the different substances in the mixture. They also help him determine what the substances are.

John Persaud, a chemical research analyst

Reading Check

Is gold a pure substance or a mixture? Is gold ore a pure substance or a mixture?

Solid Mixtures Underground

Most underground materials are solid—solid rocks. Most rocks are mixtures. For example, the rock shown in Figure 5.22 is a mixture of two pure substances: white-coloured quartzite and gold. This rock is called gold ore because it can be processed to get gold. An *ore* is a rock in the ground that contains one or more valuable substances. Gold ore is an example of a raw material. A **raw material** must be processed to obtain useful products.

The discovery of a large deposit of gold ore is exciting. Gold is valuable because it is beautiful and rare. It is also valuable because of its properties. Gold does not corrode easily, and it can be worked into jewellery because it is soft.

Gold is so valuable that people are willing to go to a lot of trouble to get it. In the nineteenth century, reports of newly discovered gold deposits led to gold rushes. The first gold rush in North America started in 1848 in California. There was another famous gold rush in Canada's Yukon Territory. British Columbia had its own gold rushes. The first was in 1858 on the Fraser River. The second was in 1862 in the Cariboo region.

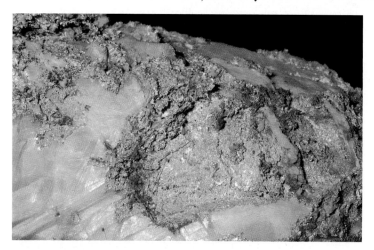

Figure 5.22 The white part of this rock is quartzite. The yellow part is nearly pure gold.

The Cariboo Gold Rush

People came from all over the world for the chance to find gold in British Columbia. They travelled many kilometres by boat and foot to get to a Cariboo mining town, such as Barkerville, shown in Figure 5.23. There, they could stake their claim and start looking for gold.

The simplest way to find gold in streams was by panning. A miner dug out material from the stream bed and swirled it in a pan with plenty of water. Gold is a very dense element. Even tiny pieces of gold are have more mass than pieces of sand or gravel. As the miner swirled the mixture from the stream bed, the pieces of sand, gravel, and mud were washed away. If the miner was lucky, a nugget or two of pure gold would be left at the bottom of the pan. In the next activity, you will make a model to illustrate panning for gold.

Figure 5.23 Barkerville was named after a miner named Billy Barker. Barker found enough gold in the creek nearby to make him very, very wealthy.

Find Out ACTIVITY 5-H

Panning for "Gold"

Make a working model to show how panning for gold works.

How does panning work?

What You Need

about 1 L Styrofoam™ packing "peanuts"
about 1 L marbles
large paper bag
large basin
hair dryer

Safety Precautions

Turn off the hair dryer when you are not using it.

What to Do

1. Pour the "peanuts" and marbles into the large paper bag, and mix them thoroughly.

2. Place the mixture in a large basin. Use a circular motion to swish and swirl the basin steadily. Have a partner aim a hair dryer across the top of the basin. Your partner should aim away from you for safety.

What Did You Find Out?

1. Identify which part of your model represents
   (a) gold nuggets
   (b) gravel
   (c) running water

2. (a) How is your model similar to the method that was used to pan for gold?
   (b) How is your model different from panning?

Chapter 5 Classifying Matter • MHR 153

Section 5.4 Summary

In this section, you learned the following methods for separating mixtures:

- filtration
- sifting (using sieves)
- magnetism
- evaporation
- flotation
- dissolving
- panning

In Chapter 6, you will learn about more methods for separating solutions. You will also learn how some of these methods are used to purify one of our most important resources—water.

Key Terms
raw material

Check Your Understanding

1. Look at the list of separation methods above. Do these methods involve chemical changes or physical changes? How do you know?

2. A mixture contains marbles of three different sizes. How could you use two different sieves to separate the mixture? Draw diagrams to illustrate your method.

3. A mixture contains Styrofoam™ "peanuts" and marbles. What are two different ways to separate this mixture?

4. How would you separate each mixture below?
   (a) wood chips and pieces of granite rock
   (b) iron filings and dirt
   (c) salt and pepper

5. **Apply** The small pitcher shown here is used to separate fat from gravy juices.
   (a) What property of fat and gravy juices allows you to separate them?
   (b) The pitcher is designed so that gravy juices, not fat, pour out of the spout. Gravy juices are mostly water. Why does the spout come from the bottom of the pitcher?
   (c) If you were going to use the gravy shown in the photo, what would you do first? Explain.
   (d) If you did not have a gravy pitcher, how could you separate the fat from the gravy juices?

6. **Thinking Critically** Petroleum (also called crude oil) is a liquid mixture that is found underground. Petroleum is processed to make useful materials such as gasoline, jet fuel, waxes, and diesel oil.
   (a) How is petroleum similar to gold ore?
   (b) How is petroleum different from gold ore?

This pitcher allows you to enjoy gravy with less fat.

CHAPTER at a glance

Now that you have completed this chapter, try to do the following. If you cannot, go back to the sections indicated in brackets after each part.

(a) In your classroom, identify something made of a material that is heterogeneous. List the properties of each type of matter in the material. (5.1)

(b) In your classroom, identify something made of a material that might be homogeneous. Explain why you would need to investigate further to find out whether or not the material is homogeneous. (5.1)

(c) Brainstorm a list of ten foods and drinks that are mixtures. Classify each mixture as precisely as possible. (5.2)

(d) A mixture contains sand, iron filings, and wood chips. Draw a flowchart to show how you would separate this mixture. (5.3)

(e) How are mixtures and compounds the same? How are they different? (5.3)

(f) Draw an atom of oxygen and a molecule of water. How are these models different from real atoms and molecules? Why are models like these useful? (5.3)

(g) What mixture is the colander shown below designed to separate? What other methods for separating mixtures do you use in your daily life? (5.4)

(h) Why are natural materials often processed before they are used to make products? (5.4)

This colander is a type of sieve. How does it work?

Prepare Your Own Summary

Summarize this chapter by doing one of the following. Use a graphic organizer (such as a concept map), create a poster, or write a summary to include the key chapter ideas. Here are a few ideas to use as a guide:

- Illustrate some ways of telling whether or not a substance is a mixture.
- List some ways to identify a pure substance.
- Create a poster to explain the difference between elements and compounds.
- Create flowcharts to show methods for separating different mixtures. What method is shown in the photograph below?

CHAPTER 5 Review

Key Terms

mixture
heterogeneous
pure substance
homogeneous
mechanical mixtures
suspension
emulsion
solutions
alloys
dissolves
solute
solvent
element
periodic table
compound
theory
atoms
molecule
raw material

Reviewing Key Terms

If you need to review, the section numbers show you where these terms were introduced.

1. In your notebook, match each description in column A with the correct term in column B. Use each description only once.

   Column A
   (a) Dalton's theory says they cannot be destroyed
   (b) type of matter that is made of only one kind of atom
   (c) type of matter that cannot be separated by physical changes
   (d) type of matter that must be processed before being used
   (e) mixture such as paint
   (f) pizza is an example
   (g) another name for salt when it is dissolved in water
   (h) particle that is made of two or more atoms linked together

   Column B
   • pure substance (5.1)
   • mechanical mixture (5.2)
   • suspension (5.2)
   • emulsion (5.2)
   • solute (5.2)
   • solvent (5.2)
   • elements (5.3)
   • atoms (5.3)
   • molecules (5.3)
   • compounds (5.3)
   • raw materials (5.4)

2. For each of the following, what is the relationship between the two terms?
   (a) pure substance, homogeneous (5.1)
   (b) solution, homogeneous (5.2)
   (c) mixture, solution (5.2)
   (d) suspension, emulsion (5.2)
   (e) mechanical mixture, properties (5.2)
   (f) element, atom (5.3)
   (g) compound, element (5.3)
   (h) atom, molecule (5.3)
   (i) theory, atom (5.3)
   (j) raw material, mixture (5.4)

Understanding Key Ideas

Section numbers are provided if you need to review.

3. The properties of a sample of matter can help you decide whether the material is homogeneous or heterogeneous. (5.1)
   (a) What is an example of matter that is clearly heterogeneous?
   (b) What is an example of matter that is clearly homogeneous?
   (c) Easily observed properties can be misleading. What is an example of a material that looks homogeneous but is actually heterogeneous?

4. Examine the model below. (5.3)
   (a) What does the model represent?
   (b) How is this model useful?
   (c) How does the model differ from the thing it represents?

156 MHR • Unit 2 Chemistry

5. How would you define the terms "mixture" and "pure substance?" Use ideas from the particle model of matter in your definitions. Give two examples for each definition. (5.1)

Developing Skills

6. How does panning separate gold from gravel? Draw a diagram to help explain.

7. How would you separate each mixture below? Outline your method in a flowchart.
   (a) oil and water
   (b) paper clips and pennies
   (c) sawdust and sugar

8. Models are useful when you are visualizing molecules.
   (a) How could you use Styrofoam™ balls and toothpicks to make models of molecules?
   (b) How would your models be better than the two-dimensional models in this textbook?
   (c) Would there be any drawbacks to your models? Explain.

Problem Solving

9. A liquid that looks like water boils at 106°C.
   (a) Is the liquid pure water? Explain how you know.
   (b) How could you change the boiling temperature of pure water but not its appearance?

10. Suppose that your friend gives you a bracelet that she says is pure copper. You decide to find out for yourself. Using a balance, you find that the mass of the bracelet is 30 g. Using displacement, you find that its volume is 3.5 mL. Is the bracelet made of pure copper? Explain your answer. (Hint: You will need to refer to Table 4.3 on page 119.)

Critical Thinking

11. One type of iron ore is composed of rock and a compound called magnetite. Magnetite is made from the elements iron and oxygen.
    (a) Magnetite is a magnetic compound. How could you separate magnetite from the rock it is mixed with?
    (b) Once you have pure magnetite, can you use physical changes to get pure iron? Explain your answer.
    (c) Magnetite and iron share the property of magnetism. How are they different?
    (d) How is iron ore different from gold ore? Explain your answer.

Magnetite

Pause & Reflect

Go back to the beginning of this chapter on page 130, and check your original answers to the Getting Ready questions. How has your thinking changed? How would you answer these questions now that you have investigated the topics in this chapter?

CHAPTER 6
Exploring Solutions

Getting Ready...

- How is "weak tea" different from "strong tea"?
- How can you make solutes dissolve faster?
- What are acids and bases?
- How can you separate solutions?

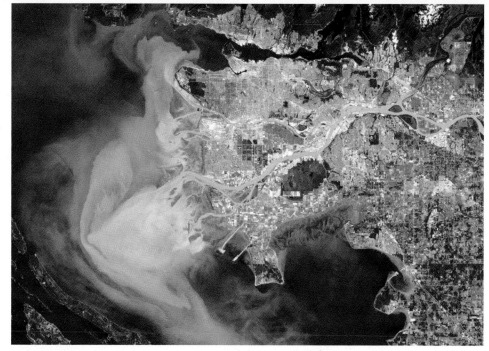

The Fraser River and the Pacific Ocean are both mixtures. The area where they meet is called the Fraser Estuary.

The mighty Fraser River empties its waters into the Pacific Ocean, south of Vancouver. This is a place where the river meets the ocean, and fresh water meets salt water. It is home to many fish and other living organisms. The river water is a complex mixture that carries mud, branches, and lumber, as well as bottles and other garbage. It also contains much smaller particles—dissolved substances that you cannot catch in a net or even see. The dissolved substances include fertilizers and other chemicals from fields along the river, industrial wastes from factories, and wastes from towns and cities.

The Fraser River began its journey to the ocean on the east side of British Columbia, high in the Rocky Mountains of Mount Robson Provincial Park. How is the clear water that trickles into the river from melting glaciers in the park different from the river water that flows into the ocean? How is the river water different from the ocean water? In this chapter, you will focus on the properties of solutions such as river water and ocean water. You will discover that some substances dissolve in water and others do not. You will learn how to classify solutions and how to separate the parts of a solution. For example, how might you get salt out of ocean water?

What You Will Learn

In this chapter, you will learn

- how concentrated solutions and dilute solutions are different
- how to identify acids and bases
- how to separate the parts of a solution

Why It Is Important

- Water dissolves a wide variety of substances. Some substances are healthful, some do not affect you, and some are harmful. Knowing about solutions helps you understand the meaning of "safe drinking water."
- You use acids and bases to clean your home, your clothes, and yourself. Acids and bases affect the taste of your food and the well-being of the environment.

Skills You Will Use

In this chapter, you will

- observe how different variables affect how fast a solute dissolves
- make observations to identify acids and bases
- classify substances as acids, bases, or neutral
- predict whether common solutions are acids, bases, or neutral

Water from melting glaciers forms the beginning of the Fraser River.

Starting Point ACTIVITY 6–A

Cold Tea, Hot Tea

Can you make iced tea with cold water?

What You Need

cold tap water
hot tap water (*not* boiling water)
2 transparent containers, such as glass mugs
3 tea bags

What to Do

1. **Observe** a dry tea bag. (You may open the bag.) Record your observations.

2. Half fill one container with cold water. Place the second tea bag in the water.

3. **Observe** the water and tea bag for a few minutes. Sketch your observations. Let the water and tea bag sit for 1 h.

4. Meanwhile, half fill the second container with hot water. Place the third tea bag in the hot water.

5. **Observe** the hot water and tea bag for a few minutes. Sketch your observations. Let the hot water and tea bag sit undisturbed for 1 h.

6. After 1 h, **observe** the contents of both containers. Record your observations.

What Did You Find Out?

1. Is the dry tea a pure substance or a mixture? Give reasons for your answer.

2. How do you know that something escaped from the tea bags?

3. What difference did you notice when you used hot water instead of cold water?

4. What is the best way to make iced tea? Explain your answer.

Chapter 6 Exploring Solutions • MHR **159**

CHAPTER 6

Section 6.1 Solutes and Solvents

READING check

What is the difference between a substance that is soluble and a substance that is insoluble?

To keep cut flowers fresh longer, some people like to add a little sugar to the water in the vase. As you know, when you mix sugar and water, the sugar dissolves in the water. Another way of saying "the sugar dissolves in the water" is to say "the sugar is soluble in the water." **Soluble** means able to dissolve in a certain solvent.

What about substances that do not dissolve? For example, glass does not dissolve in water. This is one reason why glass is a suitable material for making a vase, like the vase in Figure 6.1. Another way of saying "glass does not dissolve in water" is "glass is insoluble in water." **Insoluble** means unable to dissolve in a certain solvent. Gold, diamonds, and sand are all insoluble in water.

People sometimes call water the "universal solvent" because many different solids, liquids, and gases are soluble in water. For example, the vinegar shown in Figure 6.2 is a solution of ethanoic [eth-a-NO-ick] acid in water. Lemon-lime soda is a solution of carbon dioxide gas, sugar, and flavourings in water. About half of your blood is a solution of essential substances (such as dissolved glucose, vitamins, and minerals) in water.

Some substances are insoluble in water, however. For example, you cannot remove grease from your hands using water alone. The compounds in grease are insoluble in water. You need to use a different solvent, such as soap, to help you remove the grease. In the next activity, you will observe which solvents dissolve two different solutes.

Figure 6.1 Glass is insoluble in water. What would happen if the vase were made of sugar?

Figure 6.2 Many people enjoy a solution of ethanoic acid in water on their fries. You will learn more about acids in section 6.3.

Find Out ACTIVITY 6-B

Does It Dissolve?

In this activity, you will investigate whether salt and flour are soluble or insoluble in two different solvents.

What You Need

4 small transparent plastic cups
4 labels
4 stir sticks
water
vegetable oil
salt
flour
measuring spoons

What to Do

1. Copy the following table of observations into your science notebook. Give your table a title.

Cup	Name of solvent	Name of solute	Observations
1			
2			
3			
4			

2. Label the four cups 1, 2, 3, and 4.

3. Use a measuring spoon to pour about 2 mL of water into cups 1 and 2. Pour about 2 mL of vegetable oil into cups 3 and 4.

4. Record the solvent.

5. **Predict** whether each of the two solutes (flour and salt) will dissolve in one, both, or neither of the solvents. Record your predictions.

6. Using the stir sticks, add a little salt to cups 1 and 3 and a little flour to cups 2 and 4. Record the names of the solutes in your table.

7. Stir each mixture. Carefully **observe** the contents of each cup to determine whether the solutes have dissolved. Record your observations in your table.

8. Clean up the materials as instructed by your teacher. Wash your hands thoroughly.

What Did You Find Out?

1. Summarize your results, using complete sentences.

2. Were your predictions correct?

3. Solutes may be soluble, insoluble, or *partly* soluble. What observations might indicate that a solute is partly soluble?
   (a) Name two common substances that are soluble in water.
   (b) Name two common substances that are insoluble in water.

What Is Insoluble?

Very few materials are completely insoluble. Substances may be insoluble in one solvent but soluble in another solvent. For example, green grass stains are caused by the chemical chlorophyll, which is found in grass leaves. Grass stains are difficult to wash out of clothing because chlorophyll is insoluble in water. Even laundry detergent does not help. You can remove grass stains, however, using rubbing alcohol. Chlorophyll is soluble in rubbing alcohol.

Did You Know?

Nail polish remover is a useful solvent for dissolving nail polish. It can also dissolve the plastic on counters and tabletops, however. Like many solvents, nail polish remover is toxic.

Pause & Reflect

Diagram C in Figure 6.3 below shows that a solution consists of single particles of solute evenly scattered through the solvent. How does this model help to explain the properties of solutions?

Soluble Substances and the Particle Model of Matter

How can you model what happens when a soluble substance dissolves in a solvent? Go back to the particle model of matter in Chapter 4, section 4.3. Do you remember the following point?

- The particles are always in motion. They vibrate, rotate, and (in liquids and gases) move from place to place.

Figure 6.3 shows a model of sugar dissolving in water. The model is based on the particle model of matter. Notice that dissolving happens at the edge of the sugar crystal. First, the water particles pull one sugar particle off of the other sugar particles in the crystal. Then, the motion of the water particles carries the sugar particle away. This makes room for more water particles to move in and pull off another sugar particle. The process continues until all the sugar is dissolved. The sugar particles move around, among the water particles, until the two kinds of particles are evenly mixed.

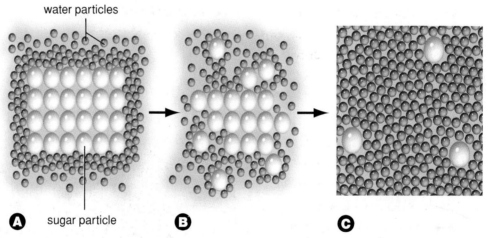

Figure 6.3 The particle model of matter provides a useful way to visualize dissolving. Remember that a grain of sugar contains not 24 but 1 800 000 000 000 000 000 sugar particles—too many to draw!

Recall that water particles are actually water molecules, according to the atomic-molecular theory. As well, sugar particles are actually sucrose molecules. In Figure 6.3, these molecules are shown as spheres. Why? To represent dissolving, you do not need to see the atoms in the molecules. A simple model is sometimes more useful than a more complex model. You will use the model of dissolving shown in Figure 6.3 several times in this chapter.

Dilute or Concentrated?

Figure 6.4 shows two cups of tea. Both cups have the same volume of tea, and both tea bags are the same size. You can see, however, that the tea on the left is darker in colour than the tea on the right. How would you explain this difference in colour?

Look at the light-coloured tea on the right. Because of its light colour, you would probably guess that the tea bag has not been sitting in the water for very long. Only small quantities of the substances in the tea bag have dissolved in the water.

Now look at the dark-coloured tea on the left. The tea bag has probably been sitting in the water for several minutes or longer. The substances in the tea bag have had more time to dissolve in the water. The dark-coloured tea contains more dissolved substances than the light-coloured tea.

The darker tea is a concentrated solution of tea and water. A **concentrated solution** has a large mass of dissolved solute for a certain quantity of solvent.

In comparison, the lighter tea is a dilute solution of tea and water. A **dilute solution** has a small mass of dissolved solute for a certain quantity of solvent.

READING check

You and a friend are making instant hot chocolate. Your friend mixes four spoonfuls of hot chocolate powder into a mug of hot water. You mix two spoonfuls of hot chocolate powder into another mug of hot water. Whose hot chocolate is more concentrated, yours or your friend's?

Concentration

The quantity of solute that is dissolved in a certain quantity of solvent is the **concentration** of the solution. You can describe concentration with the terms "concentrated" and "dilute." You can also express concentration quantitatively, in a number of ways. You will learn some of these ways in the next section.

Figure 6.4 The tea on the left is a concentrated solution. The tea on the right is a dilute solution.

You can change the concentration of a solution by
- adding more solute to make the solution more concentrated
- adding more solvent to make the solution more dilute

Suppose that you make a fruit drink using drink crystals. If you find that the taste is too strong, you can add water to make the solution more dilute. If you find that the taste is not strong enough, you can add more drink crystals to make the solution more concentrated.

Section 6.1 Summary

In this section, you explored different ways to talk about solutes, solvents, and solutions.

- A substance is soluble in a solvent when it dissolves in this solvent. For example, salt is soluble in water.
- A substance is insoluble in a solvent when it does not dissolve in this solvent. For example, diamonds are insoluble in water.
- A concentrated solution has a large mass of solute for a certain volume of solvent.
- A dilute solution has a small amount of solute for a certain volume of solvent.

Is there a limit to how concentrated a solution can be? In other words, can you add so much salt to a solution that no more salt will dissolve? You will explore this question in the next section.

Key Terms

soluble
insoluble
concentrated solution
dilute solution
concentration

Check Your Understanding

1. (a) What are two examples of substances that are soluble in water?
   (b) What are two examples of substances that are insoluble in water?
2. Why do you need to use soap to wash grease off your hands?
3. Your friend is drinking apple juice. He finds the taste too strong. What could he do to make the apple juice more dilute?
4. **Apply** The diagrams below model two different solutions.
   (a) Which diagram models a more concentrated solution? Explain your answer.
   (b) Which diagram models a more dilute solution? Explain your answer.

A   B

5. **Thinking Critically** "There is a solvent for every solute." Do you agree or disagree with this statement? Explain why.

CHAPTER 6

Section 6.2 Dissolving

If you put a spoonful of salt in a glass of water, the salt will dissolve. Suppose that you put a second spoonful of salt in the water. Then you add a third spoonful, and a fourth. The solution is becoming more and more concentrated. Will each spoonful dissolve like the spoonful before it? Do you think there is a limit, when a spoonful of salt will no longer dissolve? If so, what determines this limit? You will answer these questions in the next activity.

Find Out ACTIVITY 6-C

How Much Is Too Much?

How much salt can dissolve in a flask of water?

What You Need

graduated cylinder
tap water
250 mL beaker
2.5 mL measuring spoon
salt
stirring rod
balance

What to Do

1. Using the graduated cylinder, measure 100 mL of cold tap water. Pour the water into the beaker.

2. Using the measuring spoon, add 2.5 mL of salt to the water. Stir the mixture until all of the salt is dissolved.

3. Repeat step 2 until some of the salt will no longer dissolve. Keep track of the volume of salt as you add it. Record the total volume of salt you added.

4. Using a balance, measure and record the mass (in grams) of the total volume of salt you added. (Use fresh salt.)

5. Record your results in a class table.

The Dead Sea contains so much dissolved salt that no plants or fish can survive. The dissolved salt increases the density of the water. Thus, it is very easy to float on the Dead Sea.

What Did You Find Out?

1. (a) What was the mass of water in your beaker at the beginning? (Hint: 1 mL of water has a mass of 1 g.)

   (b) What total mass of salt did you dissolve in this mass of water?

2. Did different students get different results? If so, suggest reasons for the different results.

Chapter 6 Exploring Solutions • MHR 165

A Limit to Dissolving

A **saturated solution** forms when no more solute will dissolve in a specific amount of solvent, even with plenty of stirring. For example, a saturated solution of salt is formed when 35.7 g of salt is dissolved in 100 g of water at 0°C. If more salt is added to this solution, it will not dissolve. Figure 6.5 shows a saturated solution. The lowest part of the Dead Sea, shown on the previous page, is a saturated solution. The floor of the Dead Sea is covered in undissolved salt.

An **unsaturated solution** is a solution that is not yet saturated. Therefore, more solute can dissolve in it.

Figure 6.5 This bluestone solution is saturated. Notice the lump of undissolved solid that remains on the spoon.

DidYouKnow?

It is sometimes possible to pass the saturation limit. A solution that contains more solute than would normally dissolve at a certain temperature is called a *supersaturated solution.* You can prepare a supersaturated solution by making a saturated solution and then cooling the solution without stirring. The solute stays dissolved if the solution is not disturbed. If you add a small crystal of solute (A), however, the extra solute in the solution quickly becomes solid (B). Only some solutes can be used to make a supersaturated solution.

READING Check

What is the difference between a saturated solution and an unsaturated solution?

Solubility

Solubility refers to the mass of solute that can dissolve in a certain volume or mass of solvent, at a certain temperature. For example, scientists have determined that 35.7 g of salt will dissolve in 100 g of ice-cold water. In other words, the solubility of salt in water at 0°C is 35.7 g/100 g. You will soon find out why temperature is important.

Table 6.1 lists the solubilities of several common substances in water. Notice that the solubilities are given as mass of solute (in grams) that will dissolve in 100 g of water at 0°C to form a saturated solution.

Table 6.1 Solubilities of Some Common Substances at 0°C

Substance	State	Solubility (g/100 g water)
baking soda	solid	6.9
bluestone	solid	31.6
canola oil	liquid	insoluble
carbon dioxide	gas	0.34
Epsom salts	solid	70.0
ethyl alcohol	liquid	unlimited
limestone	solid	0.0007
nitrogen	gas	0.003
oxygen	gas	0.007
salt	solid	35.7
sugar	solid	179.2

Off the Wall

Some solids are so soluble that they will "grab" water molecules from the air to dissolve themselves! Over time, they become liquid solutions. These solids are used in devices that remove moisture from the air (dehumidifiers). The scientific name for this property is *deliquescent* [deh-lih-KWE-sent].

Look at Table 6.1 again. Compare the masses of different solutes that dissolve in water at 0°C. The most soluble solid substance in the table is sugar. Amazingly, you can dissolve nearly 180 g of sugar in only 100 g of water! The least soluble solid substance in the table is limestone. Limestone is commonly used to make buildings and statues. If limestone did not have low solubility, the buildings and statues would quickly dissolve in the rain.

Just like density and boiling point, solubility is a property of a pure substance. Thus, you can use solubility to identify a substance. Do you think you could change the solubility of a substance? You will find out in the next activity.

Pause & Reflect

The solution in the hummingbird feeder in Figure 6.6 is made by dissolving 30 g of sugar in 100 g of water at 0°C. Is the solution saturated or unsaturated? If it is unsaturated, how much more sugar could you dissolve in it?

Figure 6.6
Hummingbird feeders contain a sugar solution that is made by dissolving 30 g of sugar in 100 g of water. In other words, the concentration of the sugar water is 30 g/100 g.

THINK & LINK
INVESTIGATION 6-D

SKILLCHECK
- Interpreting Data
- Communicating
- Predicting
- Hypothesizing

How Does Temperature Affect Solubility?

Think About It
What happens to the particles in a solvent as the solvent is heated? Do you think heating a solvent will affect the amount of solute that will dissolve in it?

The data table below shows the solubility of three different solutes at various temperatures. In this investigation, you will examine these data and develop a hypothesis about how temperature affects solubility.

Question
How does temperature affect the solubility of a solid solute in a liquid solvent?

What to do

1. Draw the axes for a graph. Label the *y*-axis "Solubility (g/1000 g)." Label the *x*-axis "Temperature (°C)." Mark the scale for the *x*-axis to go from 0 to 100.

2. Plot the data in the following table. Use a different colour for each solute. Include a key to show the solute that each colour represents.

Temperature versus Solubility

Temperature (°C)	Solubility in water (g/1000 g) (grams of solute that will dissolve in 1000 g of water)		
	sodium chloride	potassium chlorate	ammonium chloride
10	370	50	320
20	373	70	370
30	378	110	410
40	380	150	460
50	382	210	500
60	387	270	550
70	390	340	600

3. Connect the points for each solute by drawing a line of best fit.

4. Use dashes to extend the line for each solute so that it crosses 100°C.

5. Give your graph a title.

Analyze
1. Are the graphed lines straight or curved?

2. What happens to the lines as temperature increases?

3. Predict the solubility of each solute at 95°C, using your extended line.

4. Which solute has the highest solubility at each of the following temperatures?
   (a) 20°C
   (b) 60°C
   (c) 95°C

Conclude and Apply
5. What happened to the solubilities of the solid solutes as the temperature of the water increased?

6. (a) Based on the evidence in this investigation, how does temperature affect solubility for a solid solute and a liquid solvent? Write a hypothesis.
   (b) How could you test your hypothesis further?

Skill POWER

For tips on drawing graphs, turn to SkillPower 5.

Rate of Dissolving

As you have discovered, the mass of a solute that will dissolve in a solvent depends on temperature. What factors do you think might affect *how fast* a particular mass of solute will dissolve? When you measure how fast a solute dissolves, you are measuring the **rate of dissolving**.

You have probably already tried one way of making a solute dissolve faster: stirring. For example, have you ever made a fruit drink by adding cold water to drink crystals (like the student in Figure 6.7)? If so, you probably stirred the mixture to make the solute dissolve faster. Why do you think stirring or shaking makes a solute dissolve faster?

Figure 6.8 shows the particles of a sugar crystal being dissolved in water. Each sugar particle is carried away by nearby water particles. Before the mixture is stirred (A), the sugar particles move into the solution at a rate that is determined by the movement of the nearby water particles. So, the sugar and water particles slowly move away from the crystal. The solution that is close to the crystal is concentrated, while the solution that is farther away from the crystal is more dilute.

When you stir the mixture (B), you push some of the concentrated solution away from the crystal. At the same time, you push dilute solution closer to the crystal. As a result, the sugar particles spread through the water much faster.

Figure 6.7 How is this student affecting the rate of dissolving?

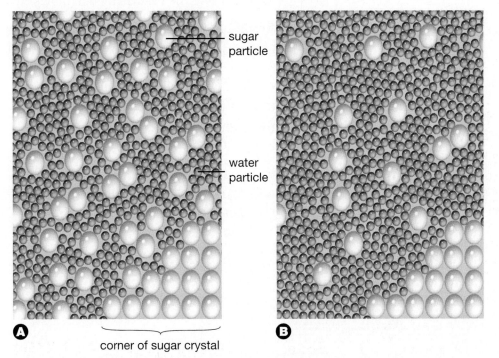

Figure 6.8 How does stirring affect the rate of dissolving? How is the arrangement of the particles in a mixture before stirring (A) different from the arrangement of the particles after stirring (B)?

In the next investigation, you will observe how variables such as stirring affect the rate of dissolving.

DESIGN YOUR OWN
INVESTIGATION 6-E

SKILLCHECK
- Designing Experiments
- Controlling Variables
- Measuring
- Interpreting Data

Changing the Rate of Dissolving

In this investigation, you will observe how the rate at which Epsom salts dissolve in water is affected by changing one of the following variables: temperature, speed of stirring, or size of the crystals.

Question

What tests would you use to find out how the rate of dissolving is affected by changing one variable, as indicated below?
- changing the temperature of the water **(Part 1)**
- changing the rate of stirring **(Part 2)**
- using small or large crystals of Epsom salts **(Part 3)**

Skill POWER

For tips on designing investigations, turn to SkillPower 6.

Safety Precautions

Do not drink any solution you make in your science classroom. Clean up any spills immediately.

Apparatus
graduated cylinder or measuring cup
2 beakers (250 mL)
2 measuring spoons (10 mL)
2 stirring rods
timer (for measuring seconds)
sieve
bowl

Materials
Epsom salts
water (hot and cold)

Procedure

1. Copy the following table into your science notebook to record your observations. Give your table a title.

2. In a small group, discuss which variables you will need to keep constant in each part of the investigation. (Note: You will have to do some stirring in Part 1. How will you ensure that your results are due to temperature, not stirring?)

3. **Predict** the effect of each variable you will change on the rate of dissolving. Explain your predictions using the particle model of matter.

4. Plan how you will carry out your tests in each part of this investigation. Each part must show how changing one variable affects the rate of dissolving, as outlined in the Question. (Hint: Use the same quantities of water and Epsom salts in all your tests. Carry out all your tests in the 250 mL beakers.)

5. Give each member of the group a task, such as timing, recording, sifting, mixing, or stirring. (Some tasks may be shared.)

6. Carry out your tests for each part of the investigation. **Record** your observations in your table.

Part	Variable	Beaker 1		Beaker 2	
		Description	Time to dissolve	Description	Time to dissolve
1	temperature				
2	rate of stirring				
3	size of crystals				

170 MHR • Unit 2 Chemistry

Analyze

1. (a) Was your prediction for Part 1 correct? Explain how the particle model helped you make your prediction.
   (b) Was your prediction for Part 2 correct? Explain how the particle model helped you make your prediction.
   (c) Suggest a reason why the size of the salt crystals in Part 3 affected the rate of dissolving.

2. (a) In your group, list at least ten more variables you could test, such as time of day, amount of light, and shape of the beaker. Be creative. Record anything that varies and that may have an effect on the rate of dissolving, however unlikely it is.
   (b) Discuss why each new variable on your list might or might not affect the rate of dissolving.

3. (a) Based on your results for Part 3, what additional variable could you have controlled in Parts 1 and 2?
   (b) Why would this have been an improvement in the design of the investigation?

Conclude and Apply

4. Summarize what you discovered in each part of your investigation.

5. You learned in Chapter 5 that mining for gold was once as simple as sifting through sand in a river. Most gold that is mined today is found as tiny flakes of gold metal embedded in gold ore. One or more tonnes of ore may contain a total of only one or two grams of gold. To extract the gold, the ore is placed in a solvent that dissolves gold but not rock. Suggest a procedure that would increase the rate at which the gold dissolves.

Extend Your Skills

6. (a) Measure the time for Epsom salts to dissolve in water at three or more different temperatures. Then graph your results.
   (b) Measure the time for three or four different speeds of stirring, and graph your results.

Part 1 These students are comparing the rate of dissolving in solvents at different temperatures.

Part 2 The students are comparing the rate of dissolving when they stir at different rates.

Part 3 The students are using a sieve to separate small crystals of Epsom salts from larger crystals.

Reading Check

Describe two ways to make a sugar cube dissolve more quickly in a glass of water.

Temperature and Rate of Dissolving

You have learned how the particle model of matter explains the effect of stirring on the rate of dissolving. Does the particle model also explain the effect of temperature on the rate of dissolving? Remember that the particles in matter are always moving. They move faster, on average, at higher temperatures. So, at higher temperatures, the particles in the concentrated solution near the solute move away from the solute more quickly. The particles in the dilute solution farther away from the solute move in quickly to take their place. In this way, the solute particles are spread throughout the solvent particles more quickly.

Size of Solute and Rate of Dissolving

Why do small pieces of solute dissolve faster than larger pieces? Dissolving a solid in a liquid takes place on the surface of the solid. By breaking up a large solid into several smaller pieces, you produce more surface for dissolving to occur.

Figure 6.9 shows a whole sugar cube and a sugar cube that has been broken into smaller pieces. The surfaces of both cubes were coloured red by a felt pen before one cube was broken. You can see that breaking the cube created new surfaces, which are white in the photograph. The total surface of the pieces is greater than the surface of the unbroken cube. Since more surface is in contact with the solvent, the broken cube will dissolve faster than the unbroken cube—even though both have the same total mass of sugar.

Did You Know?

Some candies are made with confectioner's sugar. This is the same compound as regular table sugar, but it has smaller particles. If you put a crystal of each type of sugar on your tongue, the confectioner's sugar will taste sweeter. This is because more of it has dissolved in the same time. (You can only taste dissolved sugar.)

Figure 6.9 Two identical sugar cubes were coloured red on the outside. Then, the sugar cube on the right was broken into small pieces. Which sugar cube will dissolve faster? Why?

Section 6.2 Summary

In this section, you learned how to change the mass of solute that will dissolve in a solvent. You also learned how to change the rate at which a solute dissolves.

- When a solution is unsaturated, more solute can dissolve at the same temperature. When a solution is saturated, no more solute will dissolve at the same temperature.
- Different solutes have different solubilities. The solubility of a solid in a liquid can be increased by increasing the temperature.
- The rate at which a solid dissolves in a liquid can be increased by stirring the solution, increasing the temperature of the solution, or increasing the area of the solid by breaking the solid into smaller pieces.

In the next section, you will learn a way to classify solutions.

Check Your Understanding

Key Terms
saturated solution
unsaturated solution
solubility
rate of dissolving

1. Suppose that you add some solid detergent to the water in a washing machine. Then you decide that your clothes are really dirty, so you add more detergent. Is the solution of detergent and water now more concentrated or more dilute? Explain.

2. What happens to the particles of sugar when a sugar cube is placed in water?

3. What are three ways to make a solute dissolve faster?

4. **Apply** Suppose that you are conducting a test for saturation. You add a small amount of solute to three different solutions, labelled A, B, and C. Based on the observations given, is each solution saturated, unsaturated, or supersaturated? Explain your answers.
   (a) Solution A: The crystal of solute dissolves.
   (b) Solution B: The crystal of solute does not dissolve.
   (c) Solution C: Many more crystals form suddenly.

5. **Thinking Critically** To keep roads safer in the winter, highway crews often spread salt on them to melt the ice and snow. Which form of salt would you expect to stay on a road longer, rock salt (large chunks) or table salt (fine grains)? Give reasons for your answer. (Hint: Both types of salt are shown on the right. Which would dissolve more quickly?)

CHAPTER 6

Section 6.3 Acids and Bases

What do you think of when you read the word "acid"? Do you think of any of the things shown in Figure 6.10? An **acid** is a compound that dissolves in water to form a solution with properties such as these:

- An acidic solution, such as lemon juice, tastes sour.
  (**Caution:** Remember that you must *never* taste anything in the science classroom.)
- An acidic solution corrodes metals, such as lead and iron.
- An acidic solution can cause severe burns to skin.

Figure 6.10 Lemons, vitamin C, and ant venom are acidic.

What do you think of when you read the word "base"? Do not be surprised if you cannot answer this question. Most people have heard of acids, but many people are less familiar with bases. Even so, you come in contact with bases more often than you might think. Figure 6.11 shows some examples. A **base** is a compound that dissolves in water to form a solution with properties such as these:

- A basic solution, such as tonic water, tastes bitter.
  (**Caution:** Remember that you must *never* taste anything in the science classroom.)
- A basic solution breaks down oils and fats.
- A basic solution has a slippery texture.
- A basic solution can cause severe burns to skin.

READING check

What are two examples of acids? What are two examples of bases?

Figure 6.11 Soap, oven cleaner, and antacid tablets are all basic.

Identifying Acids and Bases

Most acidic solutions and basic solutions are colourless and transparent. Therefore, you cannot tell the difference between an acid and a base by appearance alone. One safe way to determine whether something is an acid or a base is to use an indicator. An **indicator** is a compound that turns one colour in an acid and a different colour in a base. There are many types of acid-base indicators. One of these is litmus paper. As you can see in Figure 6.12, litmus paper may be blue or red. Figure 6.13 shows what happens when an acidic solution and a basic solution come in contact with litmus paper. An acid causes blue litmus paper to become red. A base causes red litmus paper to become blue.

Figure 6.12 Litmus is a compound that is extracted from plant-like organisms called lichens. Litmus paper is made by dipping a small strip of paper into a diluted solution of litmus and letting the paper dry.

A Scale for Classifying Acids and Bases

Scientists describe the acidity of a solution by using the **pH scale**. As you can see in Figure 6.14, the pH scale tells you whether a substance is an acid or a base. It also tells you how acidic and how basic a substance is. The pH scale has numerical values from 0 to 14.

- Acids have a pH that is less than 7. Substances with a lower pH are more acidic than substances with a higher pH. A substance with a pH of 0 is *very* acidic.
- Bases have a pH that is greater than 7. Substances with a higher pH are more basic than substances with a lower pH. A substance with a pH of 14 is *very* basic.
- Substances that fall in the middle of the scale, with a pH of 7, are not acidic and not basic. They are **neutral**.

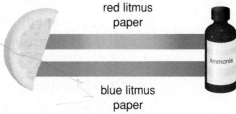

Figure 6.13 Here is a good way to remember the litmus test: Acids turn blue litmus red. Bases turn red litmus blue.

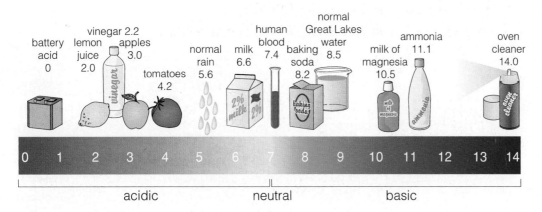

Figure 6.14 The pH scale classifies substances as acids or bases. It runs from 0 to 14. What is the pH of vinegar? Is vinegar more acidic or less acidic than milk?

CONDUCT AN INVESTIGATION 6-F

SKILLCHECK
- Predicting
- Observing
- Classifying
- Communicating

Acid, Base, or Neutral?

In this investigation, you will use litmus paper to classify common solutions as acidic, basic, or neutral. Then you will use universal pH paper to determine the pH of each solution.

Question

Which common household solutions are acids? Which are bases? Which are neutral solutions?

Some common household acids and bases

Safety Precautions

- Never taste anything in the science classroom.
- Never touch an unknown substance with your bare hands.
- Follow your teacher's instructions for disposing of solutions.
- Inform your teacher about any spills immediately.
- When you have finished the investigation, wash your hands thoroughly. Clean up your work area as instructed by your teacher.

Apparatus

stirring rod
colour key for universal pH paper

Materials

universal pH paper
paper towels
water

as many of the following solutions as you have time to test: household bleach, club soda, tonic water, milk of magnesia, liquid soap or shampoo, household window cleaner, black coffee, distilled water, rainwater, vinegar, lemon juice, milk

Procedure

Part 1: The Litmus Test

1. In your science notebook, make a table of observations like the one below. Give your table a title.

Substance	Prediction	Colour change (if any) of red litmus paper	Colour change (if any) of blue litmus paper	pH observed	Acidic, basic, or neutral?

176 MHR • Unit 2 Chemistry

② In the first column, **record** the name of your first test solution.

③ **Predict** whether the solution is acidic, basic, or neutral. **Record** your prediction in the second column.

④ Use the stirring rod to place one drop of the solution on a piece of red litmus paper. Wait until you see a colour change. Then **record** your observations in the third column.

⑤ Repeat step 4 with blue litmus paper.

⑥ Thoroughly clean the stirring rod with water and paper towels.

⑦ Repeat steps 2 to 6 for each test solution.

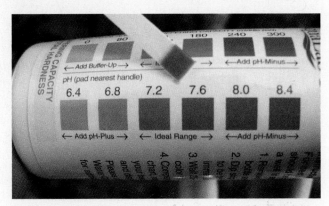

You can determine the pH of a solution by matching the colour change it causes on a piece of universal pH paper with the colour key that comes with the paper.

Part 2: What Is the pH?

1. Using the clean stirring rod, put one drop of your first test solution on a piece of universal pH paper.

2. Compare the colour change of the paper with the colour key provided. **Estimate** and **record** the pH of the solution in your table.

3. Repeat steps 1 and 2 with the remaining samples.

4. Discard the solutions and indicators as instructed by your teacher. Wash your hands thoroughly.

Analyze

1. Which of your predictions were accurate?

2. Did any results surprise you? If so, which ones? Why?

Conclude and Apply

3. Why did you need to test with both red and blue litmus paper to classify substances as acidic, basic, or neutral?

4. What colour do you predict universal pH paper would become if you tested each substance below? Give reasons for your prediction.
   (a) ant venom (pH 3.0)
   (b) rainwater (pH 5.6)
   (c) blood (pH 7.4)

Figure 6.15 Marble monuments contain a compound called calcium carbonate. Acids react easily with calcium carbonate, wearing it away.

Acids and Bases in Action

The pH of "normal" rainwater is about 5.6. Rainwater is weakly acidic. It becomes acidic because carbon dioxide gas in the air dissolves in water vapour that is also in the air. A far more acidic solution is produced when some air pollutants combine with the water vapour. For example, gases from smokestacks, fires, and automobiles dissolve in the water vapour in air. When the water vapour condenses and falls as rain, the rainwater is a diluted solution of two powerful acids: sulfuric acid and nitric acid. You probably know the common name for this rain: acid rain.

Acid rain harms buildings and statues by reacting with the rock from which they are made. Figure 6.15 shows an example. Acid rain also harms plants and animals that live on land and in lakes, rivers, and streams.

Find Out ACTIVITY 6-G

Home-Grown Indicators

In this activity, you will use the juice of a red cabbage to test the pH of common household chemicals.

What You Need

9 small test tubes or jars
test tube rack (if using test tubes)
concentrated red cabbage juice in a medicine dropper bottle
samples of the following solutions: vinegar, household ammonia, lemon juice, powdered Aspirin™, baking soda, powdered antacid, tonic water, soda water, distilled water

Safety Precautions

What to Do

1. Read through steps 2 to 6, and design a table to record your observations.

2. Label each test tube (or jar) with the name of the solution that will be poured into it.

3. Half fill each test tube with a different solution.

4. Your teacher may ask you to help prepare the cabbage juice indicator. Otherwise, obtain a dropper bottle of the indicator.

5. Add ten drops of cabbage juice indicator to each test tube.

6. **Observe** the colour change that occurs in each test tube. Record your observations.

What Did You Find Out?

1. Use table below to classify the solutions you tested as acidic, basic, or neutral.

Colour after addition of cabbage juice indicator	red	light purple or pink	purple	blue	bluish-green	green
Approximate pH value	2	4	6–7	8	10	12

2. Which test solution was most acidic? Which was most basic?

Section 6.3 Summary

Materials can be classified as acids or bases according to their properties. An indicator is a natural compound that changes colour in the presence of an acid or a base. The pH of a substance tells you whether the solution is acidic, basic, or neutral.

- Neutral solutions have a pH of 7.
- Acids have pH values that are less than 7.
- Bases have pH values that are greater than 7.

In the final section of this chapter, you will learn how and why people process solutions.

Check Your Understanding

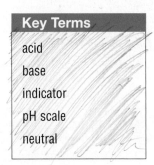

Key Terms
acid
base
indicator
pH scale
neutral

1. What are three properties of acids?

2. (a) What is one substance that you could use as an indicator?
   (b) Suppose that you used the indicator you named in part (a) to test an unknown solution. What colour change would occur if the solution is an acid?
   (c) What colour change would occur if the solution is a base?

3. An unknown substance was tested with litmus paper. Red litmus turned blue, and blue litmus stayed blue. Is the substance an acid or a base?

4. Is each of the following liquids acidic, basic, or neutral?
   (a) household bleach (pH 12.4)
   (b) urine (pH 6.0)
   (c) black coffee (pH 5.0)
   (d) sugar water (pH 7.0)
   (e) egg white (pH 7.8)

5. Arrange the solutions in question 4 in order of increasing acidity.

6. **Apply** Cheese and yogurt are made by the action of certain bacteria on milk. The pH of fresh milk is about 6.5. The pH of cheese is usually about 5.5. The pH of yogurt is about 4.5.
   (a) Which is most acidic: milk, cheese, or yogurt?
   (b) How does the taste of milk, cheese, and yogurt support your answer to part (a)?

7. **Critical Thinking** You have decided to go into the indicator business! You will sell paper indicator strips made with cabbage juice rather than litmus. How will you prepare your indicator paper? In your answer, consider what you know about the properties of matter and solutions. How has your understanding of properties helped you design your indicator paper?

CHAPTER 6

Section 6.4 Processing Solutions

Boiling a solution is a common method for separating a solute from a solvent. As the liquid evaporates, the solute remains. Therefore, the solution becomes more and more concentrated. You have probably eaten one solution processed by boiling—maple syrup.

Long ago, Aboriginal peoples in parts of North America learned how to make syrup from the sap of eastern maple trees, as shown in Figure 6.16. Maple sap is a dilute solution of sugar, water, and small amounts of other substances needed by the trees. In the spring, the sap is collected by driving a small tube into the trunk of each maple tree. The sap flows out of the tube and into a bucket that is hung below it. Buckets of sap, collected from several maple trees, are poured into large boilers and then heated over a fire for several hours. Most of the water (the solvent) is removed by boiling, leaving behind a more concentrated solution of syrup. About 30 to 40 L of sap are needed to make 1 L of maple syrup.

Figure 6.16 Buckets of sap that have been collected from maple trees are poured into a boiler. Then the sap is heated over a fire to make maple syrup.

Separation by Crystallization

Is that snow in Figure 6.17? No, it is salt. Many of the lakes in the Thompson, Okanagan, Cariboo, and Chilcotin regions of British Columbia contain high concentrations of salt. When the water evaporates from these saltwater lakes during the hot, dry, summer months, the salt is left behind. This process is called **crystallization**. In the next investigation, you will model the process of crystallization.

DidYouKnow?

On the West Coast, a similar method is used to process birch syrup. Birch syrup is a rich source of vitamins (such as vitamin C) and minerals (such as potassium, calcium, and thiamin).

Figure 6.17 What process produced these salt deposits?

180 MHR • Unit 2 Chemistry

CONDUCT AN INVESTIGATION 6-H

SKILLCHECK
- Predicting
- Observing
- Interpreting Data
- Communicating

Separating Salt from Salt Water

Solutions can be more difficult to separate than heterogeneous mixtures. You cannot use a filter or a magnet to get the salt from salt water, for example. What physical change can help you get salt from a salt water solution?

Question

How can you separate all the salt from a sample of salt water?

Safety Precautions

- Use oven mitts or tongs to handle hot glassware.
- Wear a visor when near the evaporating dish as it is heating.
- Unplug the hot plate at the end of the investigation, and let it cool before putting it away.

Apparatus
evaporating dish
50 mL graduated cylinder
hot pad
hot plate
tongs

Materials
50 mL salt water

Procedure

❶ **Predict** whether salt can be separated from a salt-water solution by heating the solution. **Record** your prediction.

❷ Pour 50 mL of salt water into the empty evaporating dish. Cover the evaporating dish with the watch glass.

❸ Put the evaporating dish on the hot plate, and very gently heat the solution. **Observe** the solution as it heats. **Record** your observations.

❹ When all the water has evaporated, remove the evaporating dish from the hot plate. Place the evaporating dish on the hot pad to cool.

❺ When the evaporating dish has cooled, remove the watch glass. **Observe** the residue in the evaporating dish. **Record** your observations.

❻ Clean the evaporating dish and watch glass with tap water.

❼ Clean up any spills, and wash your hands thoroughly.

Analyze

1. Describe the appearance of the solution in step 2.

2. Describe the residue that remained in the evaporating dish in step 5.

Conclude and Apply

3. What happened to the water in the solution?

4. What substance remained in the evaporating dish?

5. Did your observations support your prediction?

Extend Your Skills

6. What method could you use to collect the water that evaporated?

INTERNET CONNECT

www.mcgrawhill.ca/links/BCscience7

Earthquakes, tornadoes, and other natural disasters often destroy the pipes that bring drinking water to a community. One of the biggest risks to survivors is drinking contaminated water. How might you purify water in an emergency, using only energy from the Sun? To explore this question, go to the web site above. Click on **Web Links** to find out where to go next.

Separation by Distillation

Suppose that you are lost in a desert and the only liquid you have is a salt solution. You need to separate the salt from the water so that you have something to drink. If you let the water evaporate, it will be lost into the air as water vapour. You will be left with only the salt. Distillation is another method for separating solutions. Distillation lets you recover both the solvent and the solute. During **distillation**, the solution is heated until the solvent changes to a gas. Then the gas is condensed back to a liquid by cooling. The solute does not change state and is left behind. Figure 6.18 shows the apparatus that is commonly used for distillation in a laboratory. Figure 6.19 shows an apparatus that you could use to collect drinking water from salt water in a desert.

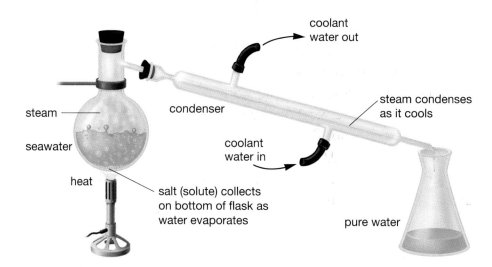

Figure 6.18 During distillation, the solvent is first evaporated to separate it from the solute. Then the gas is condensed to produce a liquid.

Pause & Reflect

Copy Figure 6.19 into your science notebook. Add labels and arrows to show how the distillation process works. Also add the following terms: liquid, vapour, evaporate, condense.

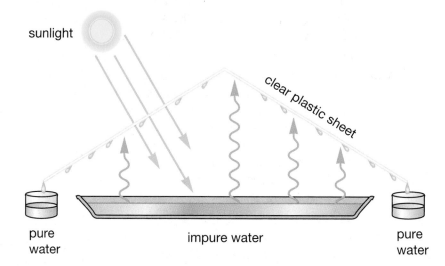

Figure 6.19 This method for distilling drinking water from salt water requires simple materials.

182 MHR • Unit 2 Chemistry

Purifying Water

In nature, fresh water is rarely pure. It contains many dissolved substances. The types of solutes, their concentration in the water, and the pH values of the water vary from lake to lake and river to river. Fish and plants that live in the water are adapted to the conditions where they live and may not be able to survive in a different location with different conditions. People who keep aquariums, like the one in Figure 6.20, know how important the quality of the water is to the health of their fish.

Water quality is also important to human health. Where does your drinking water come from? Depending on where you live, your source of water may be a lake, a reservoir, or a well. No matter what the source, water usually needs some sort of treatment to make it drinkable before it reaches the taps in your home.

Water is purified for drinking using both physical and chemical methods. For example, flotation and filtration are used to remove solid matter. After solid matter has been removed, chemicals are added to the water to kill bacteria and balance the pH.

Figure 6.20 Fish can only survive if the chemistry of their water is suitable.

Career CONNECT

Testing the Waters

When you visit an aquarium, you can expect to see everything from beluga whales to star fish and sea mollusks, all living in their own special tanks. Keeping these tanks filled with clean water is more complicated than just turning on a tap. Mackenzie Gier is an aquarist at the Vancouver Aquarium. She looks after 30 tanks that house animals such as the octopus, jellyfish, and cold-water coral.

Part of Mackenzie's job involves testing the quality of the water in the tanks. She checks the salt concentration and temperature of the water once a day. She checks the dissolved oxygen content and the pH of the water once a week. Mackenzie uses her knowledge of chemistry when she decides what kinds of chemicals need to be added to the water in the tanks. "We can predict what a certain chemical will do if we add it to the tanks simply by knowing the properties of the chemical and the chemical changes."

Mackenzie says her job provides constant variety. "On any day, I can be designing a new tank, trying to raise a new species of fish, or be out on a collecting trip in the Queen Charlotte Islands. It's pretty easy to love your job when you get to come to work and interact with wolf eels."

Mackenzie Gier

Section 6.4 Summary

Dissolving is a physical process, so the parts of a solution can be separated by physical methods.

- Crystallization occurs when a solvent evaporates from a concentrated solution, leaving crystals of solute behind.
- In distillation, a solution is heated until the solvent changes to a gas. Then the gas is condensed back to a liquid by cooling.
- Water can be purified for drinking using both physical and chemical methods.

Key Terms
crystallization
distillation

Check Your Understanding

1. How would you describe maple sap and birch sap? How would you describe the method that is used to process these solutions? Use the terms "dilute" and "concentrated" in your answers.

2. What is crystallization? Give an example to explain your answer.

3. What is the purpose of distillation? Use the terms "solvent" and "solute" in your answer.

4. **Apply** The following diagram shows how you could obtain clean drinking water from impure water if you were lost in a desert. What process is used to separate pure water from the matter that is mixed with it?
   (a) In your own words, describe *how* this process works.
   (b) Use the particle model to explain *why* the process works.

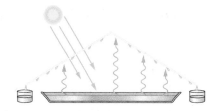

5. **Thinking Critically** Mining is an important industry in British Columbia. Mining companies dig important and useful metals such as copper, zinc, and iron from the province's ground. However, mining waste can also contribute to acid pollution.
   (a) Name one use for copper, one use for zinc, and one use for iron. Use a search engine on the Internet if you need help.
   (b) One way to deal with acidic mining waste is to make it less acidic. To do this, a basic substance called lime is added to the waste. When the acids and base are mixed, they form neutral substances. Do you think this is a chemical change or a physical change? Explain your answer.
   (c) Mining is an example of a human activity that has both risks and benefits. What is one risk and one benefit of mining?

Can you catch water with a net? A Canadian scientist, Dr. Bob Schemenaur, helps people who live in dry regions collect water from the air by hanging large nets. When fog blows into a net, tiny water droplets that are suspended in the fog catch on the fine plastic mesh of the net. These droplets then trickle down the net, into a collecting tube.

CHAPTER at a glance

Now that you have completed this chapter, try to do the following.
If you cannot, go back to the sections indicated in brackets after each part.

(a) Use the particle model of matter to explain how sugar dissolves in water. How is this model of a sugar crystal different from a real sugar crystal? (6.1)

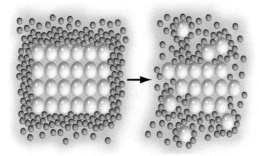

(b) Explain how you could make a solution more concentrated. Explain how you could make a concentration more dilute. (6.1)

(c) List three solids that are soluble in water and three that are not. (6.1)

(d) List three things you could do to make a solid dissolve more quickly in a liquid. (6.2)

(e) Describe how the solubility of a solid in a liquid can be affected by temperature. (6.2)

(f) Explain how you can tell if a solution is saturated or not. (6.2)

(g) Arrange the following substances in order from least soluble to most soluble in ice-cold water: limestone, sugar, salt, bluestone, canola oil, carbon dioxide. (6.2)

(h) List three examples of acids and three examples of bases. (6.3)

(i) Describe two different ways to tell an acid from a base (do not include tasting). (6.3)

(j) Describe how to use distillation to get pure water from a water solution. (6.4)

(k) Describe the process used to make maple syrup from sap. (6.4)

Prepare Your Own Summary

Summarize Chapter 6 by doing one of the following. Use a graphic organizer (such as a concept map), create a poster, or write a summary to include the key chapter ideas. Here are a few ideas to use as a guide:

- Use drawings or differently coloured marbles to model the following:
  – a dilute solution
  – a concentrated solution
  – dissolving
  – the effect of stirring on the rate of dissolving
- Create a pH scale poster. Cut pictures of foods and household products out of magazines of different acids and bases to include on the scale.
- Create a flow chart to show how you would separate the following mixtures:
  – salt, water, and marbles
  – sugar and pepper
  – Epsom salts, iron filings, and sand
- Make a comic strip that explains the terms dilute, concentrated, saturated, and unsaturated.
- Write a story about making maple syrup.
- Write an announcement for radio explaining why it is important to know about solutions.

Chapter 6 Review

Key Terms

soluble
insoluble
concentrated solution
dilute solution
concentration
saturated solution
unsaturated solution
solubility
rate of dissolving
acid
base
indicator
pH scale
neutral
crystallization
distillation

Reviewing Key Terms

If you need to review, the section numbers show you where these terms were introduced.

1. Describe the difference between:
   (a) a substance that is soluble in water and a substance that is insoluble in water (6.1)
   (b) a saturated solution and an unsaturated solution (6.2)
   (c) a dilute solution and a concentrated solution (6.1)
   (d) an acid and a base (6.3)
   (e) crystallization and distillation (6.4)

2. In your notebook, match each description in column A with the correct term in column B. Use each description only once.

   Column A
   (a) turns blue in the presence of bases
   (b) describes how quickly a solute dissolves in a solvent
   (c) has a pH lower than 7
   (d) a solution that does not contain very much solute
   (e) describes sand in water
   (f) describes sugar in water
   (g) has a pH of 7
   (h) baking soda is an example

   Column B
   - acid (6.3)
   - base (6.3)
   - universal indicator paper (6.3)
   - rate of dissolving (6.2)
   - solubility (6.2)
   - neutral (6.3)
   - litmus paper (6.3)
   - concentrated solution (6.1)
   - dilute solution (6.1)
   - insoluble (6.1)
   - soluble (6.1)
   - distillation (6.4)

Understanding Key Ideas

3. Why do you need to include temperature when you state the solubility of a substance?

4. Use the particle model to help you explain each of the following observations.
   (a) stirring helps solutes dissolve faster
   (b) solutes dissolve faster at higher temperatures
   (c) solutes in smaller pieces dissolve faster

5. Neatly sketch a graph showing what happens to the concentration of a solution (y-axis) as you add solute to it (x-axis).

6. Copy the following table into your notebook. Complete the chart by putting check marks into the appropriate boxes in each of the three blank columns.

	True for acids only	True for bases only	True for both acids and bases
(a) sour taste			
(b) bitter taste			
(c) soaps are an example			
(d) can be classified with an indicator			
(e) have a pH lower than 7			

7. Why do scientists use indicators to identify acids and bases?

8. Why do people heat maple or birch sap before using it? Why not just use the sap as it is?

Developing Skills

9. If you did not have a commercial indicator such as litmus paper or pH paper, what could you use to classify a substance as either an acid or a base? (Tasting is not an option!)

10. Copy this table into your notebook. Complete the last column.

Substance	pH	Acid or Base?
(a) tomatoes	4.2	
(b) lye (sodium hydroxide)	13.8	
(c) stomach acid	2.0	
(d) bananas	5.2	
(e) blood	7.4	
(f) ammonia	11.1	
(g) eggs	7.8	

11. Draw a pH scale, adding labels to show where acids, bases, and neutral substances are located on the scale.

12. Water from the Dead Sea is about eight or nine times as salty as ocean water. If you had a sample of water from the Dead Sea, how could you find out whether it was unsaturated or saturated with salt? Write a brief procedure for your test and explain why it would work.

13. You have a mixture of clean sand and pure sugar. How could you get pure sugar from this mixture? Describe your method in detail and list the apparatus you would need.

Problem Solving

14. Suppose you mix 65.0 g of Epsom salts with 100 g of water at 0°C.
    (a) Is the solution saturated or unsaturated? Explain your answer.
    (b) If the solution is saturated, how much solute is undissolved? If the solute is unsaturated, how much more solute could you dissolve?

15. Your friend tells you she was able to dissolve 39 g of salt in 100 g of water. She says there was no undissolved solute at the bottom of the container. But according to Table 6.1, the solubility of salt in 100 g of water is 35.7 g. Could your friend be telling the truth? Explain your answer.

Critical Thinking

16. (a) Describe the steps you would take to recover pure water from dirty dishwater. What is the name of this process?
    (b) Would this process be practical for conserving and reusing water in your home? Explain your answer.
    (c) What are three simple steps you could take to reduce the quantity of water you use at home?
    (d) Why should we conserve water?

Pause & Reflect

Check your original answers to the Getting Ready questions on page 158 at the beginning of this chapter. How has your thinking changed? How would you answer those questions now that you have investigated the topics in this chapter?

UNIT 2

Ask an Elder

Elizabeth Gravelle

Few modern fabrics can compete with leather or buckskin for warmth and comfort. These materials are also very quiet to wear, so they work well for hunters. For these reasons, Aboriginal peoples in many parts of North America have traditionally used the tanned hides of animals such as deer, elk, moose, and buffalo to make clothes. The process of tanning changes the properties of animal hides. A tanned hide resists decay and is soft and supple enough to be made into clothing.

Elizabeth Gravelle is a respected Ktunaxa [Too-NA-ha] Elder of the Tobacco Plains Band. She is an expert on many areas of Ktunaxa culture, including the Ktunaxa language, hide tanning, beading, and traditional plant knowledge. Elizabeth lives in Grasmere in southeastern British Columbia.

Q. What method do you use to preserve a hide?

A. First you skin the animal and soak the hide. Then you hang the hide on a scraping pole and scrape the back side to clean off any flesh. Then you scrape off the hair. The hair side is the good side of the buckskin when it is finished. When you have finished scraping, you wash the hide and let it dry.

When the hide is nice and clean you put it in a brain solution to soften and whiten the hide. You work it with your hands in the solution. When you remove it you wring out the solution by throwing the hide over a pole and twisting it. Then you hang it to dry. You keep working it as it dries. If it is not soft when it is dry, you put it back in the solution and continue working it. Then you put the hide on a rack to stretch it, and scrape off any loose material. Then it is ready for smoking.

Q. What is a brain solution?

A. You make the solution by simmering the brain of the animal in a pot of water. The brain has proteins that help make the hide soft and white. You can also use egg yolks for the same effect. When I don't have a brain solution I use about nine egg yolks to a hide.

Q. What materials do you use for scraping poles and tools?

A. For poles, we use white poplar because it has no pitch (sticky resin) in it. You strip off the bark and dry the pole. If you do not dry the wood first, you cannot re-use the scraping pole as it will damage the colour of the next hide.

People used to make the scraping tool out of a shin bone or a rib bone. Rocks could be used too. More recently, people use metal scythes or knives to scrape hides.

Q. How do you stretch the hide?

A. First you make a wooden rack to fit the hide. You need to use a pitch-free wood. Then you collect rawhide trimmings from around the edge of the hide. You use these to string the damp hide tightly onto the rack. As the hide stretches you tighten the strings. You leave the hide on the rack until it is dry and ready to be smoked.

Q. For how long is the hide smoked?

A. It depends what you want the hide for. The more you smoke the hide, the tougher and darker it gets. A dark hide is more weatherproof. If you do not smoke the hide at all it stays white. A white buckskin is very soft, but it cannot be washed. However, you can clean it by rubbing it with something rough.

The best wood for smoking hides is dry rotten Douglas fir. Other woods like ponderosa pine have too much pitch in them, and will blacken the buckskin. To smoke the hide you sew it together with the hair side in. You sew it onto a canvas so the smoke does not touch the good side of the hide directly. When the hide is the colour you want you remove the sewing and hang the hide in the shade until most of the smell of the smoke is gone.

Q. What are some uses of tanned hides?

A. Clothing is the main use. Animals like elk and moose have thick hides that are good for moccasins, vests, and jackets. Deer have thinner hides that are good for dresses and gloves. Tanned buffalo hides were once used as bedding and as the floor in tepees. Hides can also be tanned with the fur on to make warm winter clothing.

We always used every part of the animal. For example, we made thread from the sinews. People used hooves in ceremonial clothing and to make glue, and bones and antlers to make tools. Nothing was wasted.

Sheryl Stevens of the Ktunaxa Nation uses a drawknife to scrape a hide. Scraping tools were traditionally made from bone or stone.

EXPLORING Further

Traditional Technologies

Chokecherries (Prunus virginiana) are an important traditional food for First Nations in many parts of British Columbia. The seed of the cherry is a nutritious part of the fruit but it also contains the toxic compound hydrogen cyanide. A traditional First Nations method of preparing the fruit is to pound the whole berries into a pulp that is sun-dried. The pounding shatters the seeds. Once the seeds are broken, the hydrogen cyanide reacts with oxygen to produce a harmless compound. The pounding and drying releases the poison but keeps the fruit's high nutritional value.

Working with a partner, research a traditional First Nations technology for the preparation of food or clothing. Does the process involve physical changes, chemical changes, or both? How can you tell? Prepare a brief report that explains how the method works. How does the traditional method compare with a more recently developed technology?

UNIT 2
Ask an Oil Spill Adviser

Hesham Nabih

The words "chocolate mousse" and "blowout" may make you think of a decadent dessert and how your stomach feels after eating too much of it. These terms are also used to describe toxic oil spills. Oceanographer Hesham Nabih knows all about oil spills. He is the president of a company in Vancouver that helps government agencies and oil companies prevent, manage, and contain them.

Q. How did you become interested in cleaning up oil spills?

A. I started my career as an oceanographer. An oceanographer measures water currents and studies the chemistry and ecology of bodies of water. I studied the Red Sea and Nile River in Egypt, and the Georgia Strait and Fraser River in British Columbia. Twenty years ago, while vacationing at a Red Sea resort, I stepped on a ball of tar on the beach. After that, I decided to focus my efforts on fighting oil pollution. Now I develop computer software that is used to train people on how to clean up oil spills.

Q. What causes oil spills?

A. Oil spills occur for many different reasons both on land and water. They can be caused by accidents involving oil tankers, barges, pipelines, refineries, and storage facilities. Most accidents happen while the oil is being transported. Some oil spills are caused by human error. Other spills are natural occurrences. They are caused when underground springs of oil seep into the ocean from the sea floor. Offshore blowouts can also cause oil spills.

Q. What is an "offshore blowout"?

A. An offshore blowout occurs at sea when gas or oil, under high pressure, bursts out of an underwater reservoir during oil drilling or production. Oil rigs have control valves known as "blowout preventers" to stop the flow of oil if problems occur during drilling. Blowouts still happen, though. In 1979, an oil well blew out in the Bay of Campeche, off the coast of Mexico. By the time it was brought under control in 1980, about 140 million gallons of oil had spilled into the bay. That was one of the biggest oil spills of all time.

Q. How do oil spills affect the environment?

A. Oil spills can have a serious impact on ecosystems. Oil can contaminate and smother both plants and animals. The chemical components of oil are toxic and can taint future generations of both plants and animals in an ecosystem. The environment can also be physically damaged by the methods used to clean up the spill.

Q. Cleaning up oil spills, especially cleaning up oil after it has been spilled in the water, is an example of separating a mixture. What are some of the methods used to clean up oil spills?

A. You can divide cleanup techniques into two general categories: mechanical and chemical. Mechanical devices include pressure washers, used on land, and skimmers, used on water. Skimmers remove oil from the water's surface. The two major types of skimmers are suction skimmers and adhesion skimmers. Suction skimmers float on the surface of the water and use a vacuum to suck up the oil. Adhesion skimmers soak up the oil using drums, belts, or ropes.

Q. What about chemical cleanup?

A. With chemical cleanup, chemicals are applied directly to the oil to break it up into tiny droplets. Micro-organisms such as bacteria can more easily convert the tiny droplets of oil into less harmful substances. Breaking up the oil into droplets also prevents the oil and water from forming "chocolate mousse," the term given to emulsions of oil and water that look like oily reddish-brown pudding.

Q. How long can an oil spill affect an area even after it has been cleaned up?

A. On water, an oil spill can affect an area for a few hours or a few days, or longer. On land, the after-effects can last for many years. For example, in 1989 the oil tanker *Exxon Valdez* struck Bligh Reef. The tanker spilled nearly 11 million gallons of crude oil into the waters of Alaska's Prince William Sound. Even now, while the vast majority of the spill appears to have been recovered, pockets of crude oil remain.

This harlequin duck was rescued from the effects of the Exxon Valdez oil spill. Many other animals, however, did not survive.

EXPLORING Further

Single-Hull vs. Double-Hull

The *Exxon Valdez* oil spill is still considered to be one of the most environmentally damaging oil spills in the world. Experts now estimate that if the tanker had been a double-hulled ship instead of a single-hulled ship, the amount of oil spilled would have been reduced by more than half. Shipping organizations and governments are working to eliminate the use of single-hulled ships.

1. Why are double-hulled ships safer than single-hulled ships?
2. What else is being done to make the shipping industry safer?
3. Where do the majority of oil spills involving oil tankers occur?
4. What is a "contingency plan?" How does it work?

The *Exxon Valdez*

UNIT 2 Project

Purifying Mixtures

SKILLCHECK
- Designing experiments
- Observing
- Classifying
- Communicating

From the air you breathe to the water you drink to the soil underfoot — mixtures are everywhere! You have learned that mixtures contain two or more pure substances that are combined together *physically*. One way to identify a mixture is to separate it into its parts. The parts of a mixture are also called *fractions*. Some methods include magnetism, flotation, sifting, evaporating, and filtration. Many industries rely on these methods to isolate the fractions in as pure a state as is necessary. For example, our drinking water is cleaned and purified before it runs out of our taps for drinking and cooking. Tap water still contains some minerals, but these do not harm us (some are actually good for us!). Therefore, it is not necessary to purify water any further. If pure water is needed, it must be distilled, which requires much more time and energy to accomplish. Each new step of a separation may cost a company more money. This is why ingredients and materials that are almost 100 percent pure can be costly to buy. As companies develop methods and technologies that can separate a mixture or purify a contaminated substance more efficiently, the cost can come down.

This giant magnet is being used to lift bits of iron. How could this method be used to purify materials in a mixture?

Challenge

Separate a dry mixture into the pure substances that make it up.

Materials
- variety of mixtures prepared by your teacher
- several beakers (250 mL to 600 mL), bottles, or cups
- magnet
- plastic wrap
- variety of sieves (fine to coarser mesh)
- small funnels
- filter paper
- hot plate (optional)
- evaporating dishes
- tap water
- paper
- labels or grease pencil

Safety Precaution

- Be careful if you use the hot plate. Unplug it when it is not in use.
- Be sure to wipe up any spills as wet floors are slippery.

Design Criteria

A. In a small group, separate a mixture into its fractions.

B. Include the smallest number of separation steps as is necessary to produce pure fractions.

C. Design a flowchart showing the steps and separation methods that you used in your final separation procedure.

Plan and Construct Group Work

1. Obtain a dry mixture from your teacher. With your group, try to identify the different substances you can see in the mixture. **Record** these in a list. Then suggest any substances that might be present but not visible (for example, salt and sand can look very similar in a mixture and, therefore, may be difficult to tell apart). **Record** these substances in a separate list.

2. Brainstorm the types of methods you might be able to use to separate your mixture. Make a list of your ideas. Examine the list and make a group decision on which methods you will use and in which order.

3. Outline the steps in your method of separation and the fraction isolated at each step. You may need to revise your outline a few times before you come up with the best sequence of steps.

4. Separate your mixture. Collect each fraction in a separate, labelled container.

5. If your fractions look contaminated (for example, sand stuck to pieces of gravel), try to purify them by using one of the separation methods.

6. After several trials, decide on the method that will give you the best results in the fewest number of steps.

7. Make a flowchart to reflect your final method.

Evaluate

1. How pure were your fractions? How did the order of steps affect how pure you were able to get them?

2. Which methods did you use to separate substances that were not clearly visible? How did you know that each method worked?

3. (a) Did you identify all the parts of your mixture correctly when you first examined it? Explain your answer.
   (b) How did separating your mixture help you to identify its parts?

4. How could you improve your methods of separation? What other equipment or methods might have helped you to improve the quality and efficiency of your separations?

Extend Your Skills

5. Paper chromatography is a special method that can be used to separate mixtures such as ink and various drinks made from flavour crystals. This method is particularly useful in forensic science. Conduct research on the Internet or at your library about how mixtures can be separated using paper chromatography. With a partner, use this method to separate the substances in blue or black ink. Present your results in a poster with labels. Include a brief explanation of the procedure and practical uses for it.

UNIT 3

Earth's Crust

Planet Earth can be an exciting place to live. The ground beneath your feet is not as solid as it feels. Molten rock sometimes finds its way to Earth's surface and erupts with tremendous force. The people in the truck opposite narrowly escaped the eruption of Mount Pinatubo in the Philippines in 1991. When Mount Pinatubo awoke from its 600-year slumber, it spewed out more gas and debris than most other volcanic eruptions over the last hundred years. Fortunately, earthquakes provided a warning of the eruption, and many lives were saved.

British Columbia has more earthquakes and volcanoes than any other province in Canada. What forces unleash the incredible power of volcanoes and earthquakes? What processes make and change the rocks and minerals? What evidence do we have of activity deep inside our planet?

The answers lie in and under Earth's crust—the thin, ever-changing outer layer of Earth. In Unit 3, you will see how theories about Earth's crust were developed. You will learn how scientists made new observations and found new connections to past observations. You will also learn how recent technology has allowed scientists to study the sea floor and gain an understanding of the moving crust. Enjoy your tour of Planet Earth.

Unit Contents

Chapter 7
The Composition
of Planet Earth   196

Chapter 8
Earth's Moving
Crust   226

Chapter 9
Earth's Changing
Surface   256

Getting Ready...

- What is inside Earth?
- How is the surface of Earth made?
- How has Earth changed over long periods of time?

CHAPTER 7

The Composition

Getting Ready...

- What does the inside of Earth look like?
- What is a mineral?
- How are rocks made?

Flowing lava from the Kilauea Volcano in Hawaii. Volcanic eruptions can also make volcanic glass, such as the obsidian used in the arrowhead in the smaller photograph on the next page.

Imagine you could travel to the centre of Earth. What would you see on your journey from the surface to the centre? Would your journey take place in solid rock, or would you travel through some layers of liquid or gas? What temperature changes might you notice?

Humans are not able to travel very far under the surface of our planet. Scientists can predict, however, what Earth might be like at deeper levels. Some clues have been found in the rocks and minerals on Earth's surface. Other clues have been found in the deepest parts of the oceans. The presence of lava and volcanoes shows there is tremendous heat and pressure beneath Earth's surface. All scientific clues indicate that Earth is not a quiet, solid ball of rock.

Some of the rocks on Earth's surface show evidence of being melted and cooled deep below the surface. Other rocks show evidence of being formed and reformed at Earth's surface. In addition, rocks contain minerals that can be mined and used in our daily lives.

In Chapter 7, you will learn about the layers deep inside Earth. You will also learn how to identify minerals and how rocks form.

# of Planet Earth

What You Will Learn

In this chapter, you will learn

- what the inside of Earth looks like
- how to identify rocks and minerals
- how minerals and mineral resources can be used
- how rocks and fossils are formed

Why It Is Important

- Knowledge of Earth's internal structure can help scientists predict earthquakes and volcanic eruptions. Rocks and minerals are a source of many valuable substances, such as metals, oil, and coal.
- What you learn in Chapter 7 will help you appreciate the mineral resources that are mined in British Columbia.

Skills You Will Use

In Chapter 7, you will:

- make inferences about a hidden substance
- use models to help you understand theories about Earth's layers
- research the minerals found in British Columbia
- make observations about rocks and minerals
- model the formation of a fossil
- design a procedure for classifying rocks

Starting Point ACTIVITY 7-A

Examine the Evidence

How well can you describe something you cannot see? Earth scientists face the same problem when they examine evidence of Earth's inner layers. Imagine that you are an Earth scientist, and that the containers you examine represent Earth.

What to Do Group Work

1. Your teacher will give your group several containers. Each container will have a marble inside it. Some containers will have another substance, as well. Examine the containers. Without looking inside, try to figure out the contents.

2. As a group, make a table to record your observations. **Record** what you think is inside each container. Describe the evidence you used to make your decisions.

3. Discuss your observations and decisions with the other groups. Discuss: How is this activity similar to how scientists determine what is deep inside Earth? How is this activity different?

What Did You Find Out?

1. Which were the easiest contents to identify? Why?

2. Which were the most difficult contents to identify? Why?

3. Which evidence was confusing or misleading? (For example, the contents sounded like one item but were actually something else.)

Chapter 7 The Composition of Planet Earth • MHR

CHAPTER 7

Section 7.1 The Structure of Earth

What is deep inside Earth? Scientists have many answers to this question, but there are still mysteries waiting to be solved.

Think about the Starting Point Activity. Since you could not look inside the containers, you had to use clues, such as sound and mass, to figure out what was inside. Scientists use clues to predict what the inner layers of Earth might be like. The clues include data from earthquake waves and satellite images, which help scientists learn about the layers they cannot see. Clues about something that cannot be seen directly are called *indirect evidence*.

Scientists also use *direct evidence* to learn about Earth's inner layers. Geologists obtain direct evidence by studying rock samples from volcanoes, from mountain ranges, and from deep within the sea floor. It is not possible to gather direct evidence from Earth's deepest layers. The centre of our planet is over 6000 km below the surface. The deepest any drill has gone is about 12 km. Perhaps, during your lifetime, scientists will develop a technology for collecting direct evidence from deeper inside Earth.

Why are scientists not sure what is inside Earth's inner layers?

DidYouKnow?

Geo refers to Earth. A geologist studies the rocks on Earth's crust. A geographer may study how climate affects Earth's crust. A geochemist studies the chemical elements that are found in Earth's crust. What do you think a geophysicist studies?

A Model of Earth

Scientists have used direct and indirect evidence to make a model of Earth's structure. Figure 7.1 is a model of the four main layers of Earth: the crust, the mantle, the outer core, and the inner core. Notice that some layers are solid and some are liquid. Why do you think the layers are different states?

A Inner Core
The **inner core** is the deepest layer of Earth. Many scientists believe the inner core is made of iron and possibly other elements, such as silicon and carbon. The inner core is a solid ball, with a temperature of about 5000°C–5700°C.

D Crust

On the outside of Earth is a thin layer of solid rock called the **crust**. Earth's crust extends under the oceans (the *oceanic crust*) as well as the continents (the *continental crust*). The depth of the crust varies from less than 5 km to over 50 km thick.

Pause & Reflect

Suppose that you could invent a way to gather direct evidence from Earth's inner core. Use a diagram to show how your invention would work.

B Outer Core

The **outer core** is probably made of liquid iron and nickel. Sulphur and oxygen could also be present. The temperature in the outer core is over 4000°C. The outer core is so hot that the iron and nickel are melted.

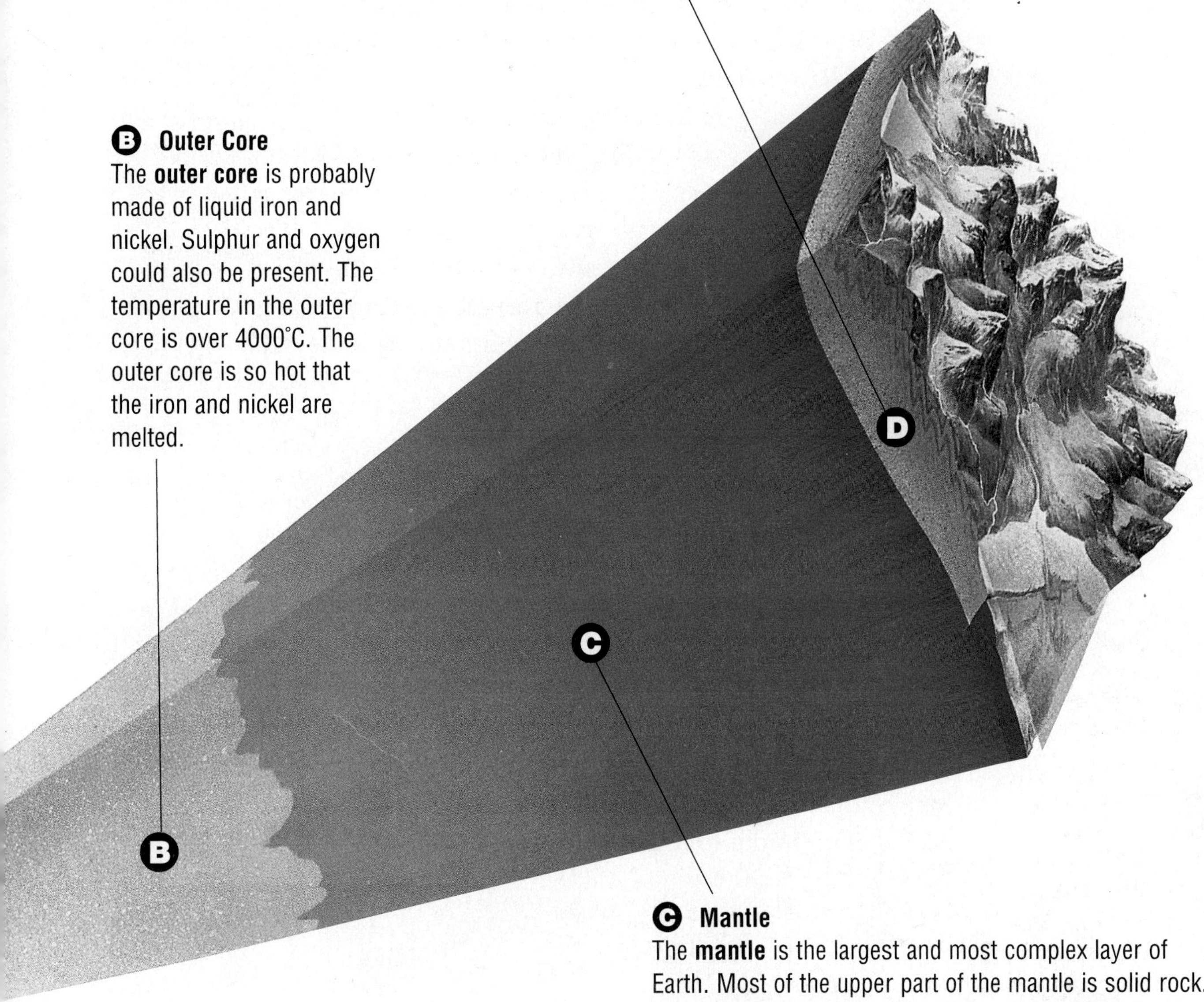

C Mantle

The **mantle** is the largest and most complex layer of Earth. Most of the upper part of the mantle is solid rock. The temperature in the upper layer of the mantle is approximately 1000°C.

The temperature in the lower area of the mantle is approximately 2500°C. Scientists picture the lower mantle as being plastic, like taffy or caramel.

The boundary separating the crust from the mantle is called the **Moho**.

Figure 7.1 The inner core, outer core, and mantle are shown at the correct scale. The crust, however, is shown much thicker than it actually is.

Chapter 7 The Composition of Planet Earth • MHR 199

Find Out ACTIVITY 7-B

A Model Planet

In what ways is Earth like an apple? The skin of the apple could represent Earth's crust. The seed core of the apple could represent Earth's outer and inner core. What part of the apple could represent Earth's mantle? When you use an apple to represent Earth's crust, you are using it as a model of Earth. What else could you use, draw, or make to model Earth's layers?

What to Do *Group Work*

1. **Make a model** of Earth's layers, or find something that you can use as a model. Think about how your model represents the different layers.

2. Present your model to the class. Explain how your model represents Earth's layers.

What Did You Find Out?

1. How did you choose the model?

2. How could you make a better model of Earth's layers?

3. How did making and using models help you understand Earth's layers?

Skill POWER

For help in modelling in science, turn to SkillPower 4.

The Plates of the Lithosphere

The crust and upper mantle make up the rigid layer of Earth called the **lithosphere** (see Figure 7.2). The semi-molten zone of the mantle just below the lithosphere is the **asthenosphere** [as-THEN-os-fear].

The lithosphere consists of about a dozen very large pieces, called **plates.** The plates fit together like the pieces of a jigsaw puzzle. Unlike the pieces of a jigsaw puzzle, however, the plates can move. How do the plates of the lithosphere move? This is one of the mysteries of Earth's structure. You will investigate one explanation in the following investigation.

Use a drawing to show the meaning of the words "lithosphere" and "asthenosphere."

Figure 7.2 The plates of the lithosphere float on the asthenosphere.

200 MHR • Unit 3 Earth's Crust

CONDUCT AN INVESTIGATION 7-C

SKILLCHECK
- Modelling
- Observing
- Inferring
- Interpreting Data

Movement in the Mantle

The asthenosphere is made from rock that is partly melted by the intense heat of the outer core. You can use hot water and cold water to model the movement of rock in Earth's mantle.

Question
What happens when hot water and cold water meet?

Safety Precautions

Apparatus
2 large (750 mL) glass or plastic jars that are exactly the same size

Materials
hot tap water
cold tap water
wax paper
food colouring

Procedure

Part 1

1. Fill one jar with very hot tap water. Make sure that the water goes up to the very top of the jar.

2. Add a few drops of food colouring to the hot water. CAUTION: Be very careful with hot water.

3. Fill the second jar with very cold water. Again, make sure that the water goes up to the very top of the jar.

4. Cut out a square piece of wax paper, 10 cm by 10 cm. Place the piece of wax paper on top of the jar of cold water.

5. Hold the wax paper tightly on the jar of cold water as you carefully invert the jar and place it on top of the jar of hot water. (If you do not hold the wax paper tightly, you may get wet!)

6. Once the cold-water jar is on top of the hot-water jar, carefully pull out the wax paper from between the jars.

7. **Observe** what happens. **Record** the results.

Part 2

1. Repeat the Procedure for Part 1, with two important differences. Place the wax paper on top of the hot-water jar. Place the cold-water jar on the bottom and the hot-water jar on the top.

2. **Observe** and **record** the results.

Analyze
1. What happened to the hot water in Part 1 of the investigation?

2. What happened to the hot water in Part 2 of the investigation?

3. Why was food colouring added to the hot water?

Conclude and Apply
4. Why were the results so different in the two parts of the investigation?

5. What conclusion can you make about what happens when a hot fluid and a cold fluid meet?

6. How might this investigation model the movement in the asthenosphere?

7. Suppose you were to repeat this investigation using different temperatures of water. Formulate a question and hypothesis you could test.

Convection Currents

In Conduct an Investigation 7-C, Movement in the Mantle, you observed the movement of hot and cold water. The movement you observed is called a convection current. A **convection current** is a movement produced by the rising of a warm material and the sinking of a cool material. You can observe convection currents in your home or at school in the following activity.

At Home ACTIVITY 7-D

Convection Currents at Home

Other fluids, such as air, move in convection currents like water. Can you detect their movement?

What to Do

Find out whether you have both warm and cold air vents in any rooms in your home or school.

What Did You Find Out?

1. When you hear the furnace fan working, put your hand near both types of vents. What do you feel?

2. **Record** your explanation of how convection heating works. Share your explanation with a classmate.

INTERNET CONNECT

www.mcgrawhill.ca/links/BCscience7

Draw your ideas of how convection currents might look deep inside Earth. Then go to the web site above to view some models of convection currents. Click on **Web Links** to find out where to go next.

Scientists have suggested there are convection currents in the upper layer of the mantle. The rock in the asthenosphere is melted by the intense heat of the inner core. Convection currents in the melted rock transfer heat to the lithosphere, as shown in Figure 7.3. Hotter rock rises and the cooler rock sinks. The plates of the lithosphere are moved as the melted rock flows. The movement of the plates results in the formation of earthquakes, mountains, and volcanoes.

Where is the hotter rock in Figure 7.3? Where is the cooler rock?

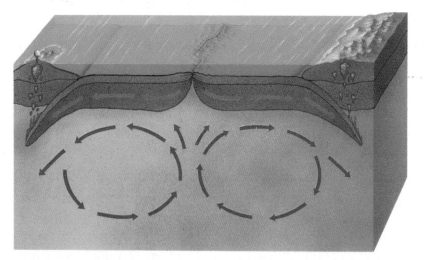

Figure 7.3 Scientists hypothesize convection currents are moving the plates of the lithosphere.

Earth's Magnetic Field

The movement of rock by convection currents in the outer core can help to explain Earth's magnetic field. A **magnetic field** is a region of force around a magnet. When you observe a compass needle pointing toward the north, you are seeing evidence of Earth's magnetic field. The magnetic field of Earth is much like the magnetic field of a bar magnet. Like a bar magnet, Earth has a north magnetic pole and a south magnetic pole. Earth's magnetic field helps to protect Earth from dangerous radiation from the Sun by deflecting cosmic rays into outer space.

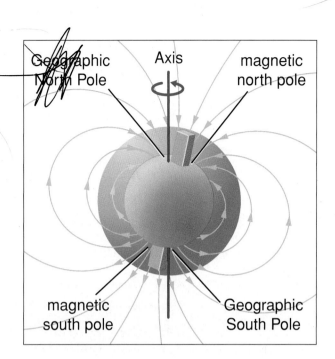

Figure 7.4 Earth's north and south magnetic poles are angled slightly differently than Earth's geographic poles.

Section 7.1 Summary

Scientists use direct and indirect evidence to model Earth's crust, mantle, outer core, and inner core.

- The lithosphere is broken into about a dozen large plates that can move.
- Convection currents may be the reason for the movement of the lithosphere.

In the next section, you will look more closely at Earth's crust.

Check Your Understanding

1. What are the four layers of Earth? Write them in order, from the centre of Earth outward, or make a labelled diagram.

2. What two types of evidence do scientists use to determine what is inside Earth? Give an example of each type of evidence.

3. Draw a Venn diagram to compare and contrast the lithosphere and the asthenosphere.

4. Develop a model to describe convection currents to someone else.

5. **Apply** How can you demonstrate convection currents? Describe an activity you could use.

6. **Thinking Critically** Could one plate in the lithosphere move by itself, or do all the plates need to move together? Explain your answer.

Key Terms

crust
mantle
Moho
outer core
inner core
lithosphere
asthenosphere
plates
convection current
magnetic field

CHAPTER 7

Section 7.2 Minerals and Mineral Resources

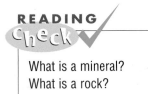

READING check

What is a mineral?
What is a rock?

How many minerals have you used today? A **mineral** is a chemical element or compound that is naturally occurring and has a crystal structure. Minerals and rocks are important resources in British Columbia. A **rock** is a natural material that is made of one or more minerals. Sometimes plant and animal remains are found in rocks. You will learn more about rocks in section 7.3.

Sulfur, gold, and diamond are examples of minerals that are a single chemical element. Quartz is an example of a mineral that is a chemical compound. Quartz is made of the elements silicon and oxygen. Most minerals are made of more than two elements.

Minerals and mineral resources have been extracted from Earth's crust for thousands of years. A **mineral resource** is a rock or mineral that can be mined and used for a specific purpose. Precious minerals like gems are polished and used to make jewellery. Metals, which are separated from minerals that have been mined, can be used to make tools. Petroleum (crude oil) is refined to make gasoline, oil, gas, plastic, petrochemicals, and many other useful substances. Oil and coal are examples of mineral resources that are made from plant materials. Coal, oil and gas are known as **fossil fuels**. Look at Figure 7.5 to discover some common uses of minerals and mineral resources.

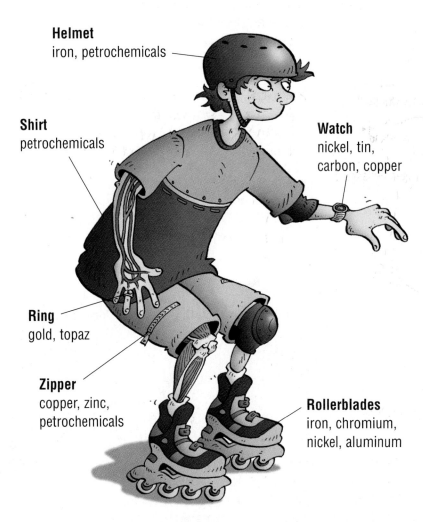

Helmet iron, petrochemicals

Shirt petrochemicals

Ring gold, topaz

Zipper copper, zinc, petrochemicals

Watch nickel, tin, carbon, copper

Rollerblades iron, chromium, nickel, aluminum

Figure 7.5 Think about the clothes you are wearing today. Which parts of your clothes come from mineral resources? Which minerals do you think your body needs to be healthy?

Find Out ACTIVITY 7-E

Research the Resource

Many minerals and mineral resources are found in British Columbia. Jade is found along the Fraser River and at Dease Lake. Zinc is found near Kimberley, and copper is mined at Logan Lake. Every year, geologists identify new mineral deposits in British Columbia. In this activity, you will discover what mineral resources are found near your community.

What You Need
map of British Columbia mines
red and black pens
research materials
map of British Columbia

Mineral Resources	Names of Mines
coal, copper, silver, gold, zinc, fireclay, magnesite, molybdenum, gypsum, limestone, magnetite, silicon, pumice	Eskay Creek, Nazko, Huckleberry, Monteith Bay, Quinsam, Myra Falls, Texada Island, Sumas Mountain, Mount Meager, Endako, Kemess South, Bullmoose, Mount Moberly, Mount Brussilof, Elkhorn, Greenhills, Line Creek, Elkview, Canal Flats, Rock Creek, Craigmont, Highland Valley

What to Do [Group Work]

1. Your teacher will give you a map of British Columbia that has symbols for the locations of mines. Mark the community where you live on the map.

2. Divide the names of mines among your group members. Each person should have at least three mines.

3. **Research** each mine you were assigned to find out what mineral resource is mined there. **Record** the mineral resource, and give one use for it.

4. **Record** the names of your mines on your map in black pen. **Record** the names of the mineral resources and their uses on your map in red pen.

5. Share the information you collected with your group. Add the names of the other mines, as well as the mineral resources and their uses, to your map.

What Did You Find Out?

1. Which mine is closest to where you live? What resource is mined there? What is one use for the resource?

2. What are five common mineral resources that are found in British Columbia? Give one use for each resource.

Extension

3. Choose one mineral resource, and research it further. **(a)** Where, other than British Columbia, is it found? **(b)** How is it mined? **(c)** What is it used for today?

4 Examine Figure 7.6 below. What environmental concerns should be considered before harvesting methane hydrates from the sea floor?

Figure 7.6 Methane hydrates are highly flammable compounds that are trapped in a cage of ice. Methane hydrates could be used to supply the world's natural gas requirements for many years. Scientists are researching how to extract methane safely.

You can find mineral crystals in credit cards, watch batteries, and most electronic devices.

Identifying Minerals

Did you recognize the names of some of the mineral resources you researched in the previous activity? Do you think you could identify these mineral resources if you saw them? There are about 4000 minerals, and some of them look alike. How can you tell which mineral is which? There are several important clues, including crystal structure, hardness, lustre, colour, streak, and fracture.

Crystal Structure

Crystal structure is an important physical property of minerals. A **crystal** has straight edges, flat sides, and regular angles. Most of the minerals in Earth's crust develop into beautiful shapes according to the six different crystal systems shown in Table 7.1. Sometimes the crystal structure is visible in a mineral, such as the galena shown in Figure 7.7. Often, however, the crystals are too small to be seen.

Table 7.1 The Six Major Crystal Systems

Mineral Examples	Systems
halite	cubic
wulfenite	tetragonal
corundum	hexagonal
topaz	orthorhombic
gypsum	monoclinic
albite	triclinic

Figure 7.7 Galena has a cubic crystal structure.

Hardness

Hardness is another important physical property of minerals. **Hardness** is a measure of the mineral's resistance to being scratched. How can hardness be used to identify minerals? German scientist Friedrich Mohs asked himself this question in 1812. He developed a scale of ten minerals with a hardness value of 1 to 10, as shown in Table 7.2.

Table 7.2 The Mohs Hardness Scale

Mineral	Mineral hardness	Hardness of common objects
talc	1 softest	soft pencil point (1.5)
gypsum	2	fingernail (2.5)
calcite	3	copper (3.5)
fluorite	4	iron nail (4.5)
apatite	5	glass (5.5)
feldspar	6	steel file (6.5)
quartz	7	porcelain tile (7)
topaz	8	flint sandpaper (7.5)
corundum	9	emery paper (9.0)
diamond	10 hardest	carborundum sandpaper (9.5)

The higher the number on the Mohs Hardness Scale, the harder the mineral. Your fingernail can scratch gypsum. Copper will scratch calcite, but not fluorite. Diamond is the hardest mineral. Diamond-tipped drill bits can cut through steel and rock. Tiny rows of diamonds are used to edge surgical scalpels, razor blades, computer parts, and dental drills.

Lustre

Some minerals, such as gold and silver, appear shiny—another clue to their identity. Other minerals appear dull. The shininess, or **lustre,** of a mineral depends on how light is reflected from its surface. If a mineral shines like a polished metal surface, it is said to have a metallic lustre. If a mineral does not shine like a metal, it has a non-metallic lustre. Table 7.3 lists different types of non-metallic lustre.

Table 7.3 Classification of Non-metallic Lustre

Name of Lustre	Appearance
adamantine	has a hard, brilliant shine like a diamond
glassy	has a surface reflection like a piece of glass
greasy	looks as if it is covered with a thin layer of oil
waxy	looks like wax
pearly	has a sheen like a pearl
silky	has a shine like silk

Colour

Colour is one of the most attractive properties of a mineral. The colour of a mineral can also be a clue to its identity. As you can see in Figure 7.8, however, colour alone cannot identify a mineral. Two or more minerals may have the same colour. Also, one mineral may be found in several different colours. Corundum (made of aluminum and oxygen) is white when pure. When corundum contains iron and/or titanium, it is blue (and called a sapphire). When corundum contains chromium, it is red (and called a ruby).

READING Check

Why can colour not be used by itself to identify a mineral?

INTERNET CONNECT

www.mcgrawhill.ca/links/BCscience7

Find out which metals are used in Canadian coins by going to the web site above. Click on **Web Links** to find out where to go next. Make a circle graph for each coin to show the percentages of different metals it contains.

Figure 7.8 Both pyrite (on the left) and gold (on the right) are golden in colour. Which properties could you use to tell them apart?

Did You Know?

Transparency is another property of minerals. If you can see through the mineral clearly, it is transparent. If you can barely see through it, it is translucent. If you cannot see through the mineral, it is opaque.

Streak

When a mineral is rubbed across a piece of unglazed porcelain tile, as in Figure 7.9, it leaves a streak. The **streak** is the colour of the powdered form of the mineral. A porcelain tile has a hardness value of 7. Minerals that have a hardness greater than 7 do not leave a streak on the tile. Graphite is a mineral used in pencils. Pencil marks are graphite streaks that are soft enough to be left on a piece of paper.

A streak test is one way to identify whether a mineral sample is gold or pyrite. Gold leaves a golden streak. Pyrite leaves a greenish-black or brownish-black streak.

Cleavage and Fracture

The way a mineral breaks apart is another clue to its identity. If a mineral breaks along planes (smooth, flat surfaces), it is said to have **cleavage.** Mica (shown in Figure 7.10) is an example of a mineral with cleavage. Separating the layers of mica is like separating the pages in a book.

Not all minerals have cleavage. Minerals that break with rough or jagged edges have **fracture.** Quartz (shown in Figure 7.11) is an example of a mineral with fracture. In order to examine cleavage and fracture, you need to look at a freshly broken surface of a mineral.

In the following investigations you will learn more about the uses and properties of minerals and mineral resources.

Figure 7.9 Hematite can be dark red, grey, or silvery in colour. The streak of hematite is always dark red-brown.

What is the difference between cleavage and fracture?

Figure 7.10 Mica has a single cleavage direction, which allows it to be pulled apart into sheets.

Figure 7.11 Notice the crystal faces of the quartz in this photograph as well as the fracture on the bottom.

THINK & LINK

INVESTIGATION 7-F

SKILLCHECK
- Inferring
- Predicting
- Communicating
- Interpreting Data

Trade Routes

Think About It

For thousands of years, Aboriginal peoples crossed Canada to find, collect, and trade valuable mineral resources. The map below shows some of their trade routes. How do you think the value of each mineral resource was determined?

What to Do [Group Work]

1. With your group, study the map below. The map shows some of the routes that Aboriginal peoples used for trading and some of the materials they traded.

2. With your group, discuss how Aboriginal peoples traditionally used rocks and minerals. Find pictures in books or on the Internet that show Aboriginal peoples using mineral resources.

3. As a group, make a poster that shows traditional uses of mineral resources by Aboriginal peoples. Label the poster with the names of the resources.

Analyze

1. Follow the routes on the map to see how silver from eastern Canada could have travelled to Vancouver Island. Do you think the silver was more valuable for trading if it was in its natural state or if it was already made into a tool or a piece of jewellery? Explain your answer.

2. Which routes could have been used to transport iron from Greenland to northern British Columbia? How do you think the iron was transported? How do you think it was used?

3. Think about the mineral resources you showed in your poster. Which properties might have been important for each resource?

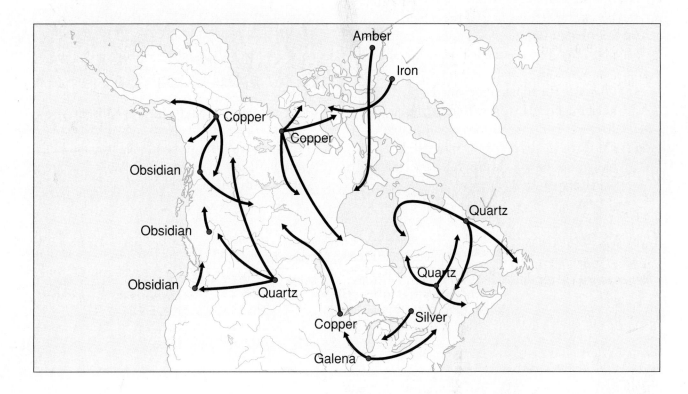

Chapter 7 The Composition of Planet Earth • MHR 209

CONDUCT AN INVESTIGATION 7-G

SKILLCHECK
- Observing
- Classifying
- Predicting
- Interpreting Data

A Geologist's Mystery

Imagine that you are a geologist. You have just received a package of minerals from your company's field team in northern British Columbia. The attached note reads, "New mine discovered. Enclosed are samples of minerals found there. Please identify." How can the Mohs Hardness Scale and the properties of minerals help you solve the mystery of the unknown minerals?

Question
How can you identify different minerals?

Safety Precautions

- Be careful when handling materials with sharp points or edges.
- Always wear safety glasses and gloves when working with acids.

Apparatus
numbered mineral samples
hand lens
iron nail
copper penny
steel file
streak plate
glass plate or jar

Materials
Tables 7.1 and 7.2 (page 206) and Table 7.3 page (207)
10% hydrochloric acid (optional)

Procedure

1. Make a table like the one below.

2. Record the number of the first mineral sample in the first column of the table.
   (a) **Record** the colour of the mineral in your table. You may use the hand lens to take a closer look.
   (b) **Observe** the mineral for any distinguishing crystal shapes. Refer to Table 7.1. Record your observations in your table.
   (c) **Observe** the lustre of the mineral. If it has a non-metallic lustre, refer to Table 7.3. **Record** your observations.

   (d) Scrape the mineral once across the streak plate. Brush off the excess powder. **Record** the colour of the streak. If the mineral is too hard to leave a streak, write "none" in the column under "Streak."

Characteristics of Some Common Minerals

Mineral number	Colour	Crystal shape (if visible)	Lustre	Streak	Hardness	Other properties	Mineral name

210 MHR • Unit 3 Earth's Crust

(e) **Predict** the hardness of the mineral sample. Scratch the sample with your fingernail. If your fingernail does not leave a scratch, use a copper penny to scratch the sample. If the penny does not leave a scratch, use the iron nail, followed by the steel file, until something leaves a scratch. Refer to Table 7.2. **Record** the hardness as a number from 1–10.

(f) **Record** any other properties, such as cleavage, fracture, and transparency under "Other properties."

(g) Your teacher may give you some 10% hydrochloric acid to test the samples. CAUTION: Be sure to wear safety glasses and gloves. After you use the acid, rinse the samples thoroughly with water and dry them. Rinse your gloves, too.

3 Repeat step 2 for the remaining mineral samples.

4 Try to identify each mineral by using a mineral identification chart provided by your teacher or your own researched information.

5 Wash your hands thoroughly after completing the investigation. Clean the streak plate. Return all the mineral samples to their proper places.

Analyze

1. Before testing, which minerals looked the same?
2. (a) Which mineral was the softest? Which was the hardest?
   (b) Were your predictions of hardness supported by your observations? Explain.
3. (a) Which minerals were the same colour as their streak or powder?
   (b) Which minerals had streaks with colours that surprised you?
4. What other features or properties helped you identify the samples?

Conclude and Apply

5. Were you able to identify all the mineral samples? If not, what other tests could you use to identify them?
6. (a) Which property was the most useful for identifying a mineral? Why?
   (b) Which properties were not very useful for identifying a mineral? Why?
7. How much does hardness seem to affect the similarity between the colour of a mineral and the colour of its streak?

Career CONNECT

Digging out the Facts

Take a look at the following occupations:

mineralogist	oil rig operator	paleontologist	geologist	paleobotanist
museum curator	miner	archaeologist	surveyor	geochronologist

With your group, produce a booklet or multimedia presentation about one of the occupations. Make sure that you describe what it has to do with the study of Earth's crust. You can use photographs and illustrations from magazines or from the Internet. If possible, interview someone working in the occupation and include your interview as an article.

Pause & Reflect

Write a poem about your favourite mineral. Be sure to include a description of each of its physical properties.

Section 7.2 Summary

British Columbia's mineral resources have been used for thousands of years. Physical properties of minerals include:

- crystal structure (six major crystal systems)
- hardness (values of 1 to 10 on the Mohs Hardness Scale)
- lustre (metallic or non-metallic lustre)
- colour (not always helpful in identification)
- streak (colour of the powdered mineral)
- cleavage and fracture (the way that a mineral breaks apart)

In the next section, you will learn how rocks are formed both on and below Earth's surface.

Key Terms

mineral
rock
mineral resource
fossil fuels
crystal
hardness
lustre
streak
cleavage
fracture

Check Your Understanding

1. In your notebook, match each description in column A with the correct term in column B.

   A
   (a) measure of how easily a mineral can be scratched
   (b) name given to the scale of hardness
   (c) one of the softest known minerals
   (d) hardest known mineral
   (e) reflection of light from the surface of a mineral
   (f) colour left by powdered mineral on unglazed porcelain
   (g) tendency to break along smooth, flat surfaces
   (h) common mineral that breaks along planes
   (i) tendency to break with rough or jagged edges
   (j) common mineral that breaks with rough or jagged edges

   B
   - cleavage
   - diamond
   - fracture
   - hardness
   - mica
   - lustre
   - Mohs
   - quartz
   - streak
   - talc

2. (a) What is the difference between metallic and non-metallic lustre?
   (b) What is an example of a mineral with each type of lustre?

3. (a) What are three minerals or mineral resources that you have used today?
   (b) How have you used them?

4. **Apply** What two properties would you use to distinguish clear, transparent quartz from clear, transparent calcite? Explain your choice of properties.

5. **Thinking Critically** In addition to identification, why might it be important to know the hardness of different minerals? Give an example.

CHAPTER 7

Section 7.3 How Rocks Are Formed

Rocks have been used since the beginning of human existence. Some of the first human shelters were made from rocks. Today, many homes around the world include rocks in their structure. The basement of your home or school may be made from concrete, which comes from a rock called limestone. Rocks such as limestone are mined for many industrial uses. Other rocks are mined for the valuable mineral resources they contain.

In section 7.2, you learned that rocks contain minerals. Take a look at the granite shown in Figure 7.12. Which four common minerals combine to form granite?

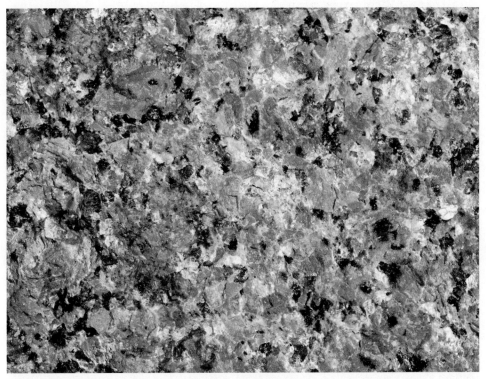

Figure 7.12 Granite is a mixture of four minerals: feldspar, quartz, mica, and hornblende. Can you find each mineral in the granite?

Three Families of Rocks

British Columbia's magnificent mountain ranges are made of many colourful rocks. Rocks can form deep inside Earth, on Earth's crust, or in the waters found on Earth's surface.

Scientists group rocks into three major families, based on how they are formed. The three families of rocks are igneous, sedimentary, and metamorphic.

The word "igneous" comes from the Latin word *ignis* meaning "fire." Write what you think the word "ignite" means. Then check a dictionary to see how close you were.

Igneous Rock

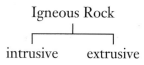

Igneous rock [IG-nee-us] forms when hot magma or lava cools and becomes solid. **Magma** is melted rock found deep below Earth's crust, where temperatures and pressures are high. Any rock that is heated at great depths can melt into magma. **Lava** is molten rock on Earth's surface.

Geologists classify igneous rock based on whether it was formed above or below Earth's surface (see Figure 7.13). Rock that cooled below Earth's surface is called **intrusive rock**. Turn back to the picture of granite in Figure 7.12. Granite is an example of an intrusive igneous rock. Granite forms very slowly, deep inside Earth.

Sometimes, hot magma pushes up to the surface through cracks in Earth's crust. Rock that forms when lava cools on Earth's surface is called **extrusive rock** (see Figure 7.14). Volcanic rocks such as obsidian, pumice, and scoria, are extrusive rocks.

Igneous rocks can cool slowly or quickly. The size of the crystals in igneous rocks differs, depending on how quickly the rocks cooled. In the next investigation, you will discover how cooling time and the size of the crystals are related.

Figure 7.13 Igneous rocks are divided into two groups, depending on where the rocks cooled.

Figure 7.14 What evidence shows that an extrusive rock cooled quickly on the surface of Earth?

CONDUCT AN INVESTIGATION 7-H

SKILLCHECK
- Observing
- Classifying
- Hypothesizing
- Interpreting Data

Cool Crystals, Hot Gems

You can use a liquid solution of copper(II) sulfate to represent melted rock. If part of the solution is cooled slowly and part is cooled quickly, the sizes of the crystals will be different.

Question
How does rate of cooling affect crystal size?

Hypothesis
Formulate a hypothesis about how rate of cooling affects crystal size.

Safety Precautions

- Do not touch copper(II) sulfate. Wash your hands thoroughly if you touch it accidentally.
- Be careful when pouring hot liquids.

Apparatus
2 test tubes
2 large (400 mL) beakers
measuring cup or graduated cylinder
250 mL beaker
stirring rod
hot plate
scoopula
2 watch glasses

Materials
masking tape
tap water
crushed ice
copper(II) sulfate

Procedure

Day 1

1. Use the masking tape to label one test tube and one 400 mL beaker "ice." Fill the beaker with crushed ice.

2. Label the other test tube and 400 mL beaker "warm water." Fill this beaker with warm tap water.

3. Pour 50 mL of tap water into the 250 mL beaker.

4. Add 40 g of copper(II) sulfate to the 250 mL beaker. Stir carefully.

5. Place the 250 mL beaker on the hot plate. Gently heat the beaker. Continue stirring until all the copper(II) sulfate has dissolved.

6. Using the oven mitts, carefully pour some of the solution into each test tube.

7. Place the test tube labelled "ice" in the beaker of crushed ice. Place the test tube labelled "warm water" in the beaker of warm water. Leave undisturbed for 24 h.

Day 2

8. Using a scoopula, gently pry the crystals loose from the test tubes. Place the crystals from each test tube on a different watch glass. **Observe** the crystals.

9. Recycle the crystals and extra solution in the container supplied by your teacher. Never wash chemicals down the sink.

Analyze

1. Which test tube had larger crystals? Which test tube had smaller crystals?

2. What was the independent variable (the feature you changed)?

3. What was the dependent variable (the feature that changed as a result of the investigation)?

Conclude and Apply

4. How did the rate of cooling affect the size of the crystals?

5. Which crystals could represent extrusive rock? Why?

6. Did your observations support your hypothesis? Why or why not?

What is the difference between compaction and cementation?

Sedimentary Rock

Sedimentary rock is made from sediment and/or by chemical reactions. **Sediment** is loose material, such as bits of rocks, minerals, plants, and animals. Figure 7.15 shows three common examples of sedimentary rocks.

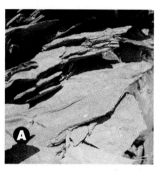
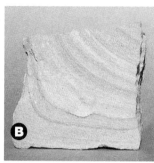
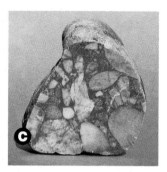

Figure 7.15 Shale is made from mud (A). Sandstone is made from sand (B). What sediment do you think is used to make conglomerate (C)?

Water, wind, and ice can move sediment to a place where it settles. Sediment slowly settles on top of other sediment, forming layers. Sedimentary rocks are often made in oceans and lakes where the larger, heavier fragments settle first. Each layer of sediment is squeezed together by the weight above it, as shown in Figure 7.16. The process of squeezing together layers of sediment is called **compaction.**

The minerals in some rocks dissolve as water soaks into the rocks. The dissolved minerals form a natural cement that sticks the larger pieces of sediment together (see Figure 7.16). The process in which pieces of sediment are held together by another material is called **cementation.**

Many sedimentary rocks are formed by compaction and cementation. Other sedimentary rocks, such as salt (halite), gypsum, and limestone are formed in water by chemical changes.

Figure 7.16 What similarities can you see between compaction and cementation? What differences can you see?

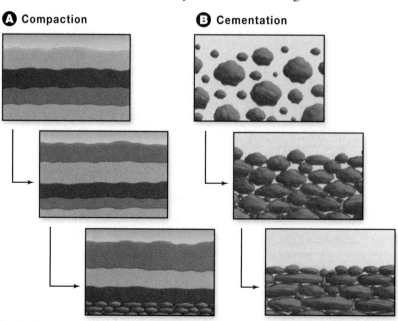

Treasures in Sedimentary Rock

Limestone, shale, and other sedimentary rocks commonly contain fossils, such as the trilobite in Figure 7.17. **Fossils** are evidence of once-living organisms. British Columbia contains a wealth of fossils. Sedimentary rocks throughout the province contain fossilized shells, bones, scales, footprints, teeth, and leaves.

Fossils provide evidence of the environment and the climate in which the animals or plants lived. Scientists know that British Columbia was once covered with tropical forests because tropical plant fossils have been discovered. Other fossil evidence shows that British Columbia was covered by ocean water in its distant past. Fossils can also be used to determine the history and age of the rocks in which they are found.

The fossils found in the sedimentary rock of the Yoho Valley, in eastern British Columbia, are so special that the United Nations has declared the Yoho Valley to be a World Heritage Site. The Burgess Shale in Yoho National Park is home to some of the most famous and unusual fossils in the world. The *Marrella* shown in Figure 7.18 is a Burgess Shale fossil. Some Burgess Shale fossils have even preserved the soft body parts of marine animals.

Figure 7.17 A fossilized trilobite. Trilobites lived in warm ocean water about 250 to 550 million years ago. Trilobites are now extinct.

Types of Fossils

Usually the remains of dead plants and animals decay quickly and are destroyed. Animals and plants can become fossilized, however, when their remains are protected from scavengers and bacteria. Soft parts (such as skin, muscle, and organs) decay rapidly and are rarely found as fossils. Hard parts like bones and shells, however, can be preserved. Figures 7.17 to 7.22 show some of the fossils that can be found in British Columbia.

Why are there not many fossils of the soft body parts of animals?

Figure 7.18 The *Marrella* is the most abundant of the Burgess Shale fossil animals.

Figure 7.19 Coiled ammonites are common fossils in British Columbia. Casts of hundreds of ammonites are sometimes found clustered together.

Figure 7.20 Shark teeth can be fossilized when water dissolves the minerals in the teeth. A harder mineral (possibly quartz) then replaces the original material.

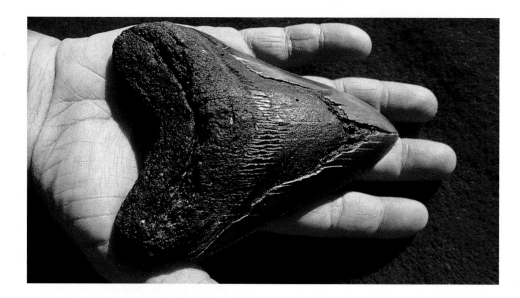

Did You Know?

- A 12-year-old girl made the first discovery of dinosaur bones in British Columbia near Courtenay, on Vancouver Island, in 1988.
- Animals and plants have been found preserved in peat bogs, tar pits, amber, and ice. When all or part of an organism is preserved, the remains are called *original remains*.

Figure 7.21 A carbon imprint may be formed when a plant is buried under many layers of sediment. Pressure and heat build up until only a thin film of carbon residue is left. The residue shows the outline of the plant.

INTERNET CONNECT

www.mcgrawhill.ca/links/BCscience7

Learn more about the discoveries of dinosaur bones and footprints in British Columbia by going to the web site above. Click on **Web Links** to find out where to go next.

Figure 7.22 Trace fossils, such as footprints, are evidence of animal activities. British Columbia is one of the best areas of Canada for discovering dinosaur footprints.

Find Out ACTIVITY 7-1

Mystery Moulds

In this activity, you will make a fossil mould of an object. Will your classmates be able to guess the identity of the object?

What You Need

small object (such as a shell, ring, or ornament)
plaster of Paris
sturdy spoon
paper plate
petroleum jelly
paper

What to Do

Day 1

1. Place a thin layer of petroleum jelly on your object and on the surface of the paper plate.

2. Mix water with the plaster of Paris to make 250 mL of the mixture.

3. Quickly spread the plaster of Paris on the greased plate.

4. Press your object in the plaster until it is almost buried.

5. Cover the plate with a piece of paper to hide the object.

Day 2

6. Gently pry your object loose from the plaster.

7. Display your mould. Try to identify the moulds that your classmates made.

What Did You Find Out?

1. Which objects made the best moulds? Why?

2. An object that is made in a mould is called a cast of the original object. Most fossils that are found on Earth are casts of living things. How could you make a cast of your object?

3. What uncertainties do scientists face when they investigate fossil evidence? Why do scientists need to investigate a variety of fossil evidence before making conclusions?

Extension

4. Read the following flow chart about how mould and cast fossils are made. Then create a similar flow chart to show the steps you followed in this activity.

The Formation of a Mould and Cast Fossil

 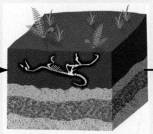 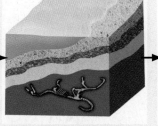 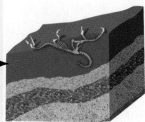

A An ancient animal dies. Its remains sink into the mud. The soft parts of the remains decay rapidly.

B The animal's hard parts are buried by sediment. Millions of years pass. The sediment slowly turns to rock.

C The skeleton gradually dissolves, leaving a mould in the rock. Other sediments fill the mould, forming a cast.

D The surface rock is eroded. The fossil is exposed.

Chapter 7 The Composition of Planet Earth • MHR 219

Metamorphic Rocks

If you find a rock with a lot of shiny mica particles and many layers, you may have found a metamorphic rock. A **metamorphic rock** is made when heat, pressure, or fluids change one type of rock into another type of rock (see Figure 7.23). Metamorphic rocks can be made from igneous and sedimentary rocks, and from other metamorphic rocks. The formation of a metamorphic rock is a long, slow process.

READING check

The word metamorphic means "changed form." How is one type of rock changed into another type?

Figure 7.23 Pressure is one condition that causes metamorphic rocks to form. When pressure is applied to granite (A), the mineral grains are flattened and aligned. This results in the formation of gneiss [nice] (B).

Metamorphic rocks are always formed below the surface of Earth, where heat and pressure are very high. Hot, watery fluids can flow into igneous, sedimentary, or metamorphic rock and change the rock both physically and chemically. The rock that has been changed into a metamorphic rock is called the **parent rock.** Depending on the amount of pressure and temperature applied, one parent rock can change into several different metamorphic rocks. A sedimentary rock called shale, for example, can change into slate (see Figure 7.24). As more pressure and heat are applied, the slate can change into schist and eventually into gneiss.

Figure 7.24 Metamorphic rock looks different from its parent rock, but the rocks have common characteristics. What characteristics of shale can you see in slate? Which properties of limestone might you find in marble?

Look back to Figure 7.23. Can you see the layers in the gneiss? Metamorphic rocks that have layers are said to be foliated. **Foliated** means having thin, leaf-like layers.

Some metamorphic rocks do not have layers. When the mineral structure changed, the rocks did not form layers. Marble and quartzite are metamorphic rocks that usually do not have layers. Metamorphic rocks that do not have layers are said to be **non-foliated.** You can see marble and its parent rock in Figure 7.24 on the previous page.

The parent rock can change so completely that the metamorphic rock no longer looks like it. The two rocks have common characteristics, however, so geologists know the two are related. For example, the limestone and marble in Figure 7.24 look different, but both are made of the mineral calcite. Also, both react with hydrochloric acid.

Find Out ACTIVITY 7-J

Find the Family

Scientists have classified all rocks into three families: igneous, sedimentary, and metamorphic. The families are not based on colour or shape, but on how the rocks are formed. How can you determine in which family a rock belongs?

Materials

a set of rocks, including rocks from your geographic area

hand lens or binocular microscope (optional)

Procedure

1. **Research** the three families of rocks and create your own classification system based on your investigations. These points may help you:
   - Sedimentary rocks are sometimes made of small particles compacted or cemented together.
   - Igneous rocks often have small crystals of minerals visible and the minerals appear to be interlocking (joined together).
   - Metamorphic rocks are created under a great deal of heat and pressure, causing some metamorphic rocks to have thin, flat layers that are easily visible.

2. You now have a good start on the classification system. Using the research you have gathered, try to **classify** each of the rocks into one of the three families. Do not be discouraged if some rocks are difficult to classify. The system you develop is more important than how accurately you sort the rocks.

What Did You Find Out?

1. Which characteristics did you use to classify igneous rocks?
2. Which characteristics did you use to classify sedimentary rocks?
3. Which characteristics did you use to classify metamorphic rocks?
4. Describe the characteristics of any rocks you were unable to classify.

Pause & Reflect

Write a story or poem about how a plant or animal becomes a fossil.

Section 7.3 Summary

Rocks are grouped into three families, based on how they are formed:
- igneous (can be intrusive or extrusive)
- sedimentary (are formed by compaction, cementation, or chemical reaction)
- metamorphic (are formed deep inside Earth; can be foliated or non-foliated)

Fossils are found in sedimentary rocks when the conditions for preservation are just right. Fossils provide evidence about the history of the rocks and the climate of the area.

In the next chapter, you will travel deeper into Earth's crust to examine the mysteries of plate movement and the recycling of rocks.

Key Terms

igneous rock
magma
intrusive rock
lava
extrusive rock
sedimentary rock
sediment
compaction
cementation
fossils
metamorphic rock
parent rock
foliated
non-foliated

Check Your Understanding

1. What are the three rock families? How are the rocks in each rock family formed?
2. What are two differences between intrusive and extrusive igneous rocks?
3. What are three examples of sedimentary rocks?
4. Describe compaction and cementation. Make a drawing of each process.
5. What are the steps in the formation of a mould and cast fossil? Use a flow diagram to help you explain your answer.
6. **Apply** Suppose that you were asked to identify whether a rock specimen was igneous, sedimentary, or metamorphic. What steps would you use to identify it?
7. **Thinking Critically** Why is it unlikely that you would find fossils in igneous or metamorphic rocks?

Did You Know?

Some rocks come from outer space. Meteorites are pieces of rock and iron that have fallen to Earth. Meteorites are often strongly magnetic. Billions of tiny meteorites have fallen to Earth.

Estimate the actual length of this tiny meteorite. What properties could help you identify what it is made of?

CHAPTER at a glance

Now that you have completed this chapter, try to do the following.
If you cannot, go back to the sections indicated in brackets after each part.

(a) List the layers of Earth from the inside of Earth outward. (7.1)
(b) Draw the internal structure of Earth, showing the relative size of the layers. (7.1)
(c) Explain the term "convection current." (7.1)
(d) Explain the difference between a mineral and a rock. (7.2)
(e) List six physical properties of minerals that might help you identify a mineral. (7.2)
(f) Compare the terms "fracture" and "cleavage" as they relate to minerals. Give an example of each. (7.2)
(g) Describe how you could determine the hardness of a mineral. Be specific. (7.2)
(h) Name the minerals in the Mohs Hardness Scale, and give the hardness of each. (7.2)
(i) Define the following terms: magma, lava, sediment, fossil. (7.3)
(j) Explain the differences among the three main families of rocks. (7.3)
(k) Draw a Venn diagram comparing intrusive and extrusive igneous rocks. (7.3)
(l) Compare the terms "compaction" and "cementation" as they relate to sedimentary rocks. (7.3)
(m) Draw a picture that shows how a carbon imprint fossil is formed. (7.3)
(n) Describe the difference between foliated and non-foliated metamorphic rocks. (7.3)
(o) Give three factors that can change a parent rock into a metamorphic rock. (7.3)

Prepare Your Own Summary

Summarize Chapter 7 by doing one of the following. Use a graphic organizer (such as a concept map), create a poster, or write a summary to include the key chapter ideas. Here are a few ideas to use as a guide:

- Use modelling clay to create a model of the inside of Earth.
- Explain why Earth has a magnetic field.
- Explain how you would identify a mineral you have never seen before.
- Using the illustration on the right, outline how the three different families of rocks are formed.
- Create a poster to show how an animal becomes a fossil.
- Write a story about the three rock families.

CHAPTER 7 Review

Key Terms

crust
mantle
Moho
outer core
inner core
lithosphere
asthenosphere
plates
convection current
magnetic field
mineral
rock
mineral resource
fossil fuels
crystal
hardness
lustre
streak
cleavage
fracture
igneous rock
magma
intrusive rock
lava
extrusive rock
sedimentary rock
sediment
compaction
cementation
fossils
metamorphic rock
parent rock
foliated
non-foliated

Reviewing Key Terms

If you need to review, the section numbers show you where the terms were introduced.

1. Which key terms describe the layers of Earth? (7.1)

2. Which key terms refer to the physical properties of minerals? (7.2)

3. Which key terms are families of rocks? (7.3)

4. Which key terms can only be used to describe metamorphic rocks? (7.3)

5. Which key terms can only be used to describe igneous rocks? (7.3)

Understanding Key Ideas

Section numbers are provided if you need to review.

6. What are the differences among the crust, the mantle, and the Moho? (7.1)

7. Draw and label the lithosphere and the asthenosphere. (7.1)

8. Copy and complete the following concept map about minerals. (7.2)

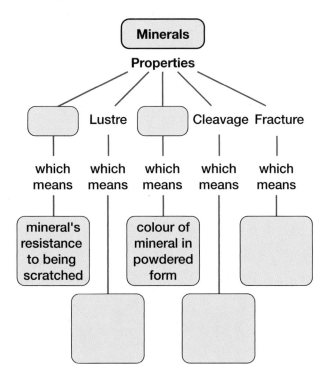

9. What does a sedimentary rock look like? Be specific. (7.3)

10. Why are metamorphic rocks created deep in Earth's crust, not on the surface? (7.3)

Developing Skills

11. What hardness does a mineral have if it can scratch an iron nail but it cannot scratch glass?

12. Use a heavy piece of paper to make a three-dimensional model of each major crystal shape. You may find some ideas on the Internet.

13. How are igneous, metamorphic, and sedimentary rocks formed? Describe the three different conditions that cause these rocks to be formed.

14. Research the Burgess Shale. Draw some of the unique fossils that are found in the Burgess Shale. Why do you think the Yoho Valley has been declared a World Heritage Site?

15. Use the following illustration to help you explain what scientists think has created Earth's magnetic field.

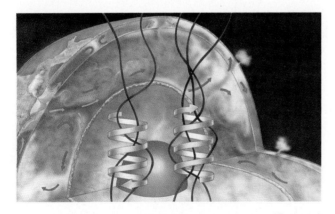

Problem Solving

16. Suppose that you had a piece of paper, a piece of steel, and a glass bottle. How could you use these items to distinguish between calcite and quartz? What other test would help you identify calcite?

17. Imagine that a new planet has been discovered in our solar system. When astronauts visit the planet, they notice that their compasses do not work. What inference could you make about the internal structure of the new planet, based on this observation?

18. What items should you take with you if you go hunting for minerals or fossils in the mountains? Explain the use and purpose of each item.

19. How could you find out the number of sides on a perfect salt crystal?

Critical Thinking

20. Why do you think some igneous rocks have bubbles or pores? Are these rocks intrusive or extrusive? Why?

21. Is ice a mineral? Explain why or why not using the definition of a mineral.

22. Diamonds are the hardest known natural substance. You can, however, easily break a diamond into several pieces if you hit the diamond with a small hammer. What does this indicate?

23. Dinosaur bones are very rare in British Columbia, but there are many dinosaur footprints. Why do you think there are so few bones but so many footprints?

Pause & Reflect

Check your orginal answers to the Getting Ready questions on page 196 at the beginning of this chapter. How has your thinking changed? How would you answer those questions now that you have investigated the topics in this chapter?

CHAPTER 8
Earth's Moving

Getting Ready...

- How do rocks change?
- How is new crust made?
- How are mountains and volcanoes formed?

Submersibles allow scientists to explore the deeper regions of the sea floor. New discoveries are made on almost every voyage.

Imagine you could travel to the bottom of the Pacific Ocean in a submersible. What features do you think you would observe along the sea floor? You might be surprised to see many mountain ranges and volcanoes. In fact, there are more mountains and volcanoes on the sea floor than on Earth's continents! The mountains of the oceans are taller than those on land, and the valleys are deeper. Continuing on your tour of the sea floor, you might also see hot-water vents, rich in chemicals and spewing black smoke. As well, you might see the creation of new crust from magma flowing through cracks in the sea floor.

Earth's crust is constantly changing, even as you read about it. Rocks are being transformed and recycled, new sea floor is being created, and the plates of the lithosphere are moving. In Chapter 8, you will learn about recent scientific discoveries that help to explain the moving crust and the changing surface of Earth.

Crust

What You Will Learn

In Chapter 8, you will learn

- how rocks are changed and recycled
- how the theory of continental drift was developed
- what major features are found on the sea floor
- what recent advances in technology have revealed about the sea floor
- how the plates of the lithosphere interact

Why It Is Important

- The movement of Earth's crust is responsible for the formation of many of Earth's features. Understanding the rock cycle may help contractors and builders make better choices about where and how to construct homes, roads, and buildings.
- Understanding how and why Earth's crust is moving may help scientists to predict earthquakes and eruptions of volcanoes more accurately.

Skills You Will Use

In Chapter 8, you will

- create colourful and tasty models of rocks
- observe the effect of chemicals on rocks
- communicate your knowledge of the rock cycle in a song
- make hypotheses about moving continents
- make models of the sea floor
- infer data about the age of rocks
- classify plate boundaries
- design a model of plate movement

Starting Point  ACTIVITY 8-A

Sediment Shake-Up

Rocks on Earth's crust are constantly being changed by natural elements, such as water and ice. How can you observe the wearing down of rocks?

Safety Precautions

What You Need

coffee can with lid
small stones collected from your schoolyard

What to Do Group Work

1. Work in a group. **Observe** the stones you collected from your schoolyard. **Draw** their shapes in your science notebook.
2. Place enough stones in the coffee can to fill one third of the can. Put the lid on the can.
3. Put on your safety glasses and apron. With other group members, take turns shaking the can vigorously from side to side for at least 10 min.
4. Remove the lid, and **observe** the stones.

What Did You Find Out?

1. What happened to the stones?
2. Describe the sediment you created.

    (a) In step 3, you shook the stones in the coffee can. How is this similar to ways that rocks are changed on Earth's surface?

    (b) How is it different from the ways that rocks are changed on Earth's surface?

CHAPTER 8

Section 8.1 Weathering and the Rock Cycle

Figure 8.1 Imagine how these rocks were carved. For millions of years, weathering has worn away softer materials, leaving behind more resistant rock.

Rocks on the sea floor and the continents of Earth are constantly being changed. Can you believe that tiny moss plants, burrowing shrews, and even oxygen in the air can affect solid rocks? Animals and plants can weaken and break apart rocks at Earth's surface. Water and ice can also break rocks into smaller pieces. **Weathering** is the breaking down of rocks into sediments. After the rocks have been broken, other forces carry the pieces far away.

Over millions of years, weathering has changed Earth's surface, as shown in Figure 8.1. The weathering process continues today. Weathering affects rocks, caves, mountains, and even buildings and streets. Over hundreds of thousands of years, weathering wears mountains down to rolling hills. Two different types of weathering—mechanical weathering and chemical weathering—work together to shape Earth's surface and create sediments.

Mechanical Weathering

Mechanical weathering occurs when rocks are broken apart by physical processes. In mechanical weathering, the overall chemical make-up of the rocks stays the same. Each piece of rock is similar to the original rock. Growing plants, burrowing animals, and expanding ice can mechanically weather rocks, as shown in Figure 8.2. Enough force is created to break the rocks into smaller pieces.

Three causes of mechanical weathering are described below.

- Water and nutrients that collect in the cracks of a rock allow plants to grow (see Figure 8.2A). As the roots of the plants grow, they enlarge the cracks. Sometimes the roots push the rock apart. You may have seen mechanical weathering in a sidewalk. The roots of a tree can break cement in a sidewalk.

- Mechanical weathering can be caused by burrowing animals, such as earthworms and moles (see Figure 8.2B). As the animals dig burrows, they loosen sediments and push the sediments to the surface. Once the sediments are at the surface, other weathering processes can act on them.

- When water enters the cracks in a rock and freezes, it expands. The cracks enlarge and the rock breaks apart (see Figure 8.2C). Then more water can enter the cracks, freeze, and expand, causing the cracks to grow even bigger. The cycle of freezing and thawing that breaks up rocks can also break up roads. When water enters cracks in the pavement and freezes, ice forces the pavement apart and causes potholes in roads.

Did You Know?

Soil is a mixture of weathered rock, organic matter, water, and air. Climate, type of rock, amount of moisture, and slope of land can influence the formation of soil.

Figure 8.2A Growing plants can push rock apart.

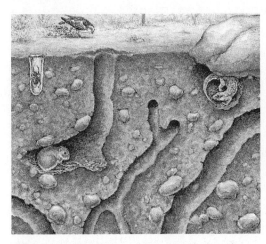

Figure 8.2B Burrowing animals break apart sediment and move it to the surface.

Figure 8.2C What effects of weathering can you see in the rock in this photograph?

Figure 8.3 The effects of chemical weathering can be seen in discoloured paint, stained statues, and rusted automobiles.

Chemical Weathering

Chemical weathering occurs when chemical changes dissolve the minerals in rocks or change the minerals into different minerals. Chemical weathering changes the chemical composition of the rocks, which can weaken the rocks.

An example of chemical weathering occurs when acid rain reacts with rocks. As you may have read in chapter 6, acid rain contains dissolved acids and other chemicals from air pollution. The acids react with some rocks, such as the rock that was used to make the statue in Figure 8.3. The rock dissolves in the acids and is washed away.

Naturally formed acids can also weather rocks chemically, as shown in Figure 8.4. When water mixes with carbon dioxide gas in the air or soil, a weak natural acid called carbonic acid forms. Carbonic acid dissolves minerals such as calcite. Calcite is the main mineral that makes up limestone. When acidic water flows through layers of limestone, rock is dissolved and redeposited to make spectacular caves.

Figure 8.4 Caves form over many thousands of years, as slightly acidic ground water dissolves large amounts of limestone.

CONDUCT AN INVESTIGATION 8-B

SKILLCHECK
- Predicting
- Hypothesizing
- Observing
- Classifying

Rocks That Fizz

Question
When acids react with certain rocks, some minerals dissolve and carbon dioxide gas is formed. Which rocks do acids affect?

Hypothesis
Formulate a hypothesis about which rocks are affected by acid rain and acid ground water.

Safety Precautions

Apparatus
watch glass
tongs or tweezers
medicine dropper

Materials
10% hydrochloric acid
small pieces of identified rock
two unidentified rock samples from your geographic area

Procedure

① Make a table like the one below, to record your observations.

Table of Observations

Name	General observations	Hydrochloric acid test
granite		
chalk		
sandstone		
shale		
marble		
limestone		
unknown rock A		
unknown rock B		

② **Observe** the physical characteristics (such as colour and texture) of one of the specimens. **Record** your observations under "General observations."

③ Put the specimen on a watch glass.

④ Put on your safety glasses. Use the medicine dropper to place a few drops of 10% hydrochloric acid on the specimen. **Observe** what happens. **Record** your observations in your table.
CAUTION: If acid gets on your skin immediately run cool water on skin and wash the area.

⑤ Rinse the specimen with water, and then dry the specimen. Return the specimen to the proper place.

⑥ Repeat steps 2 to 5 for each of the other specimens.

Analyze

1. (a) What was the independent variable (the feature you changed)?
   (b) What was the dependent variable (the feature you observed changing)?

2. Which rocks were affected by chemical weathering? How could you tell?

3. Was your hypothesis supported by the results of the investigation? Explain why or why not.

4. What new hypothesis could you make about chemical weathering? How might you test your new hypothesis?

Conclude and Apply

5. From your observations, what can you conclude about the types of rock that are most affected by chemical weathering?

6. (a) Predict the names of the unidentified rocks.
   (b) What properties did you base your prediction on?

7. Write your own definition of "chemical weathering."

Chapter 8 Earth's Moving Crust • MHR **231**

READING check

What verbs (action words) could help you describe the rock cycle?

An Endless Cycle of Change

Rocks can be broken apart, chemically altered, dissolved, and even carried away by water or ice. Once a rock has reached a new location, the rock can undergo even more drastic changes and become part of an entirely different family of rocks. Sedimentary rocks can become metamorphic rocks. Metamorphic rocks can become igneous rocks. The **rock cycle** consists of the processes in which rocks are continually changed over long periods of time.

All the rocks around you, including the rocks used to build homes and schools, are part of the rock cycle. The changes that take place in the rock cycle never destroy or create matter. The matter is just recycled in other forms. Examine the model of the rock cycle in Figure 8.5. Notice that all three families of rocks are shown, along with the processes that make them. **Erosion** is the transportation of sediments to another location. You will learn more about erosion in Chapter 9. In the following activity and investigation, you will make "sediments" and take them through some of the processes in the rock cycle.

At Home ACTIVITY 8-C

Chewable Rocks

Many foods are made by the same processes that rocks undergo in the rock cycle.

How is chocolate fudge like an igneous rock? (Hint: Think about heating and cooling.)

How is a granola bar like a sedimentary rock? (Hint: Think about different types of sediments cemented together.)

How is a grilled cheese sandwich like a metamorphic rock? (Hint: Think about how heat transforms the cheese.)

What to Do

Think of a food you could use to represent processes in the rock cycle. Make or share the food with your family. Explain the similarities between rocks and what you are eating. Discuss what steps of the food preparation represent processes in the rock cycle.

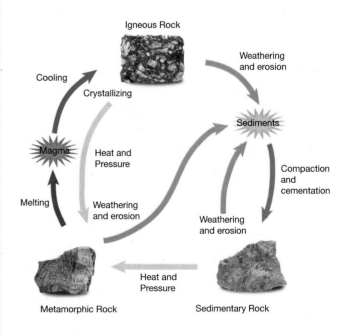

Figure 8.5 The processes in the rock cycle do not occur in a set order. Notice the various shortcuts and detours. Each of these processes can take thousands or millions of years.

CONDUCT AN INVESTIGATION 8-D

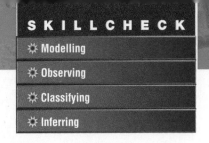

SKILLCHECK
- Modelling
- Observing
- Classifying
- Inferring

Colouring the Rock Cycle

Rocks can be weathered and eroded into sediments if they are exposed to rain, wind, and ice. The sediments can then be compacted or cemented to form new sedimentary rocks. If the new sedimentary rocks are buried deep underground, they can be changed into metamorphic rocks by the processes of heat and pressure. Rocks could also melt and become new igneous rocks. The rock cycle includes the processes of melting, cooling, crystallizing, weathering, erosion, compaction, cementation, heating, and pressure.

Question
Which of the processes of the rock cycle can you model?

Safety Precautions

- Use care when handling the hot water and when working around the heat source.
- Carefully unplug the hot plate at the end of the investigation. Let it cool completely before you put it away.

Apparatus
beaker or small glass bowl
aluminum dish
coin
rolling pin or heavy weight
hot plate or electric kettle
plastic spoon
tongs

Materials
different colours of wax crayons
aluminum foil
water

Part 1
1. Choose three to five different colours of crayons. Remove the paper coverings from the crayons.

2. Use the edge of a coin to scrape small bits ("sediments") of each crayon into a pile on a piece of aluminum foil.

Part 2
3. Place some of the coloured sediments from Part 1 in a neat pile in the centre of a 10 cm by 10 cm piece of aluminum foil. Fold the piece of foil until all the sediments are securely inside.

4. Use the rolling pin (or heavy weight) to apply mild pressure to the foil package. Roll the rolling pin back and forth several times.

5. Open the foil, and **observe** the sediments. **Record** your observations. Save the sediments and foil for use in Part 3.

Part 3
6. Fold the foil securely around the sediments once again.

7. Put on your safety glasses, apron, and oven mitts. Your teacher will supply you with a source of boiling water. Use the tongs to place the foil package in the boiling water. CAUTION: Be very careful near the boiling water. It is very hot and can severely burn your skin.

8. Carefully remove the foil package with the tongs after 20 s.

9. Use the rolling pin to apply a lot of pressure to the package quickly.

10. Open the foil, and **observe** your "rock." **Record** your observations.

continued

Part 4

11. Create more sediments by scraping the crayons with the coin.

12. Warm up a hot plate. Be sure to wear your safety glasses and oven mitts. CAUTION: Be very careful when using the hot plate.

13. Collect a spoonful of sediments. Put your sediments in a small aluminum dish. Place the dish on the hot plate.

14. Heat the sediments, just until they all melt. Do not overheat them.

15. Hold the aluminum dish with the tongs. Pour your heated sediments into a beaker of cold water. Leave the sediments in the water for 1 min.

16. Use the plastic spoon to remove your newly formed "rock." **Observe** your "rock," and **record** your observations.

17. Clean up all the apparatus and materials. Return them to the appropriate places.

Analyze

1. What processes of the rock cycle were you modelling in Part 1 of the investigation? How do you know?

2. What processes of the rock cycle were you modelling in Part 2? How do you know?

3. What process of the rock cycle were you modelling in Part 3? How do you know?

4. What processes of the rock cycle were you modelling in Part 4? How do you know?

Conclude and Apply

5. What processes in the rock cycle were you not able to model in this investigation? Why were you not able to model these processes?

6. How might you improve this investigation so that you could model all the processes in the rock cycle?

DidYouKnow?

The rock cycle takes place underwater as well as on land. Many parts of the sea floor are covered by thick deposits of sediments, up to 1 km deep. The sediments come from the continents and are brought to the sea by rivers.

Figure 8.6 Geologists study rocks in the field and in laboratories under microscopes. The geologist above is investigating how climate affects weathering. Which do you think has a slower rate of weathering, rocks in the temperate rain forests on the coast of British Columbia or the same type of rocks in the dry, hot Okanagan Valley? Why do you think so?

Find Out ACTIVITY 8-E

Real Rock and Roll

Share your knowledge of the rock cycle with your friends and family.

What You Need

recorded music
instruments
props

What to Do *Group Work*

1. With your group, brainstorm how you will use music or props to represent each process in the rock cycle.

2. With your teacher's approval, prepare your presentation. Ensure that you represent all the processes in the rock cycle.

What Did You Find Out?

1. Which process in the rock cycle was the most difficult to represent? Why?

2. Which process in the rock cycle was the easiest to represent? Why?

3. What might you do differently if you were to create another musical presentation of the rock cycle?

Section 8.1 Summary

Rocks are constantly being changed in the rock cycle.

- Sediments are formed from rocks through mechanical and chemical weathering. Soil is formed from a combination of sediments, organic matter, water, and air.
- Each process in the rock cycle can take thousands of years to occur.

In section 8.2, you will travel to the bottom of the sea to learn more about how Earth's crust changes.

Check Your Understanding

1. Where do sediments come from?

2. What are three causes of mechanical weathering?

3. When does chemical weathering occur?

4. Describe the rock cycle. You may wish to draw a diagram to illustrate your description.

5. **Apply** What evidence of weathering have you observed in your community?

6. **Thinking Critically** A hard candy dissolves much faster in your mouth if you first break it into pieces. How is this dissolving process like mechanical and chemical weathering?

Key Terms

weathering
mechanical weathering
chemical weathering
rock cycle
erosion

CHAPTER 8

Section 8.2 Clues in the Crust

Some movements of Earth's crust are slow and gradual, such as the formation of mountains on Earth's surface. Other movements of Earth's crust are rapid and violent, such as earthquakes and volcanoes. Surprisingly, scientists can explain both types of movements by studying a cross section of the sea floor. A **cross section** is a drawing of what you see when you cut through an object. If you cut through the Pacific Ocean floor, the cross section might look like Figure 8.7.

The very long mountain ranges on the sea floor are called ocean ridges. An **ocean ridge** is a raised part of the sea floor, which can become large enough to be considered an underwater mountain range. The Juan de Fuca Ridge, off the west coast of British Columbia, is one example of an ocean ridge.

A **continental slope** is a deep slope in the seabed between a continental shelf and the sea floor. A **continental shelf** is a shallow underwater ledge located between a continent and the deep ocean. Continental shelves and continental slopes are made of continental crust. The sea floor begins at the bottom of a continental slope and is made of oceanic crust.

An **abyssal plain** is a large flat area on the sea floor. Abyssal plains make up a large part of the sea floor. A **rift** is an opening in the oceanic crust, where molten materials from Earth's mantle can escape. All of these features have provided clues of how Earth's crust moves.

List and describe the features found on the sea floor.

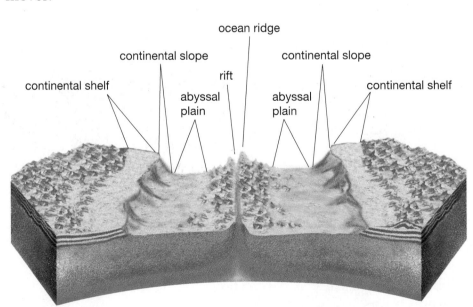

Figure 8.7 The formation of features on the sea floor is mainly due to movements of Earth's crust.

Pangaea

How does Earth's Crust move? Scientists have been investigating this question for many years. One scientist who searched for answers was a German meteorologist named Alfred Wegener [VEG-nuhr] (1880–1930). A meteorologist is a scientist who studies weather and climate. Like other scientists before him, Wegener noticed something interesting about the shapes of the continents. If you cut out the continents from a map, the continents almost fit together like puzzle pieces. The fit is even better if you include the continental shelf around each continent.

Wegener thought that the fit of the continents was more than just a coincidence. He suggested that all the continents were joined together in a huge land mass called **Pangaea** (pan-JEE-uh). He suggested that Pangaea broke apart about 200 million years ago, as shown in Figure 8.8. As well, Wegener proposed the theory of continental drift. According to the **theory of continental drift**, the continents change position very slowly, moving over the surface of Earth a few centimetres every year.

Most of the scientists of Wegener's time strongly disagreed with his theory. Wegener could not explain how the continents moved, but he collected as much data as he could to support his theory. He collected evidence from fossils and rocks. Wegener studied clues about climate change, to try to persuade the scientific world that his theory was valid. You can examine some of his evidence in Figure 8.9 on the next page, and in the following investigation.

Pause & Reflect

What do you think the continents will look like 135 million years from now? Sketch your ideas in your science notebook.

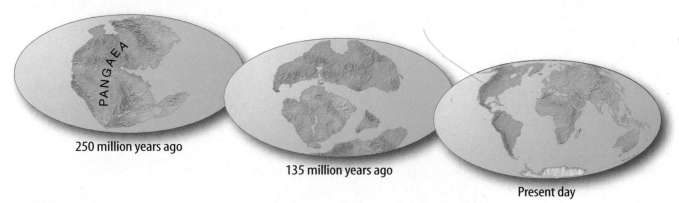

Figure 8.8 Over million of years the continents have changed positions. Find North America in each of the three models. How has the shape of North America changed?

READING check

How did fossil clues support Wegener's theory of continental drift?

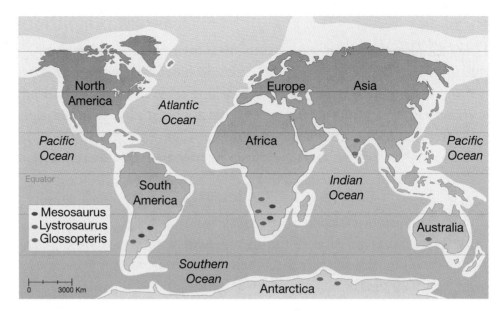

Figure 8.9A *Fossil Evidence:* Fossils of the reptile *Mesosaurus* have been found in South America and Africa. *Mesosaurus* lived in fresh water and on land and probably could not swim between continents. Wegener hypothesized that this reptile lived on both continents when they were joined.

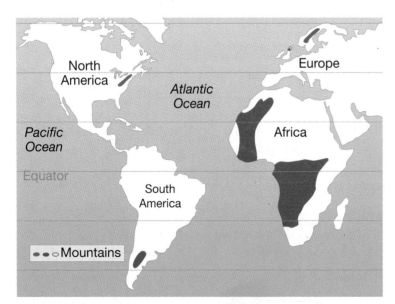

Figure 8.9B *Evidence from Rocks:* The Appalachian Mountains in eastern North America are made of the same kind of rock as a mountain range in Britain and Norway. There are similarities between rock in Québec and rock in northern Britain. There are also similarities between rock in South America and rock in Africa.

Figure 8.9C *Evidence of Climate Change:* The fossil plant *Glossopteris* [glahs-AHP-tur-us] grew in a temperate climate. Glossopteris fossils have been found in Africa, Australia, India, South America, and Antarctica.

CONDUCT AN INVESTIGATION 8-F

SKILLCHECK
- Modelling
- Classifying
- Interpreting Data
- Inferring

Give Me a Clue!

Question
Most scientists now support the idea of continents moving. In Wegener's time, however, most scientists disagreed with his theory. Imagine that you are part of a scientific panel in Wegener's time. Would you agree or disagree with his theory?

Hypothesis
Using the evidence that Wegener collected, form a hypothesis about whether continents can move.

Safety Precautions

Materials
world map with continental shelf boundaries
blue paper
coloured pencils
scissors
glue

Procedure
1. Examine Figure 8.9A. Add the fossil evidence from Figure 8.9A to your blank world map. Put a legend on your map, colour-coding each of the three fossils.

2. Examine Figure 8.9B. Add the three types of rock evidence to your map. Use a different bright colour for each type of rock evidence. Do not cover up your fossil evidence.

3. Label each of the seven continents.

4. Cut out the continents on the world map, around the edges of the continental shelves. Cut India away from Asia along the Himalayan Mountains.

5. Fit the pieces of the world map together to make Pangaea. Once the pieces are in place, glue them to a sheet of blue paper.

6. Transfer the legend to the blue paper by cutting it out or copying it.

Analyze
1. (a) Which pieces were the hardest to fit together?
   (b) How might these pieces have looked 300 million years ago?
   (c) How could you test your ideas?

2. (a) How does including the continental shelves around the continents help you fit the pieces together?
   (b) How well would the continents fit if the continental shelves were not included?

3. (a) Did the evidence you examined in this investigation reinforce your hypothesis? Explain.
   (b) What other hypothesis can you suggest to explain the evidence?

Conclude and Apply
4. Based on the evidence you examined in this investigation, do you think the theory of continental drift is a reasonable theory? Why or why not?

5. (a) As a young child, what ideas did you have that you had to change as you learned more? Give one example.
   (b) How might your example be compared with the experiences of scientists regarding the theory of continental drift?

Find Out ACTIVITY 8-G

Deep-Sea Diorama

What would the floor of the Pacific Ocean look like if all the water evaporated? In this activity, you will find out by creating a model of the sea floor. You can make a diorama, a poster, a sculpture, a computer simulation, or any other type of model you choose.

What to Do [Group Work]

You may wish to work with a partner or in a group. First, list all the processes and features you want to represent in your model. Then, decide what type of model you will make. You may find it helpful to make rough sketches of the model, to help you organize where you will place all the features. Your finished model should include labels.

What Did You Find Out?

1. In what ways is your model an accurate representation of the sea floor?

2. If you could use any building materials, how would you improve your model?

3. How did making a model help you understand the processes and features of the sea floor?

READING check

What is the cause of sea floor spreading?

The Spreading Sea Floor

In the 1960s, geologists were finally able to collect and study rock samples from the sea floor. Scientists used the data they gathered about ages of rocks on the sea floor to form a new hypothesis. They hypothesized that magma is forced upward at a ridge by convection currents. The sea floor is then carried away from the ridge by convection currents and cools. The cooler sea floor sinks back into the mantle. After studying many samples, scientists concluded that the sea floor is indeed spreading apart, as shown in Figure 8.10 below. **Sea floor spreading** is the process in which the sea floor slowly increases in size because of the formation of new crust.

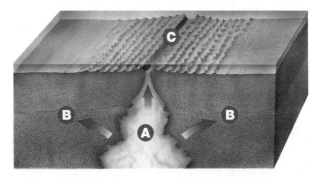

Figure 8.10 *Sea Floor Spreading:* Magma from deep inside Earth is forced upward (A). When the magma reaches the more rigid upper mantle, the magma is moved aside and the plates of the lithosphere move along with it (B). The plates of the lithosphere are forced apart. A rift forms between them (C).

Advances in Technology

As technology advanced, scientists continued to gather information about the sea floor. **Sonar** technology uses a device that sends out sound waves and then records the time that the sound waves take to bounce back. The sound waves can be sent from a ship and reflected off the sea floor, as shown in Figure 8.11A. The longer the sound waves take to return to the ship, the deeper the water is. Scientists use the sound waves to map large areas of the sea floor in detail.

Seismic waves are also used to study the sea floor, as shown in Figure 8.11B. **Seismic waves** [SIHZ-mik] are energy waves that are created by earthquakes or explosions. Air guns are used on board a ship to create seismic waves. The seismic waves can then be studied to determine the types of rocks that make up the sea floor.

In Figures 8.11C and 8.11D, samples are collected from the sea floor using a drilling rig. A long cylindrical sample of rock, called a **core sample,** is obtained from the drilling rig.

Using all three of these methods, scientists are able to create maps of the sea floor, like the map shown in Figure 8.12 on the next page.

> **Pause & Reflect**
>
> Continental shelves often hold oil deposits and other valuable resources. In addition, continental shelves are littered with shipwrecks. Make a hypothesis about why there are so many shipwrecks on continental shelves.

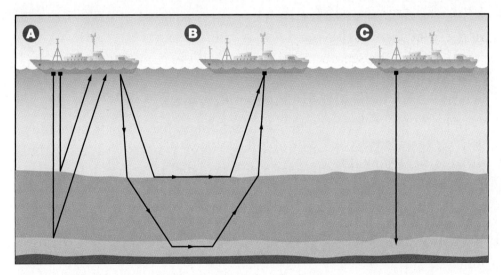

Figure 8.11 Scientists use three methods to study the sea floor. Why are all three methods necessary? What information do scientists gather from each method?

Figure 8.11D Scientists on the *Glomar Challenger* used oil drilling technology to sample the rock in one of the longest mountain ranges in the world, the Mid-Atlantic Ridge.

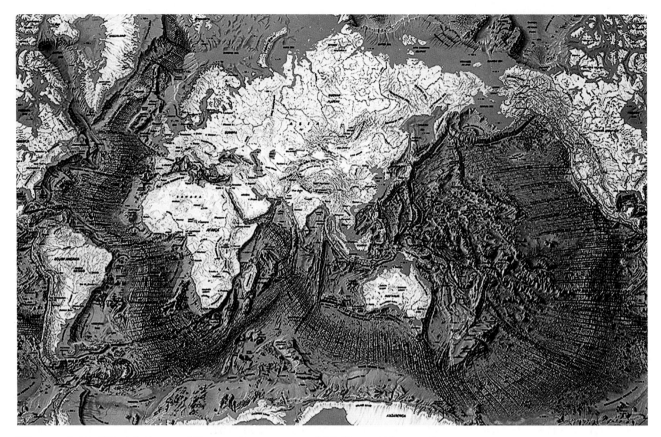

Figure 8.12 What features of the sea floor can you identify on this map?

Magnetic Reversals

Earth's magnetic field has a north pole and a south pole. During a magnetic reversal, the magnetic poles change position. The "north" arrow on a compass points south. Scientists have determined that Earth's magnetic field has reversed itself many times in the past. These reversals have occurred over periods of thousands or even millions of years.

Using a sensing device called a **magnetometer** [mag-nuh-TOM-uh-tuhr] to detect magnetic fields, scientists have found that rocks on the sea floor show evidence of these magnetic reversals. Figure 8.13 shows how the magnetic reversals in the rocks are found in strips that are parallel to the ocean ridges.

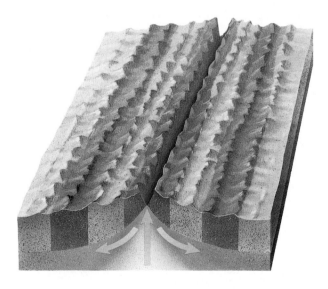

Figure 8.13 Changes in Earth's magnetic field are preserved in the rocks that form on both sides of an ocean ridge. The lighter coloured strips show periods of magnetic field reversals.

Igneous rocks provide a clue to the magnetic reversals. The magma that forms igneous rocks contains iron-bearing minerals, such as magnetite. Tiny magnetite grains line up with Earth's magnetic field just like tiny compasses. As the magma hardens at Earth's surface, the mineral particles stay in line with the magnetic field. Therefore, the reversed magnetic strips must have formed at different times—times when Earth experienced a reversal of its magnetic field. Since the strips are parallel to the ridges, the rock must have formed at the ridges. Therefore, the sea floor must be spreading.

READING check

What type of rock stores a record of Earth's magnetic field?

Recent Discoveries Along Ocean Ridges

Scientists use submersibles to explore the ocean ridges. Along many of these ridges, scientists have discovered new life forms at hydrothermal vents. A **hydrothermal vent** (Figure 8.14) is an opening in the sea floor through which mineral-rich water escapes just like hot springs on land. The water that is forced up though hydrothermal vents along ocean ridges is heated by magma. As well, the water is saturated with minerals and hydrogen sulfide. The minerals can crystallize and form tall towers on the sea floor.

Surprisingly, the organisms that live near hydrothermal vents do not get their energy from the Sun. The vents are too deep in the ocean for any sunlight to penetrate. Instead, hydrothermal organisms live on the chemicals that are dissolved in the water. Scientists have discovered so many new life forms at hydrothermal vents they have created a new classification system for these organisms. Figure 8.15 shows giant tubeworms found near hydrothermal vents.

Did You Know?

Many new and exciting features of the sea floor have been discovered using Canadian technology. One example is ROPOS, a Canadian remote-control vehicle.

Figure 8.14 Temperatures of over 300°C have been recorded at hydrothermal vents. An active hydrothermal vent is called a black smoker. Black smokers can reach heights of over 50 m.

Figure 8.15 You may recall from chapter 2 that giant tubeworms inhabit areas near hydrothermal vents along ocean ridges. Tubeworms can grow to almost 4 m long.

INTERNET CONNECT

www.mcgrawhill.ca/links/BCscience7

Imagine watching Pangaea break apart and seeing the continents drift to new locations. To watch video clips that model Pangaea's transformation, go to the web site above. Click on **Web Links** to find out where to go next.

Section 8.2 Summary

Earth's crust is in constant motion. The following ideas help to explain the movements of Earth's crust:

- Alfred Wegener proposed the theory of continental drift. Wegener collected evidence from the shape of the coastlines and from fossils, rocks, and ancient climates.
- Studies of the sea floor have shown the oceanic crust found farther away from a ridge is older than oceanic crust found closer to the ridge.
- Sonar, seismic waves, and core samples have been used to confirm Wegener's theory of continental drift.
- Magnetometers measure the reversal in Earth's magnetic field and help to support the theory of sea floor spreading.

In section 8.3, you will discover how the theory of continental drift has been refined to include movements of the oceanic crust as well as movements of the continental crust.

Key Terms

cross section
ocean ridge
continental slope
continental shelf
abyssal plain
rift
Pangaea
theory of continental drift
sea floor spreading
sonar
seismic waves
core sample
magnetometer
hydrothermal vent

Check Your Understanding

1. What evidence did Wegener use to support his theory of continental drift?

2. Draw a cross section of the sea floor from the edge of a continent to the middle of an ocean. Label the continental shelf, continental slope, ocean ridge, abyssal plain, and rift.

3. How do the ages of the rocks on the sea floor support the theory of sea floor spreading?

4. How is magnetic evidence preserved in the rocks that form along an ocean ridge?

5. **Apply** Why do you think Alfred Wegener's theory of continental drift was not accepted by most of the scientific world in his time?

6. **Thinking Critically** If new crust is being added along ocean ridges, why does Earth's surface not keep expanding?

CHAPTER 8

Section 8.3 # The Theory of Plate Tectonics

In Chapter 7, you learned that the lithosphere (Earth's crust and upper mantle) is broken into pieces called plates (see Figure 8.16). Discoveries on the sea floor showed that when magma rises from an ocean ridge, the magma produces new crust, which pushes the plates apart. As these plates are pushed apart, other plates are pushed together. Movement along any plate boundary results in changes at other boundaries. At a **convergent boundary**, plates move toward each other and collide. At a **divergent boundary**, plates move away from each other.

J. Tuzo Wilson (shown in Figure 8.17) made an important contribution when he proposed a third kind of movement along plate boundaries. He proposed that some plates slide past each other along what he called a **transform boundary**. Plates moving apart, moving together, or sliding past one another cause many of the features found on Earth.

Scientists gathered evidence showing that both the sea floor and the continents were moving. The name "continental drift" was no longer appropriate, so a new theory was formulated. The **theory of plate tectonics** [tek-TON-iks] proposes that the plates of Earth's lithosphere interact with each other. **Tectonics** is the study of the movement of large-scale structural features on Earth's crust.

Create a hint to help yourself remember the meanings of divergent and convergent. For example, "**div**ergent" might remind you of "**div**ide."

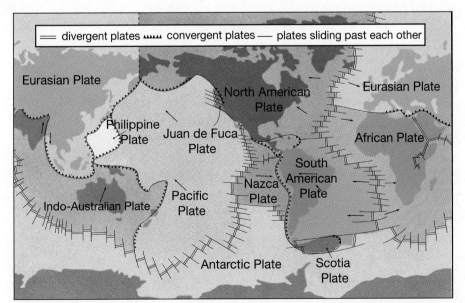

Figure 8.16 The diagram above shows the major plates of the lithosphere, their direction of movement, and the types of boundaries between them. Which types of plate boundaries are found off the west coast of British Columbia?

Figure 8.17 J. Tuzo Wilson, a Canadian scientist, was one of many scientists who have contributed to our understanding of Earth's crust.

Figure 8.18 The Himalayan Mountains are still forming today, as the Indo-Australian Plate collides with the Eurasian Plate.

Convergent Boundaries

There are three different types of convergent boundaries. The different boundaries depend on which types of plates come together.

1. *Oceanic-Continental Convergence* (Figure 8.19C) When an oceanic plate converges with a continental plate, the oceanic plate sinks under the continental plate. The area where an oceanic plate goes down under another plate into the mantle is called a **subduction zone.** Subduction zones along an oceanic-continental boundary create a deep-sea **trench**: a long, narrow, deep depression in the sea floor. Many trenches occur around the margin of the Pacific Ocean because of subduction. High temperatures cause rock to melt around the oceanic plate as it goes under the continental plate. The newly formed magma is forced upward through the upper plate, forming volcanoes. Subduction is responsible for most of the volcanoes and mountain ranges of British Columbia.

2. *Oceanic-Oceanic Convergence* (Figure 8.19E) When an oceanic plate meets another oceanic plate, the older, colder, denser oceanic plate bends and sinks into the mantle. New crust is then formed as the magma rises to make volcanoes. Oceanic-oceanic convergence also creates deep-sea trenches. The islands of Japan and the Japan Trench were formed by oceanic plates converging.

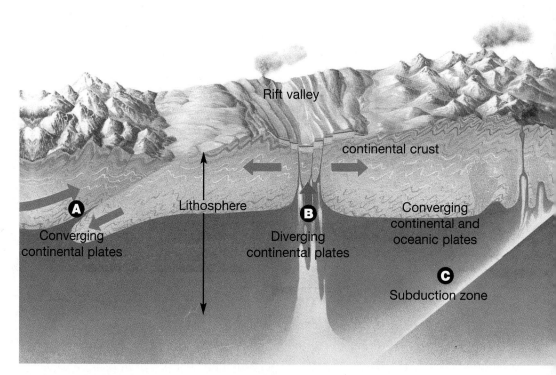

Figure 8.19 Movement along plate boundaries.

3. *Continental-Continental Convergence* (Figure 8.19A) When two plates of continental crust collide and crumple up, tall mountain ranges may form (see Figure 8.18). Earthquakes are common at these convergent boundaries. Volcanoes do not form, however, because there is little or no subduction when two continental plates collide.

Divergent Boundaries

A divergent boundary occurs when plates move away from each other, as shown in Figure 8.19B and 8.19D. Sea floor spreading occurs along a divergent boundary. The Mid-Atlantic Ridge in the Atlantic Ocean is an example of a divergent boundary.

Transform Boundaries

Transform boundaries occur where two plates slide past one another. The plates move in opposite directions or in the same direction at different rates. When one plate slips past another plate suddenly, an earthquake occurs. The Pacific Plate is sliding past the North American Plate, forming the San Andreas Fault in California.

> **READING check**
>
> For each of the sections of Figure 8.19 explain to yourself what is happening. Then review the captions and check your explanation.

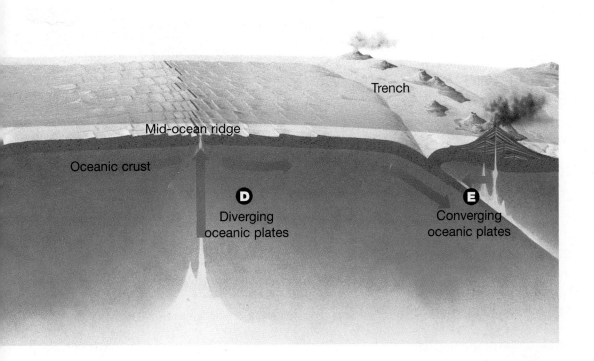

CONDUCT AN
INVESTIGATION 8-H

SKILLCHECK
- Modelling
- Classifying
- Inferring
- Interpreting Data

On a Collision Course!

Geologists often have difficulty duplicating the conditions that are found in Earth's crust. To help them, models and computer simulations have been developed. Most of the movement of Earth's crust can be explained by the theory of plate tectonics. Plate tectonics is responsible for the creation of mountains, volcanoes, and earthquakes.

Question

How can you design a model of Earth's crust that shows the movement of Earth's plates? How can you show at least one result of this movement, such as a mountain, a volcano, or an earthquake?

Safety Precautions

- Do not mix chemicals without your teacher's knowledge and approval.
- Check with your teacher before you begin work on any part of this investigation.

Materials

Work with your group to choose the materials that will be most appropriate for your model. Here is a partial list of materials that you may want to use.

1 kg cornstarch (2 boxes)
500 mL water
spoon
disposable gloves (optional)
large, clean plastic tub (approximately 35 cm by 25 cm by 15 cm)
measuring container
small items, such as puzzle pieces, and marbles

Procedure

1 With your group, brainstorm ways to set up your model. Your model should show the rigid plates of the lithosphere floating on the plastic-like layer of the asthenosphere. In addition, your model should show at least one result of the movement of the plates, such as a mountain, a volcano, or an earthquake.

2 If you make a physical model, you need a recipe for a liquid for the asthenosphere that will support your plates. For example, you could mix 1 kg of cornstarch with 500 mL of water, or you could research another recipe on the Internet. Remember that you must not mix chemicals without permission from your teacher.

3 Prepare a written plan. Include a list of materials and a neat, labelled diagram of your model. Present your plan to your teacher.

4 Divide all stages of the work among your group members.

5 Prepare a poster to accompany and explain your model.

Analyze

1. How do you think your model helps to explain the concept of plate tectonics?

2. What models in the class worked well? Why?

Conclude and Apply

3. What might you change and improve if you built another model of plate tectonics?

4. (a) What hypothesis about Earth's crust could you formulate based on your model?
   (b) How could you test your hypothesis?

Skill
POWER

For tips on modelling in science, turn to Skill Power 4.

Measuring Plate Movement

Earthquakes, volcanoes, and mountain ranges are evidence of plate movement, as shown in Figure 8.20. The theory of plate tectonics suggests that convection currents inside Earth can affect the crust differently in different locations. You have seen how the plates have moved since Pangaea separated. How do you think scientists measure the distances that plates move each year?

Scientists now use satellite and laser technology to measure the exact movements of the plates, to the nearest centimeter. Laser pulses are shot from the ground to a satellite (Figure 8.21), to determine an exact location on the ground. New data confirm that parts of the crust are moving from 1 to 12 cm each year. For example, the Atlantic Ocean is expanding. North America and Europe are being carried about 3 cm farther apart each year. Hawaii is moving toward Japan at a rate of about 8.3 cm each year.

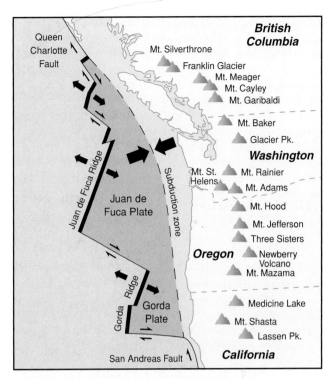

Figure 8.20 Many people in British Columbia live near the edge of two plate boundaries. The Juan de Fuca Plate is converging with the North American Plate.

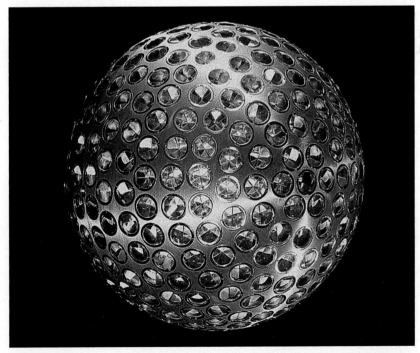

Figure 8.21 Laser pulses are shot from the ground to the satellite. The pulses reflect off the satellite and are used to determine an exact location on the ground. Records are maintained for years, and compared.

Pause & Reflect

The North American Plate is only moving about 3 cm each year. The middle of the Pacific Plate, however, is moving between 8 and 10 cm each year. Why do you think the oceanic plate is moving faster than the continental plate? Record your ideas in your science notebook.

THINK & LINK

INVESTIGATION 8-1

SKILLCHECK
- Classifying
- Interpreting Data
- Hypothesizing
- Communicating

Where Do You Draw the Line?

Think About It

In this investigation, you will plot earthquake, volcano, and plate boundary locations on a map using lines of latitude and longitude. You will look for patterns in the occurrence of earthquakes and volcanoes, and the locations of plate boundaries.

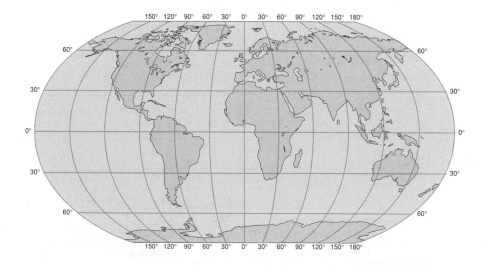

INTERNET CONNECT

www.mcgrawhill.ca/links/BCscience7

Look on the Internet for a map of world-wide earthquake activity, prepared by the Geological Survey of Canada and the United States Geological Survey. Go to the web site above. Click on **Web Links** to find out where to go next. How does the pattern of earthquake and volcano activity on this map compare with the pattern on the map you created in the investigation?

What to Do

1. Use the tables in this investigation, or obtain your own list of earthquake and volcano locations from the Internet.

2. Mark the locations of earthquakes with blue dots on the map provided by your teacher.

3. Mark the locations of volcanoes on your map with red dots.

4. Mark the locations of the plate boundaries on your map in purple.

5. Create a legend on your map to show what each symbol represents.

Earthquakes Around the World

Longitude	Latitude	Location	Year
122°W	37°N	San Francisco, California	1906
72°W	33°S	Valparaiso, Chile	1906
78°E	44°N	Tien Shan, China	1911
105°E	36°N	Kansu, China	1920
140°E	36°N	Tokyo, Japan	1923
102°E	37°N	Nan Shan, China	1927
85°E	28°N	Bihar, India	1934
39°E	35°N	Erzincan, Turkey	1939
136°E	36°N	Fukui, Japan	1948
133°W	54°N	near Queen Charlotte Islands	1949
97°E	29°N	Assam, India	1950
3°E	35°N	Agadir, Morocco	1960
48°E	38°N	Northwestern Iran	1962
147°W	61°N	Seward, Alaska	1964
57°E	30°N	Southern Iran	1972
87°W	12°N	Managua, Nicaragua	1972
92°W	15°N	Central Guatemala	1976
118°E	39°N	Tangshan, China	1976
40°E	40°N	Eastern Turkey	1976
68°W	25°S	Northwestern Argentina	1977
78°W	1°N	Ecuador-Colombia border	1979
137°E	37°N	Honshu, Japan	1983
102°W	18°N	Western Mexico	1985
45°E	41°N	Northwestern Armenia	1988
122°W	37°N	San Francisco, California	1989
135°E	35°N	Kobe, Japan	1995
122°E	47°N	Nisqually, Washington	2001
58°E	29°N	Bam, Iran	2003

Volcanoes Around the World

Longitude	Latitude	Location
122°W	46°N	Mount St. Helens, Washington
123°W	50°N	Garibaldi, British Columbia
130°E	32°N	Unzen, Japan
25°W	39°N	Fayal, Azores
29°E	1°S	Nyiragongo, Zaire
152°W	60°N	Redoubt, Alaska
102°W	19°N	Paricutin, Mexico
156°W	19°N	Mauna Loa, Hawaii
140°E	36°S	Tarwera, Australia
20°W	63°N	Heimaey, Iceland
14°E	41°N	Vesuvius, Italy
78°W	1°S	Cotopaxi, Ecuador
25°E	36°N	Santorini, Greece
123°E	13°N	Mayon, Philippines
93°W	17°N	Fuego, Mexico
105°E	6°S	Krakatoa, Indonesia
132°W	57°N	Ediziza, British Columbia
74°W	41°S	Osorno, Chile
138°E	35°N	Fujiyama, Japan
15°E	38°N	Etna, Sicily
168°W	54°N	Bogoslov, Alaska
121°W	40°N	Lassen Peak, California
60°W	15°N	Mount Pelée, Martinique
70°W	16°S	El Misti, Peru
90°W	12°N	Coseguina, Nicaragua
122°W	49°N	Mount Baker, Washington State
121°E	15°N	Mount Pinatubo, Philippines

Analyze

1. Are most of the earthquakes located near volcanoes? Explain.

2. Describe the pattern of earthquakes, volcanoes, and plate boundaries in or around the Pacific Ocean.

3. Does the pattern around the Atlantic Ocean look similar to or different from the pattern around the Pacific Ocean?

4. Where do most earthquakes occur in North America?

5. Describe any other places in the world that appear to have a large number of earthquakes.

Conclude and Apply

6. What conclusion can you make about earthquake and volcano locations, based on your observations?

7. If you were a scientist, what might you hypothesize about the areas of Earth's crust where volcanoes and earthquakes are found?

INTERNET CONNECT

www.mcgrawhill.ca/
links/BCscience7

What technologies are being used to collect data in the Neptune Project? How can classrooms become involved in this project? List questions you have about the NEPTUNE Project, and then click on **Web Links** to find out where to go next. Share the answers to your questions with your classmates.

Wiring the Juan de Fuca Plate

One of the most ambitious projects related to plate tectonics is called the NEPTUNE Project. The NEPTUNE Project is a Canadian-American joint effort, which will cost over $250 million dollars. More than 3000 km of fibre-optic/power cables are being laid over the Juan de Fuca Plate, to cover an area roughly 500 km by 1000 km in size. (This is about half the size of British Columbia!) Scientists plan to use a variety of technologies 24 h a day, 7 days a week, to record thousands of observations of the ocean. They hope to learn more about the effects of plate tectonics, underwater earthquakes, whale migrations, and changes in fish populations.

Section 8.3 Summary

According to the theory of plate tectonics, Earth's lithosphere is broken into plates that interact. Mountains, volcanoes, and earthquakes are created by the movement of the plates. Scientific data has shown that

- plates move together at convergent boundaries
- plates move apart at divergent boundaries
- plates slide past each other at transform boundaries

In Chapter 9, you will learn about both the gradual and sudden changes that affect Earth's crust.

Key Terms

convergent boundary
divergent boundary
transform boundary
theory of plate tectonics
tectonics
subduction zone
trench

Check Your Understanding

1. How is the theory of plate tectonics different from the theory of continental drift?

2. Draw a cross section of the oceanic crust. Label the following on your diagram: convergent boundary, divergent boundary, trench, and subduction zone.

3. How do converging plates form mountains?

4. At which types of plate boundaries would you expect to find volcanoes? Explain why.

5. **Apply** "Subduction is responsible for the volcanoes and mountain ranges of British Columbia." How would you explain this statement?

6. **Thinking Critically** What features caused by plate tectonics are present in or near your community?

CHAPTER at a glance

Now that you have completed Chapter 8, try to do the following. If you cannot, go back to the sections indicated in brackets after each part.

(a) Explain the difference between weathering and erosion. (8.1)

(b) Describe three types of mechanical weathering. (8.1)

(c) Describe chemical weathering. (8.1)

(d) Draw a picture to illustrate the rock cycle. (8.1)

(e) Draw a picture of Pangaea. (8.2)

(f) Explain the theory of continental drift. (8.2)

(g) List three pieces of evidence for continental drift. (8.2)

(h) Draw a cross section across the sea floor. Label the features. (8.2)

(i) Explain how an ocean ridge is formed. (8.2)

(j) Describe some forms of advanced technology that were used to support the theory of continental drift. (8.2)

(k) Explain sea floor spreading. (8.2)

(l) Describe the theory of plate tectonics. (8.3)

(m) Compare the three different types of plate boundaries. (8.3)

(n) Draw and describe a subduction zone. (8.3)

(o) Draw and describe the features created where the Juan de Fuca Plate meets the North American Plate on the west side of British Columbia. (8.3)

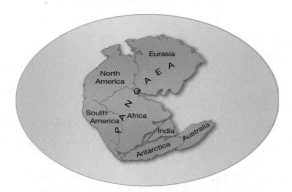

Prepare Your Own Summary

Summarize Chapter 8 by doing one of the following. Use a graphic organizer (such as a concept map), create a poster, or write a summary to include the key chapter ideas. Here are a few ideas to use as a guide:

- Describe how you would explain the rock cycle to a new student who has just transferred into your school.
- Explain the shape of the continents today, using the theory of continental drift.
- Explain the formation of mountains, volcanoes, and earthquakes, using the theory of plate tectonics.
- Copy the concept map below and fill in the missing words, based on what you have learned in this chapter.

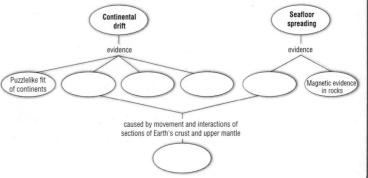

CHAPTER 8 Review

Key Terms

weathering
mechanical weathering
chemical weathering
rock cycle
erosion
cross section
ocean ridge
continental slope
continental shelf
abyssal plain
rift
Pangaea
theory of continental drift
sea floor spreading
sonar
seismic waves
core sample
magnetometer
hydrothermal vent
convergent boundary
divergent boundary
transform boundary
theory of plate tectonics
tectonics
subduction zone
trench

Reviewing Key Terms

If you need to review, the section numbers show you where these terms were introduced.

1. What is the meaning of each of the following terms? (8.2)
   (a) continental drift
   (b) continental slope
   (c) continental shelf

2. In your notebook, match each description in Column A with the correct term in Column B.

   A
   (a) area created by plates pushing together
   (b) area that contains a trench
   (c) flat, broad area of sea floor
   (d) process in which rocks are broken down
   (e) creating new crust
   (f) area created by plates pulling apart
   (g) process of transporting sediments
   (h) opening in the oceanic crust
   (i) area created by plates sliding past each other

   B
   - weathering (8.1)
   - erosion (8.1)
   - abyssal plain (8.2)
   - sea floor spreading (8.2)
   - rift (8.2)
   - transform boundary (8.3)
   - convergent boundary (8.3)
   - divergent boundary (8.3)
   - subduction zone (8.3)

Understanding Key Ideas

Section numbers are provided if you need to review.

3. Describe Pangaea. Why did Wegener believe Pangaea existed? (8.2)

4. Is chemical weathering more rapid in a cool, dry climate or in a warm, wet climate? Explain your reasoning. (8.1)

5. (a) How does sonar technology work?
   (b) What information has sonar provided about the sea floor? (8.2)

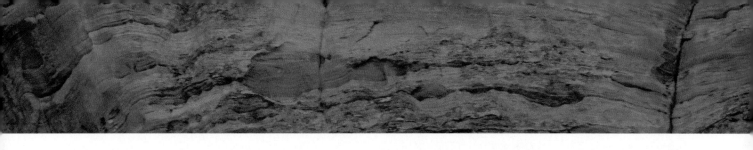

6. Describe and draw the three basic types of plate boundaries. (8.3)

7. (a) What happens when two continental plates collide?
   (b) What happens when a continental plate and an oceanic plate collide?
   (c) What happens when oceanic plates collide? (8.3)

8. Explain which type of plate boundaries produce volcanic eruptions. Be specific. (8.3)

Developing Skills

9. Study the cross section of the sea floor shown below.
   (a) What title would you give this picture?
   (b) What processes are occurring at point A, point B, and point C? Describe each process. Use key terms that you learned in this chapter.

10. Write a story or a poem that explains what can happen to a sedimentary rock as it changes through the rock cycle.

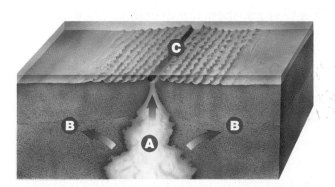

11. (a) What are the three kinds of evidence that Wegener collected to support his theory of continental drift?
    (b) Give one example for each kind of evidence. How does each example suggest that the continents moved?

Problem Solving

12. Why are there few volcanoes in the Himalayan range but many earthquakes?

13. Why are there are so many mountains, volcanoes, and earthquakes in British Columbia?

Critical Thinking

14. How can water be a factor in both mechanical weathering and chemical weathering?

15. Fossils of trilobites are common in the Himalayan Mountains. Trilobites lived in the oceans 200 million years ago. There is no ocean next to the Himalayan Mountains today. Why do you think fossils of trilobites are found there?

Pause & Reflect

Check your original answers to the Getting Ready questions on page 226 at the beginning of this chapter. How has your thinking changed? How would you answer those questions now that you have investigated the topics in this chapter?

CHAPTER 9
Earth's Changing

Getting Ready...

- How do wind and water change the surface of Earth?
- What are the effects of earthquakes?
- What happens when volcanoes erupt?

Earthquakes are destructive, but they also help scientists learn about Earth's internal structure.

Can you imagine how the students in Alaska felt when they saw how an earthquake damaged their school? Luckily, the earthquake occurred at night and no one was injured in the school. Did you know that most schools in western British Columbia have an earthquake plan for their students?

The forces of nature are strong enough to change the surface of Earth. Sometimes the changes are slow and sometimes the changes occur quickly. For millions of years, nature has been eroding the rocks, wearing away softer materials and leaving behind more resistant rock. The work of running water, ice, and wind has slowly changed the surface of rocks and carried particles far away into the ocean. Sometimes even plants and animals have helped in the process.

Volcanoes and earthquakes also change the surface of Earth. Volcanoes release water and carbon dioxide from the interior of Earth, adding important gases to our atmosphere. Volcanoes create new land and fertile soil. Unfortunately, volcanoes can also destroy homes, trees, and landscapes. Although volcanoes and earthquakes are destructive, they provide scientists with valuable information about the internal structure of Earth.

Watch for changes on the surface of Earth as you continue your tour of Earth's crust in Chapter 9.

Surface

What You Will Learn

In this chapter, you will learn

- how wind, ice, and water change the surface of British Columbia
- about the different types of faulting
- how earthquakes are recorded and studied
- how to prepare for an earthquake
- about the different types of volcanoes
- how the age of rocks is determined
- about the major events in geologic time

Why It Is Important

- Erosion can slowly or quickly change the surface of Earth. When you are familiar with the forces of nature, you can be better prepared to deal with them.
- British Columbia has many volcanoes and earthquakes. If you study and understand the strong forces associated with earthquakes and volcanoes, you will be better prepared to protect yourself and your family.

Skills You Will Use

In this chapter, you will

- model the forces that move sediments
- map the location of earthquakes
- research earthquake safety
- investigate the effects of volcanoes
- research volcanoes in British Columbia
- communicate British Columbia's geologic past
- design a model of an earthquake-resistant building

Starting Point ACTIVITY 9-A

Use the Force

Can you think of ways to move something without touching it? In nature, sediment is moved from one location to another by a variety of forces. What are some of the forces that can move sediment and rocks on Earth?

Safety Precautions

What You Need

a mixture of sand and gravel
a shoebox or metal tray

What to Do

1. Place a small pile of sand and gravel in one end of a large shoebox or metal tray.
2. Your task is to move the sediment to the other end of the box without touching the particles with your hands. You can use any force or object approved by your teacher.
3. You can touch and move the outside of the box, but you cannot touch the sediment.
4. Try to move the mixture in a number of different ways.

What Did You Find Out?

1. Describe the methods you used to move the sediment.
2. Which method was most effective?
3. Explain how your methods compare with forces of nature that move sediment.

Chapter 9 Earth's Changing Surface • MHR **257**

CHAPTER 9

Section 9.1 Changing Landforms

In Chapter 8 you learned about features found on the sea floor. There are many similar features to be found on continents (see Figure 9.1). Three of the most common landforms are plains, plateaus, and mountains.

Plains are large, flat areas often used for agriculture. Plains often have thick, fertile soils and abundant grassy meadows suitable for grazing animals. Plains near the ocean are called coastal plains.

Plateaus are flat, raised areas of land made up of nearly horizontal rocks that have been uplifted by forces within Earth. Some plateaus in British Columbia are made from lava flows.

Mountains are formed when rock layers are squeezed from opposite sides when forces within Earth push up crust. Mountains also form when some rock blocks move up and others move down. Another common landform is a delta. A **delta** is a triangular deposit of sediment at the mouth of a river.

As you learned in Chapter 8, erosion is the movement of rock particles or soil from one place to another. The major causes or agents of erosion are gravity, ice, wind, and water (see Figure 9.1).

READING check

What are two ways that mountains can form?

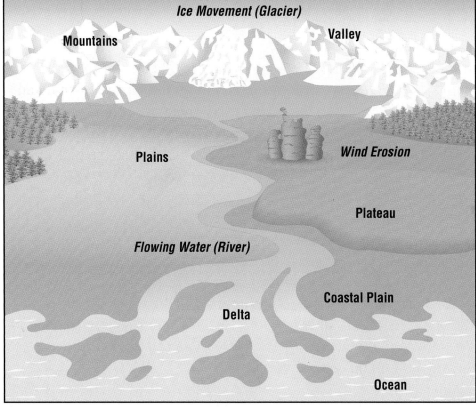

Figure 9.1 Earth's major landforms are changed by the agents of erosion.

Erosion

Have you ever been by a river just after heavy rain? A river looks muddy when there is a lot of sediment in the water (see Figure 9.2). Some of the sediment comes from the riverbank. The rest of the sediment comes from more distant sources and is carried by wind, water, and ice. Muddy water is a product of erosion. Some types of erosion are gradual, and happen over many thousands of years. Changes caused by flash floods and landslides happen suddenly (see Figure 9.3).

> **Pause & Reflect**
>
> In what ways can humans cause erosion? Make a list and record your ideas of how humans can prevent each cause.

Figure 9.2 Erosion from exposed land can cause streams to fill with unusually high amounts of sediment. How could too much sediment damage an area?

Figure 9.3 Heavy rains can cause mudslides that can destroy homes and roads.

Deposition

As you investigate the agents of erosion in Chapter 9, you will notice they have several things in common. Gravity, ice, wind, and water all wear away materials and carry them off. However, agents of erosion erode materials only when they have enough energy to do work. Air cannot erode sediment when the air is still, but once air begins moving and develops into wind, it carries dust, soil, and even rocks along with it. Strong winds can remove the soil from farmland and carry the important resource several kilometres away. Gravity is the force that pulls the sediments down toward Earth's surface. All agents of erosion deposit the sediment they are carrying when their energy decreases. The dropping of sediment is called **deposition**. Deposition is the final stage in the process of erosion. Sediment, mud, and rocks are deposited every day in British Columbia, changing the surface of the province. The same sediment may be eroded again and again over millions of years.

Water in Motion

Water in motion is one of the most powerful agents of erosion. Sudden changes can occur as rivers erode their banks and fast-moving floodwaters carry away large amounts of soil (see Figure 9.4). A **floodplain** is the flat layer of sediment deposited by rivers in flood stage. In addition to causing flooding, heavy rain can disturb the stability of a slope, causing landslides. A **landslide** is the rapid movement of rock fragments and soil. Landslides are common in British Columbia, as shown in Figure 9.5.

Figure 9.4 Sediment deposited during floods makes floodplains some of the most fertile land on Earth.

Figure 9.5 The Hope Slide in 1965 was one of the largest landslides in Canadian history.

READING Check

How does water in rivers, heavy rain, and oceans cause erosion?

Oceans, seas, and large lakes also erode their shorelines. When waves hit cliffs and shores, rocks are broken down and land is eroded. In some places erosion happens quickly. A coastline can lose several metres every year to erosion by sea water.

Running water can also create slower changes on Earth. Rivers erode material, but when they enter a calm body of water, the river deposits its sediment and forms a delta (see Figure 9.6).

Figure 9.6
Richmond, Delta, and Pitt Meadows are built on the delta of the Fraser River. Other deltas are found at Squamish and Bella Coola.

Ice in Motion

Glaciers are large bodies of ice that move slowly downhill (see Figures 9.7 and 9.8). There are many glaciers in the mountain ranges of Western Canada. Glaciers have been formed by the weight of snow piling up year after year. Geologists study the effects of glaciers to learn more about past periods of glaciation. During the Ice Ages, sheets of ice covered much of the northern hemisphere. Remnants of these sheets are found today in Greenland.

DidYouKnow?

Sometimes glaciers carry so much material that as they melt, the material deposited forms unusual but distinctive hills called drumlins.

Figure 9.7 In what ways could you compare Bear Glacier to a river?

Figure 9.8 Some glaciers in Greenland, Antarctica, and parts of British Columbia cover whole mountain ranges.

Glaciers erode the surface of Earth and create interesting landforms. The glaciers that once covered all of British Columbia carved and eroded a great deal of material away from the mountains. Some of the eroded material was carried to the ocean. However, if you have been to Langley, Haney, Aldergrove, or other areas of the Fraser lowlands, you have walked over thousands of tonnes of glacial material deposited there. Materials deposited by glaciers, sometimes called ice age sediments, provide excellent drainage and important building materials for roads, basement walls, and concrete malls. Ice age sediments include sand, gravel, silt, and clay (see Figure 9.9).

Figure 9.9 Sand and gravel deposits left by glaciers are used to construct roads.

READING check

What are three ways that glaciers change the land?

Glaciers Shape the Land

As glaciers move over land, they act like bulldozers and giant scouring pads. Glaciers push aside and forward any loose materials they encounter. Eroded sediments pile up along a glacier's sides and in front of a glacier as shown in Figure 9.10. The large ridge of material deposited by a glacier is called a **moraine**. Sometimes moraines block water melting from a glacier and create a lake. Glaciers can also create features in the mountains such as cirques and horns. A **cirque** is a bowl-shaped basin formed by erosion at the start of a mountain glacier. A **horn** is a sharpened peak formed by the glacial action of three cirques on a mountaintop. All of these glacial features are very common in the mountains of British Columbia.

Mountain glaciers flow down mountain slopes and along valleys eroding as they go. Valleys eroded by glaciers are U-shaped because a glacier plucks and scrapes soil and rock from the sides as well as from the bottom. In contrast, a valley eroded by a river is V-shaped.

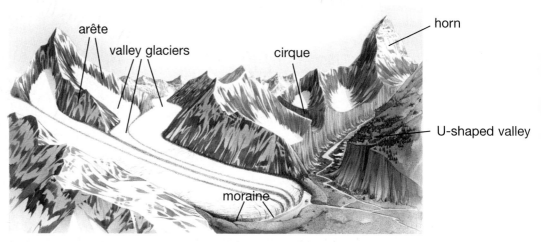

Figure 9.10 Many of the features in the mountains of British Columbia have been created by glacial erosion and glacial deposition.

Glaciers have had a profound effect on Earth's surface. Glaciers have eroded mountaintops and transformed valleys. Vast areas of the continents have sediments that were deposited by great ice sheets. Today, glaciers in polar regions and in mountains continue to change the surface of Earth. In addition to changing the appearance of Earth's surface, glaciers act as natural reservoirs of fresh water. Most of the fresh water on Earth is frozen in glaciers.

Ancient ice provides clues to what the climate was like millions of years ago. Scientists use long drills to remove cores of ice from glaciers (see Figure 9.11). Each layer of ice in a core corresponds to a single year or season. The layers include wind-blown dust, ash, and atmospheric gases. Ice layers are a record of global climate change over millions of years.

Figure 9.11 Scientists examine ice cores to discover clues about ancient climates.

Find Out ACTIVITY 9-B

How Does a Glacier Move?

Glaciers are moving rivers of ice. They move so slowly that it is difficult to observe their movement. In this activity, you can make a model glacier and "speed up" its movement.

Safety Precautions

- Wear gloves during this activity.
- Do not get the borax solution on your skin or in your eyes.
- Dispose of the mixture as instructed by your teacher.
- Wash your hands after doing this activity.

Materials

waxed paper	2 mixing bowls
large spoon	measuring cup
measuring spoons	water
sponges	white glue
borax powder	funnel

Procedure

1. Cover your work surface with layers of waxed paper.

2. Crinkle up some of the waxed paper to make a slight slope. Then cover the slope and the surrounding work surface with more waxed paper. CAUTION: Be sure that the glue and the model glacier stay on the waxed paper throughout this activity.

3. In one mixing bowl, combine 250 mL water with 15 mL borax. Stir until the borax dissolves and there is some solid material at the bottom of the bowl. Add a little more borax, if necessary.

4. In the second bowl, mix 100 mL of water with 100 mL white glue. If necessary, pour off any excess water after you stir the mixture.

5. Stir the borax solution again. Then add about half the solution to the glue-and-water mixture. Stir until the mixture sets. (Add more small amounts of the borax solution if you need to, until the mixture sets.)

6. Remove the mixture from the bowl. Have a partner help you pour it into a funnel.

7. Pour the mixture through the funnel onto the waxed paper slope. Let it pile up on the paper and then spread out. **Observe** what happens and **record** your observations.

What Did You Find Out?

1. Draw and describe a series of diagrams to show what you observed.

2. From your observations, explain how your model glacier moves like a real glacier.

Most glaciers move less than 1 m per day. A glacier in Alaska holds the record as the fastest glacier on Earth. In 1937, the Black Rapids Glacier advanced at the incredible rate of 30 m per day. Why do you think this glacier was moving so fast?

Pause & Reflect

What can be done to slow erosion? Record your ideas for each of these landforms: a steep slope, the plains, and an ocean shoreline.

Career CONNECT

When Dr. Charles Yonge joined a caving club at university, he didn't know that he would be spending much of his future life underground. Charles was a physics student, but after he experienced the beauty of cave rock, he was hooked. "Ninety-nine percent of the world's caves are limestone or dolomite," he says, "and they are the most fascinating landscapes you'll see anywhere."

After Charles switched his studies to geology, he worked with moon rocks and began investigating the rocks and minerals in caves. While he explores, Charles conducts research in climate studies,

Dr. Charles Yonge

searching for clues about the climate hundreds of thousands of years ago. "A cave might be a few million years old, but the parent rock could be 320 million years old or more. The rock can help us understand cycles of glaciation. It can also tell us about the past rain and snowfall, soil, forest cover, and even solar events."

What other ways do scientists study ancient climates? List your ideas, then research with a partner to find answers. Share your results with another pair of students or make a class list.

Section 9.1 Summary

- Erosion is the movement of rock particles or soil from one place to another.
- Deposition is the dropping of sediment.
- The agents of erosion, including gravity, water, wind, and ice, constantly change the surface of Earth.
- Changes caused by erosion can be gradual or sudden.

In Section 9.2 you will learn more about sudden change.

Key Terms
plains
plateaus
delta
deposition
floodplain
landslide
glaciers
moraine
cirque
horn

Check Your Understanding

1. Explain the difference between erosion and deposition.

2. List three agents of erosion.

3. What is the main agent of erosion on the surface of Earth and why is it so common?

4. Explain some of the features created by a glacier as it moves over the land.

5. **Apply** Describe some local features around your city or town that might be formed by glaciers.

6. **Thinking Critically** What would happen to the level of water in oceans if all the glaciers on Earth melted? Why?

Section 9.2 Earthquakes

Earthquakes have occurred in British Columbia for millions of years. An **earthquake** is the shaking of the ground caused by the sudden release of energy stored in rocks. Valuable records of earthquakes have been kept by Aboriginal peoples. Devices for measuring earthquakes have been used for thousands of years (see Figure 9.12).

The rock in Earth's crust is under pressure from internal forces. Tremendous amounts of energy build up in the rocks of Earth. The stresses of moving tectonic plates can cause the rock to bend and stretch. When the pressure is too great, the rock breaks suddenly, creating an earthquake. A **fault** is the surface along which rocks break and move.

Figure 9.12 An ancient earthquake detector from China. The eight dragons on this urn have little balls inside them. Earthquake movements shake the balls into the toads' open mouths. The direction of the earthquake is determined by which toad swallows the ball.

Rocks on the Move

Rock layers can move differently depending on the forces that act upon them. Movement along a fault can spread more than a kilometre in a second. Fault zones exist where tectonic plates meet. Three types of faults are shown in Figure 9.13.

Figure 9.13 Three types of faults

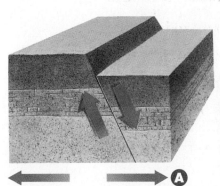

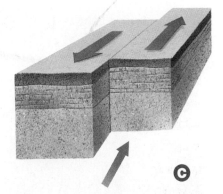

Normal Fault: Tension is the force that causes stretching. In places where plates are moving apart, the tension can pull rocks apart and create a normal fault. In a normal fault, rock above the fault moves downward.

Reverse Fault: Compression is a force or stress that squeezes or compresses. In places where the rock is squeezed by the movement of plates, the compression can cause rocks to bend and break. In a reverse fault, rock above the fault is forced up and over rock below the fault.

Strike Slip Fault or Transform Fault: Shear is a force that causes slipping. Some plates move sideways past each other. As the plates try to move, the forces build up until the rocks break and an earthquake occurs.

READING check

Which type of seismic waves causes the most damage during an earthquake?

Types of Seismic Waves

You can recall from Chapter 8 that seismic waves are energy waves released by an earthquake or explosion. When an earthquake happens, three kinds of seismic waves are released.

Primary waves or P waves travel the fastest of all three types of waves. P waves can pass through solids, liquids, and gases. Primary waves cause a slight vibration (compression) that might rattle dishes on the shelves. P waves provide a warning that an earthquake is happening, and can give people a few seconds to prepare for the waves that will follow.

Secondary waves or S waves travel more slowly than P waves. S waves can pass only through solids, not through liquids or gases.

Surface waves are the slowest of the three waves, but their rolling motion breaks up roads and buildings (see Figure 9.14). Surface waves cause the most damage during earthquakes. You have probably thrown a small stone into water and watched the ripples spread out from the point where the stone entered the water. Surface waves travel through Earth in just the same way. Surface waves cause parts of a building to move up while other parts move down. Rigid structures will collapse if the movement is too great.

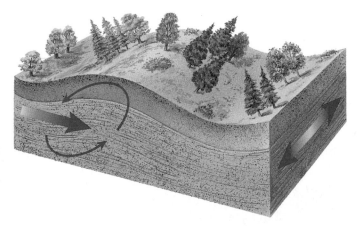

Figure 9.14 Surface waves move rock particles in a backward, rolling motion, and a side-to-side swaying motion.

Evidence of Earth's Layers

Seismic waves provide some evidence of what might be inside Earth. The earthquake that happened in Kobe, Japan, in 1995 was measured at the University of Manitoba. The P waves from Japan had travelled right through the centre of Earth. S waves did not register. We know that P waves can travel through liquid and that S waves cannot. Therefore, we can hypothesize that Earth's outer core must be liquid. Using information from earthquake waves, scientists have developed the model of Earth shown in Figure 9.15.

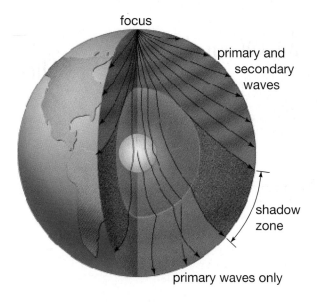

Figure 9.15 Primary waves are refracted (bent) as they travel. There is an area where primary waves do not come through on the other side of Earth. The area is called a *shadow zone*.

Locating an Earthquake

You can use the arrival time of seismic waves to determine earthquake locations. P waves travel faster than S waves. Therefore, it is possible to determine the location of an earthquake by measuring the time interval between the P and S wave. The farther apart the P and S waves are, the farther away is the earthquake.

The place deep in the crust where an earthquake begins is called the **focus** of an earthquake. The primary and secondary waves come from the focus of an earthquake. The surface location directly above the focus is called the **epicentre**. Surface waves travel out from the epicentre (see Figure 9.16).

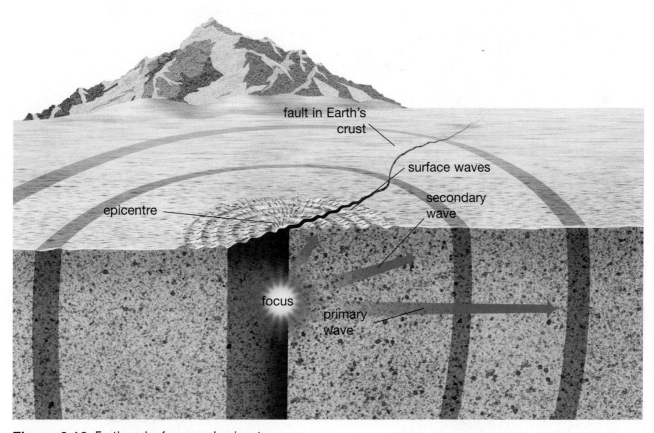

Figure 9.16 Earthquake focus and epicentre

A Sudden movement in Earth's crust releases energy that causes an earthquake.

B P and S waves originate at the focus and travel outward in all directions.

C When the P and S waves reach the surface, they generate surface waves, which are the slowest kind of waves.

D Surface waves travel outward from the epicentre along Earth's surface.

Figure 9.17 Seismologists study earthquakes by reading seismograms.

Measuring Earthquakes

Scientists called seismologists use a special machine called a **seismograph** to measure earthquakes (see Figure 9.17). Seismographs must be attached to **bedrock**, the solid rock that lies beneath the soil, in order to detect the vibrations that result from an earthquake. Inside the seismograph, a marking pen hangs over a rotating drum, just touching the drum. The drum is covered with paper to record the vibrations marked by the pen. When an earthquake strikes, the pen tip moves, making a jagged line. The record of seismic waves detected by a seismograph is called a **seismogram** (see Figure 9.18). Most modern seismographs are electronic, but they are based on the same principles.

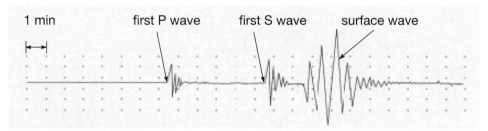

Figure 9.18 The jagged line on the seismogram represents the three types of seismic waves.

Seismologists use a method of measurement called the **Richter scale** to describe the magnitude (strength) of an earthquake (see Table 9.1). The scale starts at zero and can go as high as necessary. The amount of energy released increases greatly as the numbers increase. An earthquake that registered 7 would be about 30 times stronger than one that registered 6, and about 900 times stronger than one that registered 5. Most earthquakes that cause damage and loss of life register between 6 and 8 on the Richter scale. The Alaskan earthquake (shown in the Chapter 9 opener) registered 9.2.

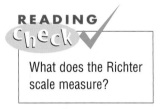

What does the Richter scale measure?

Table 9.1 Richter scale

Richter magnitudes	Earthquake effects	Estimated number per year
< 2.0	generally not felt, but recorded	600 000
2.0 – 2.9	felt by few	300 000
3.0 – 3.9	felt by some	49 000
4.0 – 4.9	felt by most	6200
5.0 – 5.9	damaging shocks	800
6.0 – 6.9	destructive in populated regions	266
7.0 – 7.9	major earthquakes, which inflict serious damage	18
≥ 8.0	great earthquakes, which produce total destruction to communities near the source	1.4

Find Out ACTIVITY 9-C

Waiting for the Waves

Seismologists study the arrival times of primary and secondary waves to determine the distance to an earthquake epicentre. The farther apart the waves arrive, the farther away the location of the epicentre.

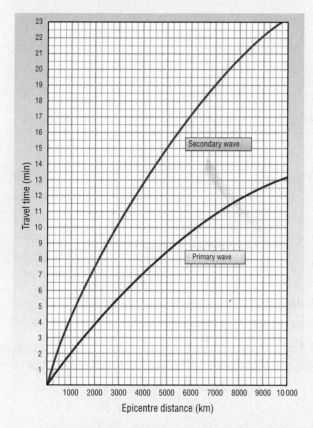

Time Travel Graph for P and S Waves

What to Do

1. Make a data table like the following one:

Arrival Times

Epicentre Distance (km)	Difference in Arrival Time
1500	2 min; 50 s
2250	
2750	
3000	
4000	5 min; 45 s
7000	
9000	

2. Use the Time Travel Graph for P and S Waves to determine the difference in arrival times for primary and secondary waves. Measure the two lines on the graph at the epicentre distances in the table. For example, for an epicentre 4000 km away, the difference in arrival time would be 5 min and 45 s.

3. Use the same procedure to determine the difference in arrival time for the remaining distances.

What Did You Find Out?

1. What happens to the difference in arrival times as the distance from an earthquake increases?

2. Why do seismologists measure arrival time differences?

3. If you were watching a seismogram and the arrival time between the P and S waves was less than 10 s, what should you do?

INTERNET CONNECT

www.mcgrawhill.ca/links/BCscience7

You can compare arrival times of seismic waves and locate the epicentre of earthquakes on-line. Visit the web site above to find out where to go next.

Chapter 9 Earth's Changing Surface • MHR

INTERNET CONNECT

www.mcgrawhill.ca/links/BCscience7

Did you know that Dancing Elephants have been pounding the surface of Canada for the past 20 years? The elephants are actually large trucks that send out seismic waves. The trucks are part of Project Lithoprobe, and the project's results have changed many ideas about the ancient geology of Canada. Find out more by visiting the web site above.

Epicentre Location

If seismic wave information is recorded from at least three seismograph stations, the location of the epicentre can be determined, as shown in Figure 9.19.

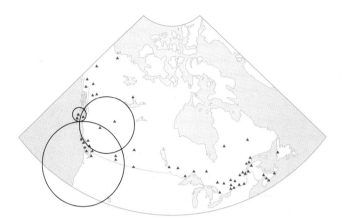

Figure 9.19 The map above shows seismograph stations in Canada. To locate an epicentre, scientists draw circles with radiuses equal to that station's distance from an earthquake epicentre. The point where all three circles intersect is the location of the earthquake.

There is usually an earthquake of very low magnitude every day in British Columbia. Most earthquakes in British Columbia originate beneath the ocean floor. Some earthquakes have their focus in the crust at depths of 20 km or less. Major earthquakes occur within the Juan de Fuca plate and the subduction zone shown in Figure 9.20. Only earthquakes larger than magnitude 4 are shown on the diagram below.

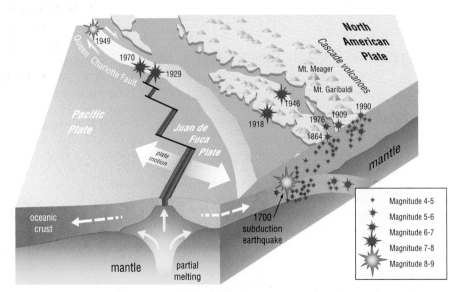

Figure 9.20 Epicentre locations of earthquakes in the Juan de Fuca plate. Why do you think there are so many earthquakes in this area?

THINK & LINK

INVESTIGATION 9-D

SKILLCHECK
- Classifying
- Interpreting
- Inferring
- Communicating

In the Zone

Think About It

How can earthquake damage such as the highway at right be prevented? People who live in earthquake zones learn how to prepare for earthquakes. Schools have earthquake drills. In some homes, furniture and water heaters are attached to the walls so they will not fall over during an earthquake. Engineers try to make buildings and roads strong enough to withstand the shaking of the ground that occurs during an earthquake. The office building in Vancouver (shown below) is specially built so it will not collapse during an earthquake. The floors are suspended from the central core of the building by huge cables that are visible at the top.

In what other ways can safety be increased before, during, and after an earthquake?

Procedure Group Work

① With your partner or group, create a chart with three headings: Before an Earthquake, During an Earthquake, and After an Earthquake. Discuss and list safety points for each heading. Some ideas to get you started are the following:

- How can the danger of injuries from falling objects and buildings be reduced?
- Where should you seek shelter if you are in a building during an earthquake?
- What should you avoid if you are outdoors during an earthquake?
- What should you do if you smell gas after an earthquake?

② Research earthquake safety information by using the Internet or library.

③ Create a brochure of what to do before, during, and after an earthquake. Share your brochure with other groups.

Analyze

1. What are five types of damage an earthquake can cause?

2. What can you do to make your home safer during an earthquake?

Find Out ACTIVITY 9-E

Pass the Wave

How can you model all three types of seismic waves?

What You Need

soft cotton rope, assorted coiled springs, Slinky™ (optional)

Safety Precaution

What to Do *Group Work*

Brainstorm with your group how to create a model of the three types of seismic waves created in an earthquake. It will help you to know that primary waves are compression waves where energy travels through earth by compression and expansion. Secondary waves are shear waves where energy travels through Earth with a series of vibrations perpendicular to the direction of the wave.

What Did You Find Out?

1. Describe and draw a picture of the types of waves you were able to create.

2. Describe the type of energy movement created by the rope and the Slinky™.

3. Which waves were the most difficult to create with the materials you were given? Why?

Tsunamis

Some earthquakes occur under the sea. The water displaced by landslides, volcanoes, or earthquakes can become huge waves called tsunamis. Tsunamis can travel across oceans and cause great damage when the waves break on the shore.

DidYouKnow?

A large tsunami struck the west shore of Vancouver Island in 1700. All the Huu-ay-aht people of the village of Anacla, south of Bamfield, were killed except for the daughter of the chief. She had recently married and moved to a neighbouring village that was not destroyed. She is the ancestor of all the Huu-ay-aht chiefs dating back to 1700.

Figure 9.21 Tsunamis are common along Japan's coastline. The painting above by artist Katsushika Hokusai shows a huge ocean wave near Japan, with Mount Fuji in the background. Tsunami is a Japanese word meaning "harbour wave."

Liquefaction

The shaking of an earthquake can cause wet soil to act like a liquid (liquefaction). When liquefaction occurs in the soil under buildings, the buildings can sink into the soil and collapse. One of the most damaging earthquakes happened about 350 km east of Mexico City in 1985. When the surface waves from an earthquake reached Mexico City, the sandy base of the city turned into quicksand. Many buildings fell over.

Section 9.2 Summary

- Earthquakes are caused when stress is released along a fault.
- During an earthquake three types of seismic waves are produced: primary, secondary, and surface.
- If you have information from three different seismographs you can determine the epicentre of an earthquake.
- The west coast of British Columbia has many earthquakes because of the location of plate boundaries.
- Buildings can be made more earthquake-resistant and people can be prepared for an earthquake.

In Section 9.3 you will learn about the fiery strength of volcanoes.

Figure 9.22 The middle building is in the process of falling over during an earthquake. The sediment underneath it is acting like quicksand.

Check Your Understanding

1. Describe the three types of seismic waves. Use coloured diagrams in your answer.
2. What is the difference between an epicentre and a focus?
3. Explain how earthquakes are measured.
4. **Apply** Imagine waking up in the middle of the night to find an earthquake occurring. When the shaking stops, you notice cracks in the walls of your home. Should you stay inside or leave? Explain.
5. **Thinking Critically** Why is it easier to predict *where* an earthquake will occur than it is to predict *when* an earthquake will occur?

Key Terms
earthquake
fault
seismic waves
primary waves
secondary waves
surface waves
focus
epicentre
seismograph
bedrock
seismogram
Richter scale

Section 9.3 Volcanoes

Figure 9.23 The volcano Paricutin erupted in a farmer's field in 1943.

Imagine watching a volcano appear right before your eyes. A farmer in Mexico had this experience when the ground in his cornfield suddenly sent clouds of ash, smoke, and glowing cinders into the air. The volcanic activity continued for several days. The ash and cinders cooled, fell to the ground, and eventually formed a volcano (see Figure 9.23).

A **volcano** is an opening in Earth's crust that releases lava, smoke, and ash when it **erupts** (becomes active). The ash can travel great distances (see Figure 9.24). Volcanoes that are not active are called **dormant**. The hole from which magma emerges is called a **vent**. Scientists try to predict when volcanoes will erupt so that the people living near them can leave the area to avoid injury.

Types of Volcanoes

A **cinder cone volcano** is made of cinder with steep sides and a bowl-shaped crater at the top (see Figure 9.25). Cinder cones are the smallest of the three types of volcanoes. Paricutin, the volcano that formed in a Mexican farmer's field, is only several hundred metres high.

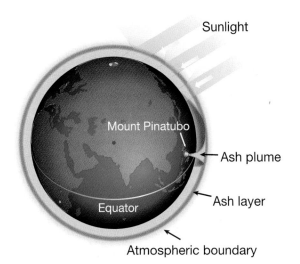

Figure 9.24 Mount Pinatubo erupted in the Philippines in 1991. The huge amount of ash blown out of the volcano formed an ash layer within the atmosphere. The ash layer circled the globe and cooled temperatures around the world.

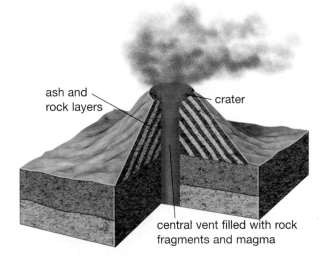

Figure 9.25 Cinder cones are built from layers of ash and small volcanic rocks. Cinder cones are usually less than 300 m high. Why do you think cinder cones are so small?

Shield Volcanoes

Mount Edziza (Figure 9.26) is an example of the second type of volcano, the shield volcano. A **shield volcano** is made of gently sloping layers composed entirely of cooled lava (see Figure 9.27A). Shield volcanoes are the largest (in area) of the three main types of volcanoes. Shield volcanoes do not erupt violently as do the cinder cones.

The Hawaiian Islands are shield volcanoes that grew up from the bottom of the Pacific Ocean floor (Figure 9.27B). The chain of islands is located above an area of Earth's crust known as a hot spot. A **hot spot** is an area of volcanic activity produced by magma rising from the mantle. The temperature under the crust at a hot spot is much higher than elsewhere. As a result, magma is forced upward through cracks in Earth's crust. In central British Columbia, the Nazko volcanic cone forms the eastern end of a possible hot spot volcanic field that can be traced back 14 million years. Many of the Nazko volcanoes are now eroded and covered with trees.

Figure 9.26 Mount Edziza, near Telegraph Creek, British Columbia, was once a large shield volcano. Glaciers have heavily eroded the volcano.

What is a hot spot?

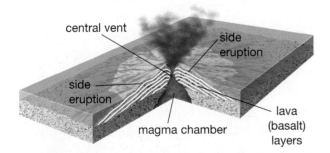

Figure 9.27A A shield volcano consists of layer upon layer of cooled lava. Ash layers are rare in shield volcanoes.

Figure 9.27B Hot spots are found below the Hawaiian Island chain stretching north in the Pacific Ocean. The northern islands in the chain no longer have any volcanic activity. Which islands in the chain do you think are the oldest?

Chapter 9 Earth's Changing Surface • MHR **275**

Composite Volcanoes

Some of the volcanoes in British Columbia are formed from magma that is pushed upwards when the plates of Earth diverge or converge. Many volcanoes are formed at subduction zones, as shown in Figure 9.28. The volcanoes of the Cascade Range on the west coast represent the third type of volcano known as a composite or stratovolcano.

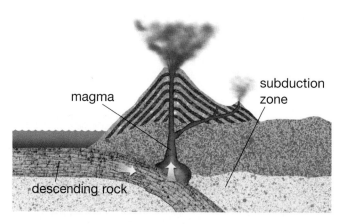

Figure 9.28 In a subduction zone, the descending rock moves deeper and deeper until it melts into magma. The magma rises up through cracks in the rock until it exerts enough pressure to cause the volcano to erupt.

A **composite volcano** is made of alternating layers of lava and cinder or ash. The eruption of a composite volcano may be quiet, as in a shield volcano, or it may be violent, as in a cinder cone. Some of the volcanoes in the Cascade Range have erupted so violently that ash was carried across great areas by the wind.

Ring of Fire

Think back to the map you made in Chapter 8 that showed the locations of volcanoes (see also Figure 9.29). Volcanoes often form along plate boundaries. The volcanoes encircling the Pacific Ocean are called the Ring of Fire.

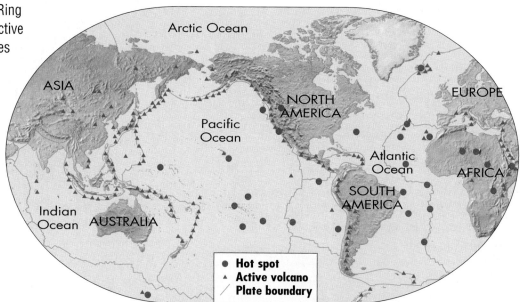

Figure 9.29 The Ring of Fire is a belt of active volcanoes that circles the Pacific Ocean.

276 MHR • Unit 3 Earth's Crust

Find Out ACTIVITY 9-F

Ash in the Area

An erupting volcano can produce huge clouds of ash that extend hundreds of kilometres. The ash can cause breathing problems and damage buildings, crops, electronics, and machinery.

Are parts of British Columbia at risk from volcanic ash?

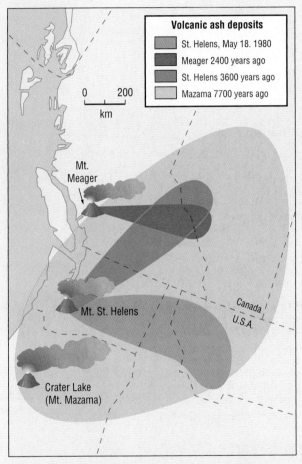

Map of Historical Volcanic Ash Deposits in Western North America

What to Do

Using the map of historical ash deposits answer the following questions. You can refer to Figure 9.30 for the names of other volcanoes on the west coast of North America.

What Did You Find Out?

1. What is the name of the volcano that created the largest ash fall? In what year did the volcano erupt?

2. Which volcano created the smallest ash fall? In what year did the eruption occur?

3. Formulate a hypothesis about the general direction of the wind when the volcanoes erupted.

4. Most scientists believe that Vancouver is more at risk from Mount Baker in Washington state than from any of the volcanoes in British Columbia. Why do you think this is so? (See the map on page 278.)

5. Why are cities in the interior of British Columbia more likely than Vancouver to have an ash fall from the volcanoes shown?

Mount Baker is a large, active, composite volcano visible from many parts of southwestern British Columbia. On a clear day, steam can be seen rising from the crater.

Chapter 9 Earth's Changing Surface • MHR **277**

Pause & Reflect

There are many volcanoes in British Columbia. Why do volcanic eruptions result in so little property damage in the province?

Volcanoes and People

Take a look at Figure 9.30 and determine which volcanoes are close to residential areas in British Columbia. What happens when a volcano erupts near a residential area?

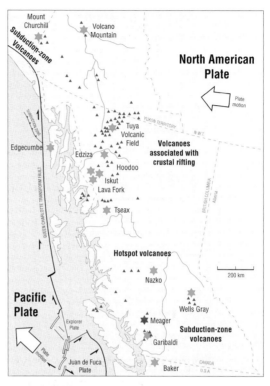

Figure 9.30 Volcanoes 10 000 years old and younger are shown as stars. Other volcanoes are shown as triangles. Can you infer what caused each volcano to form?

Around the world people live close to volcanoes because the soil is very fertile. Volcanic ash is rich in nitrogen, which makes the soil ideal for growing a variety of plants, fruits, and vegetables. However, sometimes communities are wiped out by erupting volcanoes. Mount Vesuvius, a famous composite volcano in Italy, erupted several thousand years ago (see Figures 9.31A and B) Many scientists believe that Vesuvius, dormant since 1944, is due for a large eruption.

Off the Wall

Some volcanoes have flat tops because they erupted beneath a glacier or in a glacial lake. Hyalo Ridge in Wells Gray Provincial Park is a typical, flat-topped, steep-sided, sub-glacial volcano.

Figure 9.31A The photograph above shows a plaster cast of a body buried in the eruption of Vesuvius in 79 C.E.

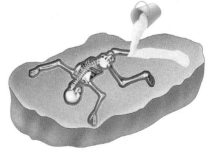

Figure 9.31B After the body decayed, plaster was poured into the cavity. When the plaster hardened, the surrounding ash was removed.

278 MHR • Unit 3 Earth's Crust

Find Out ACTIVITY 9-G

Blast from the Past

Researching the history of past volcanoes can help people prepare for future eruptions.

What to Do

1. With a partner, choose a volcano. Find out what you can from books or the Internet. See what you can learn about something called *pyroclastic flow*. Why is it one of the deadliest parts of a volcanic eruption?

2. Organize your findings in the form of an annotated poster, a storyboard, a booklet, or a demonstration (live or on video) that you can present to your class.

What Did You Find Out?

1. Explain pyroclastic flow.

2. How does knowing more about volcanoes help you to be better prepared for future eruptions?

3. What surprising information did you discover about volcanoes?

4. What additional questions do you have about volcanoes?

Intrusive Volcanic Features

You can observe volcanic eruptions because they occur at Earth's surface. However, most magma never reaches Earth's surface to form volcanoes. Instead, magma cools slowly underground and produces rock bodies that could become exposed later by erosion. Rock bodies formed by cooling magma are called intrusive volcanic features. There are several different types of intrusive volcanic features, as shown in Figure 9.32. All of these features can be found in British Columbia.

A **volcanic neck** is the core of a volcano left behind after the softer cone has been eroded. Black Tusk, a favourite hiking site in Garibaldi Provincial Park, is a famous volcanic neck. A **sill** is formed when magma flows parallel to the layers of rock and then hardens. A **dike** is formed when magma flows across layers of rock and hardens.

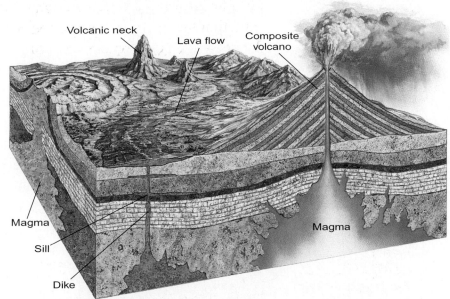

Figure 9.32 As magma intrudes into pre-existing rock, dikes and sills can be formed. What are the differences between the two features?

Energy from Inside Earth

Earth's interior contains a tremendous amount of heat. Boiling water or steam occurs naturally throughout the rock in areas where magma is close to Earth's surface. The energy harnessed from Earth's interior is called **geothermal energy**. Geothermal energy is a much cleaner form of energy than burning fossil fuels. The intense heat deep inside Earth has been harnessed in more than 20 countries, such as Iceland, Italy, and New Zealand. The large amounts of geothermal energy below Mount Meager, a dormant volcano in British Columbia, have been investigated as a possible geothermal energy source.

The Relative Age of Rocks

What information does relative dating provide?

Geologists use their knowledge of intrusive volcanic features and of faults caused by earthquakes to understand the geologic history of layers of rock. Take a look at the layers of rocks in Figure 9.33. Can you infer which layer is the oldest?

The relative age of the rock layers can be determined by using the principle of superposition. The **principle of superposition** states that for undisturbed layers of rock, the oldest rocks are on the bottom and the youngest rocks are on the top. Sediments are often deposited in horizontal beds, forming layers of sedimentary rock. The first layer to form is usually on the bottom. Each additional layer forms on top of the previous one. Unless forces turn the layers upside down, the oldest rocks are found at the bottom.

When layers have been turned upside down, geologists use other clues such as fossils and faults in the rock layers to determine the original positions and relative ages of rock layers. The **relative age** of something is its age in comparison to other things. Relative dating does not tell you the exact age of rock layers in years. Geologists determine the age of rocks and other structures by examining their places in a sequence. For example, if layers of sedimentary rock are moved by a fault, you know that the layers had to be there first before a fault could cut through them. Therefore, the relative age of the rocks is older than the relative age of the fault. If a dike or sill is present in the layer of rock, the magma that created the dike or sill must be younger than the layer of rock.

Figure 9.33 Which do you think is the oldest layer of rock? Why?

Find Out ACTIVITY 9-H

Pardon My Intrusion

You can use a variety of methods to determine the relative ages of the events in the following cross section of Earth's crust.

What to Do — Group Work

1. Discuss the relative ages of the rock layers in the illustration with your group. Study the fault and igneous intrusions. The legend will help you interpret the layers.

2. Make a sketch similar to the illustration on your own paper. Label the relative age of each rock layer on your sketch. Label the oldest layer with a 1. Mark the next oldest layer with a 2, and so on.

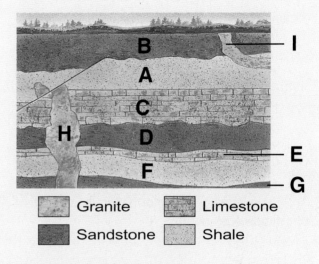

Legend: Granite, Limestone, Sandstone, Shale

What Did You Find Out?

1. Which came first, the fault or igneous intrusion H? Why do you think so?

2. Make a list of the geologic history of this area. Step 1 should be the formation of sandstone layer G.

Absolute Age

Earth is about 4.6 billion years old. Scientists found evidence that the planet is this old by using absolute dating techniques. **Absolute age** is the age, in years, of a rock or other objects. Geologists determine absolute ages by using properties of some of the atoms that make up the object. For example, fossils of mammoths have been found on Vancouver Island and near Chilliwack (see Figure 9.34). Using absolute dating, scientists have determined the mammoth fossils are about 17 000 years old.

The Geologic Time Scale

Determining the relative and absolute age of animals and plants that lived on Earth in the past has allowed scientists to create a geologic time scale. The geologic time scale divides the long history of Earth into smaller units called **eras**. The division of eras is based on geologic events and the fossil record of the appearance and disappearance of living things. In the next activity you will study the geological history of your province.

Figure 9.34
Fossils of mammoths have been found in several locations in British Columbia.

Find Out ACTIVITY 9-1

Discovering the Past

Imagine how British Columbia looked millions of years ago. Do you think there were mountains, valleys, plains, and plateaus?

What to Do *Group Work*

1. The four main divisions of the geologic time scale are called Precambrian, Paleozoic, Mesozoic, and Cenozoic. Your teacher will assign your group one of four divisions to research and the format to use in presenting your information, such as a timeline or mural.

2. Decide group roles in researching and preparing the presentation.

3. Use the Internet or library resources to discover what British Columbia was like during your time division.

What Did You Find Out?

1. Describe British Columbia during your time division.

2. Which part of British Columbia's history did you find most interesting? Why?

Section 9.3 Summary

- The three types of volcanoes are cinder cone, shield, and composite.
- Most volcanoes occur at the boundaries of the tectonic plates. Other volcanoes occur above hot spots.
- Magma can form volcanic necks, dikes, and sills.
- Relative and absolute methods can be used to determine the age of rocks.
- The geologic time scale divides Earth's long history into eras.

Key Terms
volcano
erupts
dormant
vent
cinder cone volcano
shield volcano
hot spot
composite volcano
volcanic neck
sill
dike
geothermal energy
principle of superposition
relative age
absolute age
eras

Check Your Understanding

1. Describe the differences among the three different types of volcanoes.

2. Where are most of the world's volcanoes found? Why?

3. What is the principle of superposition? How does it help geologists to determine the ages of rocks?

4. Compare absolute dating with relative dating. What does each method indicate?

5. **Apply** How could heat from inside Earth provide electrical energy?

6. **Thinking Critically** Some volcanoes erupt quietly and other eruptions are violent. What do you think causes this difference?

CHAPTER at a glance

Now that you have completed this chapter, try to do the following.
If you cannot, go back to the sections indicated in brackets after each part.

(a) Explain the differences between erosion and deposition. (9.1)
(b) List four main agents of erosion. (9.1)
(c) Draw and label a picture of a delta. (9.1)
(d) Describe features formed by glaciers. (9.1)
(e) Explain the importance of glaciers to the mineral resource industry. (9.1)
(f) Explain how seismic waves give evidence of whether layers of Earth are solid or liquid. (9.2)
(g) List the three types of seismic waves and explain the differences among them. (9.2)
(h) Explain the numbering system used in the Richter scale and describe how it is used. (9.2)
(i) Describe the difference between the focus and epicentre of an earthquake. (9.2)
(j) Explain how you can use the difference between the arrival time of primary and secondary waves to determine the distance from an earthquake. (9.2)
(k) Explain how buildings can be made earthquake resistant. (9.2)
(l) Draw and label examples of the three types of volcanoes. (9.3)
(m) Give one example of each type of volcano. (9.3)
(n) Draw a picture illustrating the intrusive features dike, sill, and volcanic neck. (9.3)
(o) Describe geothermal energy. (9.3)
(p) Explain the geologic time scale. (9.3)

Summarize Chapter 9 by doing one of the following. Use a graphic organizer (such as a concept map), produce a poster, or write a summary to include the key chapter ideas. Here are a few ideas to use as a guide:
- Use the theory of plate tectonics to explain the location of earthquakes and volcanoes.
- Draw a cross-section of an earthquake area to explain the difference between an epicentre and a focus.
- Give examples of how your family could prepare for an earthquake.
- Make a Venn diagram comparing the three types of volcanoes.

Prepare Your Own Summary

- Copy the concept map below and fill in the missing words as explained in this chapter.

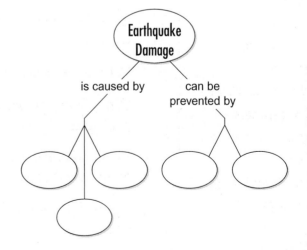

CHAPTER 9 Review

Key Terms

plains
plateaus
deposition
floodplain
landslide
glaciers
moraine
cirque
horn
earthquake
fault
seismic waves
primary waves
secondary waves
surface waves
focus
epicentre
seismograph
bedrock
seismogram
Richter scale
volcano
erupts
dormant
vent
cinder cone volcano
shield volcano
hot spot
composite volcano
volcanic neck
sill
dike
geothermal energy
principle of superposition
relative age
absolute age
eras

Reviewing Key Terms

If you need to review, the section numbers show you where these terms were introduced.

1. In your notebook, match each description in column A with the correct term in column B. Use each description only once.

   A
   (a) fastest travelling earthquake wave
   (b) second type of earthquake wave generated
   (c) opening in crust of Earth
   (d) wave that causes the most damage
   (e) encircles the Pacific Ocean

   B
   - P wave (9.2)
   - vent (9.3)
   - surface wave (9.2)
   - S wave (9.2)
   - Ring of Fire (9.3)

2. Listed below are four answers for four different questions:

   deposition (9.1)
   floodplain (9.1)
   relative age (9.3)
   absolute age (9.3)

   Write a question for each answer.

Understanding Key Ideas

Section numbers are provided if you need to review.

3. Copy each of the following sentences on your own paper. Fill in the key term that is missing.

   (a) In the spring, rivers often overflow their banks and create large flat areas of sediments called _____. (9.1)

   (b) Seismographs must be attached to the solid rock known as _____. (9.2)

   (c) Magma can flow parallel to layers of rock and solidify as a _____ or magma can cross layers of rock to become solid and form a feature known as a _____. (9.3)

Developing Skills

4. Make a concept map about where volcanoes can occur. Include the following words and phrases: hot spots, divergent plate boundaries and convergent plate boundaries. Include the names and locations of some famous volcanoes.

5. The Hawaiian Islands and Emperor Seamounts were formed when the Pacific plate moved over a fixed hot spot. The Emperor Seamounts are aligned in a different direction than the Hawaiian Islands. What can you infer about the movement of the Pacific plate?

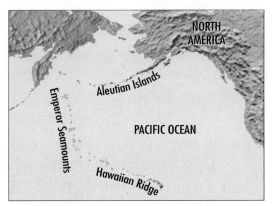

6. Explain the differences and similarities between the ways in which you would prepare for an earthquake and a volcanic eruption.

Problem Solving

7. You have discovered three rock layers that have not been turned upside down. The absolute age of the middle layer is 120 million years. What can you say about the ages of the layers above and below it?

Pause & Reflect

Go back to the beginning of the chapter on page 256, and check your original answers to the Getting Ready questions. How has your thinking changed? How would you answer those questions now that you have investigated the topics in this chapter?

8. Scientists have produced maps indicating how long a tsunami takes to travel. The map below shows tsunami travel time to Hawaii in hours. Some tsunamis travel at speeds up to 950 km/h. Use the map to infer how long a tsunami would take to reach Hawaii if the tsunami originated near Japan. New Zealand? British Columbia?

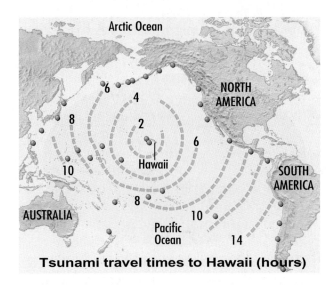

Tsunami travel times to Hawaii (hours)

Critical Thinking

9. Imagine returning to your community 100 years from now. How might the landscape be different? What sort of weathering and erosion might have happened?

10. Animals can sometimes sense when an earthquake is about to happen. How do you think they can do this?

11. Discuss an example of human action in your community that has been taken to prevent damage from one of the changes described in this chapter.

UNIT 3

Ask an Elder

Herbert Morven

The most recent volcanic eruption in Canada happened about 250 years ago near the Nass River in Northwestern British Columbia. This area is home to the Nisga'a peoples. Nisga'a stories about the volcano give a traditional perspective on the event and its causes.

Herbert Morven is a Nisga'a Hereditary Chief. He received his first Sub-Chieftainship title when he was three years old. Chief Morven lives in Gitlaxt'aamiks [Git-lak-DA-mix], also known as New Aiyansh, in the Nass Valley. His Nisga'a name is K'ee<u>x</u>kw [K-EH-uhk].

Q. What impact did the volcano's eruption have on the land and Nisga'a communities?

A. When the lava flowed down the mountain, it buried two Nisga'a villages. About 2000 people died. There was also a hut on the mountainside where young women would go to be alone when they first reached puberty; this was destroyed, too. The volcano also changed the river. Before the eruption, that part of the Nass river was slow and meandering. The lava flow altered the course of the river, pushing it out. It is swifter now.

Q. How does the traditional Nisga'a story explain the volcano?

A. Back then, the Nass River was very rich in salmon. An Elder who was walking along the bank heard children laughing with delight. He went closer and saw a group of young boys catching salmon. They were very good at it and the Elder marvelled at their skill. He thought they were enjoying themselves because they were fishing so well. But then he went closer and saw what they were doing.

The boys caught a salmon and cut a slit in its back with a piece of shale. They put pitch on the shale and set fire to it, and put the shale back into the slit in the salmon's back. Then they put the salmon back in the water. They did this over and over. The boys were laughing at the way the salmon tried to swim with the rocks and burning pitch in their backs. The Elder told them, "I appreciate your skill at catching salmon, but you are being disrespectful. The salmon will curse us, and the Creator will respond in kind." The boys did not pay any heed to the Elder's warning, and continued their game. Some time later, the people heard a rumbling from the mountain. The rumbling turned into a roar. There was a big panic when the volcano erupted, and many people died. Our Elders tell us that if the children had listened to the warning, this disaster would never have happened.

Q. What lessons are contained in this story?

A. In our Nisga'a language, our words have a meaning that respects the Spirit and the sacredness of our environment. You have to respect the environment, you have to respect the fish that sustain the people, and you have to respect yourself. This is what the young boys forgot.

Whenever a young woman went into that hut on the mountain, her grandmothers and aunts would come to her to instruct her in the changes in her life and her responsibilities. When the volcano erupted, a young woman and her grandmother were in the hut. They died in the lava. For a long time afterwards you could see their handprints and footprints on the walls of the tunnel leading to the hut. So the story also talks about the will to survive.

But you can't just ask, "What does this story mean?" Our Elders never give you the answers like that. Instead, they point you in the direction to find the answer yourself, so the answer is really yours. It is like going to school and learning how to do math. You are not given the solution, but you are given a process so you can find the solution for yourself.

Q. How do traditional stories like this fit with scientific theories?

A. What I am sharing with you is how our people took something that happened to us and used it for the good of our people. Our philosophy is that for anything bad that happens, there is an equal amount of good. This story makes something good out of a bad event. It becomes a lesson in how to treat our environment and our resources with respect.

Scientific theories are fine — but what is the Spirit of science? For something to be alive it has to have a Spirit, it has to be part of the creative process. It is not enough just to collect knowledge. You also have to apply it. Our stories and traditions apply our knowledge and technologies in ways that respect the sacredness of our land.

The Nisga'a lava beds near Gitlaxt'aamiks (or New Aiyansh) in the Nass valley. The lava flowed over an area 10 km long and 3 km wide.

EXPLORING Further

Landform Stories

Scientists researching volcanoes, earthquakes, and other geological events often study traditional aboriginal stories to find evidence that can add to scientific data. For example, these stories can help to confirm the timing, location, and impact of events that took place long ago.

Working with a partner, select a landform such as a mountain, plateau, or significant rock anywhere in British Columbia. Try to find legends or traditional stories associated with it. Research its name, including its aboriginal name. Do the stories provide information that would add to scientific evidence about the origin of the landform? Record all the information you gather in your notebook or on a computer. Then create a poster or prepare a brief presentation about the landform. Include scientific and traditional explanations about how it came to exist.

Xa:ytem (also known as Hatzic Rock) is a prominent rock in the Fraser Valley near Mission, British Columbia. According to the oral history of the Sto:lo First Nations, the rock was created when the mythical being Xa:ls transformed three Sto:lo Chiefs into stone.

UNIT 3

Ask a Geophysicist

Herb Dragert

Southwest coastal British Columbia is the most seismically active region in Canada – which makes it a great place to study earthquakes. Dr. Herb Dragert is a research scientist working with the Geological Survey of Canada at the Pacific Geoscience Centre near Sidney. He has been studying the movement of the Earth's crust along the coast for over 20 years. Together with his colleagues, Dr. Dragert has recently discovered some new clues that could someday help scientists to predict when and where an earthquake will occur.

Q. What made you decide on a career in geodynamics?

A. I carried out my undergraduate studies at the University of Toronto and one of my professors was Dr. Tuzo Wilson. It was a very exciting time in geophysics, and he really made the science come alive. I've also always liked the practical side of science. In this field you are applying physics to the Earth, so it's very real.

Q. How do you study the changes in the Earth's surface?

A. We use a set of very sensitive GPS stations to gather data about surface movement along the coast. These measurements tell us where the coast is being "squeezed," and how much. We then use a computer to model where the tectonic plates are locked together. The locked zone is where the pressure builds up. This pressure will someday be released through a big earthquake.

Q. So this information tells you where an earthquake is likely to happen. Can it tell you when?

A. This is where our most recent studies come in. We used to think that the tectonic pressure was made up of a steady "push" from the offshore ridge above the locked zone, and a steady "pull" from below the locked zone. Recently we discovered that the pull from below is not steady at all. It sticks and then slips. There is very little pull for about 14 months, and then there is a short tug or slip over a period of about two weeks. The slip puts extra pressure on the locked zone, so a big earthquake is more likely to be triggered during one of the slip periods.

Q. Can you tell when a slip is about to happen?

A. Yes. Around the same time that we were discovering the slip motion, my colleague Gary Rogers was collecting seismic recordings that showed a regular pattern of tremors from deep below the Earth's surface. Gary and I put our findings together and discovered that the tremor and the slip always happen together. That is, every 14 months or so we see a slip from our surface measurements and we detect tremors on our seismometers. So when we pick up the signal that a tremor is beginning, we know that the slip has started.

Q. How can this discovery help to predict an earthquake?

A. The tremor may give us a kind of early warning signal. It tells us that we are going into a two- or three-week period of time when a big earthquake is more likely.

But of course it is not really that simple. These slips and tremors have been going on for a long time. In the past 200 years, none of them has triggered a major earthquake. What is more important is what this tells us about our models. We see this stick-slip action in materials that we thought could not do this, so we have to revise our theories. We are being told something fundamental about the plate interactions at depths of 25 to 45 km below the Earth's surface. We just don't know exactly what it is yet. The more we find out, the closer we will be to truly being able to forecast earthquakes.

Q. What is your next step in this research?

A. The tremor and slip pattern is very regular, so we know when the next one is going to be. We will have extra instruments out to record the movement and seismic signal. The better we can monitor this, the closer we are to understanding what is going on. We are also working with researchers around the world to compare what is happening here with findings in other subduction zones.

Q. What do you like best about your work?

A. I really enjoy this kind of investigative science. I come into work thinking "what will I discover today?" Of course you don't make a big discovery every day, but it is fun to put the pieces together. If you follow your curiosity you will always enjoy the work you do.

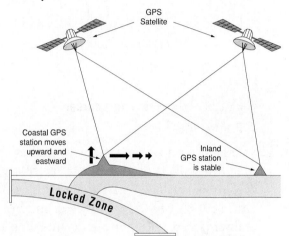

Very sensitive GPS instruments track the movements of the Earth's surface at the coast. Scientists use the data to model plate interactions.

EXPLORING Further

Tracking changes in the Earth's crust

The scientific name for the stick-slip motion Dr. Dragert discovered is "episodic tremor and slip" (ETS). "Episodic" means "happening from time to time." However, the first name given to this motion was "silent slip." Dr. Dragert and his colleagues revised the name in 2003 when they discovered that the slip movement was always linked to a seismic tremor.

The graph at right shows how Dr. Dragert and his colleagues plot the data collected by their continuous GPS stations and seismic recorders. Use a search engine to find an explanation of this graph on the internet. Working with a partner or in a small group, draw your own version of the graph. Then, in your own words, describe what the graph shows. Explain why the periods of slip and tremor could create a higher risk of a major earthquake.

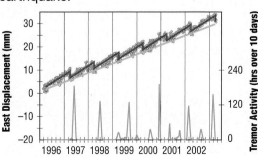

UNIT 3 Project

Shake and Quake

SKILLCHECK
- Modelling
- Observing
- Measuring
- Communicating

Millions of people in Canada live in earthquake zones. Building codes for the construction industry in earthquake prone areas are strictly enforced in these areas. Homes, office buildings, and schools must meet strict earthquake-resistant guidelines. Buildings must be built to withstand some type of vertical and horizontal motion.

Can you see any features of this building that might help it withstand an earthquake?

Challenge

You are part of a team that has been asked to build a three-story school in an earthquake prone area. Before construction of the school can begin, your group needs to develop a scale model of a school and test it on a shake table in the classroom. The team with the best design will be awarded the contract to build the new school. The scale model must be safe enough to withstand 3 strong earthquakes, each lasting 10 s. The earthquake will move the shake table very quickly in a 15 cm horizontal and vertical direction. Three plastic figures will be placed in the school before the earthquake. All of the plastic figures should be in the school after the earthquake.

Specifications

A. The school should have three stories. Each story should be at least 10 cm high.

B. The entire school should have a total height of at least 30 cm.

C. Three small plastic figures will be placed in the school. The plastic figures must be free standing, which means they cannot be taped or attached to the structure in any way.

D. Make your school as lightweight as possible, while maintaining its stability and safety.

E. You may use 15 cm of masking tape to attach your school to the shake table.

F. If you have a shake table with holes in it, part of your school can be beneath the ground.

G. All forms and types of building materials approved by your teacher are allowed, with the exception of materials from commercial construction kits.

H. The school must survive three earthquakes, each lasting 10 s.

I. All three plastic figures should still be in the school after all three earthquakes.

Plan and Construct *Group Work*

1. With your group, develop the question you will be testing in this investigation. Which variables are involved?

2. Determine the scale you will use. Discuss possible solutions based on the specifications. What are the key components of the scale model?

3. Each team member should independently draw a blueprint for the school. Plans should reflect the specifications listed for the project.

4. At a team meeting, all members can present their plans. After discussion, your team can choose a single plan or combine ideas from all team members.

5. A final blueprint accompanied by a list of materials and tools needed should be submitted to your teacher for approval. Explain how you will apply all appropriate safety measures and control variables.

6. Throughout the construction process, your team should test the structure to see if it is capable of meeting the specifications. Adjust your plan if unforeseen problems arise.

7. Draw and label a diagram of your completed school. Your diagram should include the names of all team members.

8. Once the school is complete and meets all specifications, it is ready for testing. Give your school a name and carry it to the shake table for testing.

Evaluate

1. Measure your school to make sure it has met all the specifications listed. Record your measurements on your diagram.

2. Place the three plastic figures in your school.

3. Move the shake table 15 cm both horizontally and vertically for 10 s. Repeat the shaking two more times. The teacher may choose to shake the table for you.

4. Examine your school for any signs of structural failure after each earthquake. Record any problems on your diagram.

5. After the 3 earthquakes, your teacher will score your structure based on the following evaluation scheme.

Shake and Quake Scoring Guide

Criteria	Comments	Score (Circle one score)
Meets or exceeds all specifications		Poor 0 1 2 3 4 5 Excellent
Figures in school after the earthquake		Poor 0 1 2 3 4 5 Excellent
Passing the earthquake test		Poor 0 1 2 3 4 5 Excellent
Built to scale		Poor 0 1 2 3 4 5 Excellent
Total Score		/20

6. Describe any changes your team made during the construction process. Explain why the changes were made.

7. If you were to build the structure again, what improvements would you make?

USING GRAPHIC ORGANIZERS

A **graphic organizer** is a way of arranging ideas visually. A typical graphic organizer consists of a group of boxes or circles in which key terms or concepts are written. Lines connecting the boxes show how the concepts are connected. Sometimes a short phrase is written on the lines to explain the connection. Graphic organizers are tools that help you understand and remember new ideas and information. You can often recognize new relationships when you use a graphic organizer. Several examples of graphic organizers are described below. These graphic organizers are sometimes also called *concept maps*.

Network Tree

A graphic organizer called a **network tree** starts with a description of one central concept or idea written in a box. Terms that describe characteristics or smaller parts of the central concept are written in more boxes. The lines, or branches of the tree, show how the central concept is broken down into smaller parts. The network tree below shows all of the things that make up an ecosystem.

Flowchart

A graphic organizer called a **flowchart** shows a sequence of occurrences or steps in a process. The flowchart below shows the events that follow a volcanic eruption.

Events Chain

A graphic organizer called an **events chain** shows the order in which events occur. The events chain on the right shows what happens after your alarm goes off in the morning.

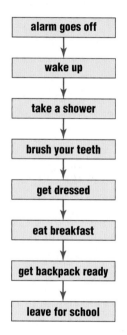

Cycle Map

A graphic organizer called a **cycle map** is like an events chain that has no beginning and no ending. The events occur over and over. Since the events are placed in a circle, you need arrows to show the order in which the events occur. The events chain below shows the life cycle of a frog.

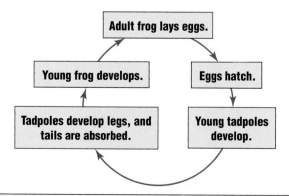

Spider Map

A graphic organizer called a **spider map** connects several objects or ideas to one central event or object. You can use a spider map for brainstorming. To draw a spider map, you place the most important event or object in the centre of the page. Then you write related ideas, events, or descriptions on lines drawn outward from the centre. These topics might not be related to each other, but they are related to the central topic. Finally, you write other ideas or descriptions on horizontal lines connected to the lines that point outward. The map soon begins to look like a spider, as shown here. This spider map describes the three states of matter. The central idea, matter, is the "body" of the spider. The related ideas, solid, liquid, and gas, are the "legs" of the spider.

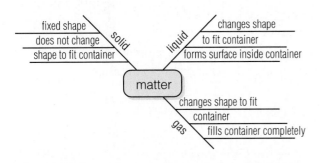

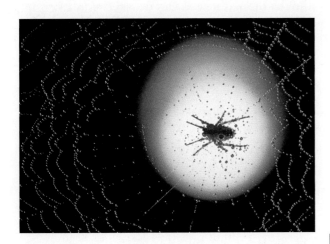

Venn Diagram

A graphic organizer called a **Venn diagram** shows the similarities and differences among objects or ideas. Venn diagrams are made of two or more overlapping ovals or circles. Each oval represents a concept, an object, or an idea. In the area in which the ovals overlap, you write the *similarities* between the objects or ideas. You write the *differences* in the parts of the circles that do not overlap. Study the following example to see the similarities and differences between a print textbook and an e-book CD-ROM.

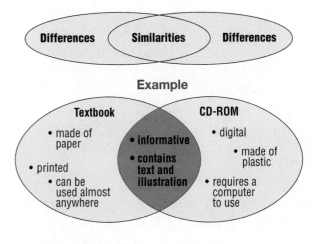

Instant Practice

1. Use these words to produce a network tree: hockey, team sports, ice, diamond, field, bat, puck, hardball, cleats, ice skates, baseball, soccer ball, stick, soccer, feet.

2. Produce an events chain that starts with lunch and ends with your return home from school.

3. Produce a cycle map using the following words: summer, winter, fall, spring.

4. Make a Venn diagram to compare and contrast a video cassette and a DVD.

5. In a group, create a spider map based on one of the following topics:
   (a) food and nutrition
   (b) music
   (c) scientific discoveries
   (d) communication

METRIC CONVERSION AND SI UNITS

Throughout history, groups of people have developed their own units of measurement. When different groups of people began to communicate, they became confused because they did not understand each other's units of measurement. For example, imagine trying to report your weight in scruples or buying a "hogshead" of strawberries! For consistency, scientists throughout the world have agreed to use the metric system of units of measurement.

The Metric System

The **metric system** is based on multiples of ten. The basic unit of length is the metre. All larger units of length are expressed in units based on metres multiplied by 10, 100, 1000, or more. Smaller units of length are expressed in units based on metres divided by 10, 100, 1000, or more. A prefix (a part of a word joined to the beginning of another word) tells you what number to use when you multiply or divide. For example, *kilo-* means "multiplied by 1000." One kilometre is one metre multiplied by 1000. You can write this relationship mathematically as shown here.

$$1 \text{ km} = 1000 \text{ m}$$

The prefix *centi-* means "divided by 100." As shown below, one centimetre is one metre divided by 100.

$$1 \text{ cm} = \frac{1}{100} \text{ m} \quad \text{or} \quad 1 \text{ cm} = 0.01 \text{ m}$$

In the metric system, the same prefixes are used with nearly all units of measurement such as units of mass, weight, distance, and energy. The table on this page lists the most commonly used metric prefixes.

Commonly Used Metric Prefixes

Prefixes	Symbol	Relationship to the base unit
giga-	G	1 000 000 000
mega-	M	1 000 000
kilo-	k	1 000
hecto-	h	100
deca-	da	10
–	–	1
deci-	d	0.1
centi-	c	0.01
milli-	m	0.001
micro-	μ	0.000 001
nano-	n	0.000 000 001

(**Note:** Time does not have a metric form of measure. Time is measured in seconds, minutes, and hours. There are 60 s in 1 min, 60 min in 1 h, and 24 h in 1 d.)

The following examples show you how to convert from one metric unit to another.

> **Problem Tip**
>
> When you are solving problems and you have the same unit in the numerator and in the denominator, you can cancel the units. Notice how the units are cancelled in the following examples.

Example 1

The length of Canada's longest river, the Mackenzie River, is 4241 km. Convert the length of the river from kilometres to metres.

Solution

$$4241 \text{ km} = \underline{} \text{ m}$$
$$1 \text{ km} = 1000 \text{ m}$$
$$(4241 \text{ km})\left(\frac{1000 \text{ m}}{\text{km}}\right) = \frac{4241 \cancel{\text{km}} \times 1000 \text{ m}}{\cancel{\text{km}}}$$
$$4241 \text{ km} = 4\ 241\ 000 \text{ m}$$

Example 2

There are 250 g of rice in a package. Express this mass in kilograms.

Solution

$$250 \text{ g} = \boxed{} \text{ kg}$$
$$1000 \text{ g} = 1 \text{ kg}$$
$$(250 \text{ g})\left(\frac{1 \text{ kg}}{1000 \text{ g}}\right) = \frac{250 \text{ g} \times 1 \text{ kg}}{1000 \text{ g}}$$
$$250 \text{ g} = 0.250 \text{ kg}$$

Quantities Described by More Than One Term

Mass

One quantity for which you will need to learn more than one unit of measurement is mass. To report an amount of mass, you would usually use the unit of grams with a prefix. For example, one thousand grams is a kilogram (kg). A special unit has been given to one thousand kilograms. That unit is the tonne (t). A tonne is defined as one thousand kilograms. When working with very large amounts of mass, many people prefer to use units of tonnes instead of grams. The relationships are summarized here.

$$1\,000\,000 \text{ g} = 1000 \text{ kg} = 1 \text{ t}$$

Area

You can also express large areas in two different units of measurement. The basic unit of area is the square metre (m^2). For example, a square with sides of 1 m has an area of $1 \text{ m} \times 1 \text{ m} = 1 \text{ m}^2$. Ten thousand square metres is given the unit hectare (ha). You usually use the hectare to describe areas of land. You can write the relationship between square metres and hectares as shown here.

$$10\,000 \text{ m}^2 = 1 \text{ ha}$$

Volume

You can express volume in cubic metres (m^3). For example, a cube that is 1 m long, 1 m wide, and 1 m high has a volume of $1 \text{ m} \times 1 \text{ m} \times 1 \text{ m} = 1 \text{ m}^3$. When you are working with fluids, however, you usually express the volume in units of litres (L) or millilitres (mL). One millilitre has the same volume as one cubic centimetre (cm^3). One litre is one thousand millilitres.

Some important relationships are listed below. Notice that one litre is NOT equivalent to a cubic metre. There are one thousand litres in a cubic metre.

$$1 \text{ cm}^3 = 1 \text{ mL}$$
$$1000 \text{ mL} = 1 \text{ L}$$
$$1000 \text{ L} = 1 \text{ m}^3$$

Metric Unit Conversions Involving Squares or Cubes

When you are converting units of area and volume, you must be very careful when you determine the number of square or cubic units that are equal to each other. For example, one hundred square centimetres is not equal to one square metre. The following examples show you how to calculate the correct relationships.

Example 1

Convert 125 cm^2 into m^2.

Solution

$$125 \text{ cm}^2 = \boxed{} \text{ m}^2$$
$$100 \text{ cm} = 1 \text{ m}$$
$$(100 \text{ cm})^2 = (1 \text{ m})^2$$
$$100^2 \text{ cm}^2 = 1^2 \text{ m}^2$$
$$10\,000 \text{ cm}^2 = 1 \text{ m}^2$$
$$(125 \text{ cm}^2)\left(\frac{1 \text{ m}^2}{10\,000 \text{ cm}^2}\right) = \frac{125 \text{ cm}^2 \times 1 \text{ m}^2}{10\,000 \text{ cm}^2}$$
$$125 \text{ cm}^2 = 0.0125 \text{ m}^2$$

Example 2

Convert 3.5 m^3 into cm^3.

Solution

$$3.5 \text{ m}^3 = \boxed{} \text{ cm}^3$$
$$1 \text{ m} = 100 \text{ cm}$$
$$(1 \text{ m})^3 = (100 \text{ cm})^3$$
$$1^3 \text{ m}^3 = 100^3 \text{ cm}^3$$
$$1 \text{ m}^3 = 1\,000\,000 \text{ cm}^3$$
$$(3.5 \text{ m}^3)\left(\frac{1\,000\,000 \text{ cm}^3}{1 \text{ m}^3}\right) = \frac{3.5 \text{ m}^3 \times 1\,000\,000 \text{ cm}^3}{1 \text{ m}^3}$$
$$3.5 \text{ m}^3 = 3\,500\,000 \text{ cm}^3$$

Use the table of prefixes on the previous page and the relationships discussed above to solve the following Instant Practice problems.

Instant Practice

1. A can contains 0.355 L of pop. How many millilitres does the can contain?
2. The height of a table is 0.75 m. How high is the table in centimetres?
3. A package of chocolate-chip cookies has a mass of 396 g. What is the mass of the cookies in milligrams?
4. One cup of water contains 250 mL. What is the volume of one cup of water in litres?
5. The distance from your kitchen to the front door is 6000 mm. How far is the front door in metres?
6. A student added 0.0025 L of lemon juice to water. How much lemon juice did the student add in millilitres?
7. A jug contains 1525 cm^3 of water. What is the volume of the water in units of m^3?

SI Units

Even when scientists began to use the metric system, some confusion still arose. Some scientists would report distance or length in centimetres while others used metres. As well, some scientists reported mass in grams while others reported it in kilograms. Finally, scientists gathered at an international conference in Paris and agreed upon a specific set of units for communicating with other scientists. In 1960, the SI system of units was accepted worldwide. The name SI is taken from the French name *Le Système International d'unités*.

There are seven base units in the SI system. Many other units consist of a combination of base units. These units are called *derived* units, which you will learn about in later science courses. You will use the following base units in this course.

Base Units:
- The SI unit of mass is the kilogram (kg).
- The SI unit of distance is the metre (m).
- The SI unit of time is the second (s).

(**Note:** There are no metric units for time. However, the second (s) is an SI unit.)

When solving problems, you should always report your answers in SI units unless the problem requires a different type of unit. Study the examples below. Then complete the Instant Practice problems that follow.

Example 1
Convert 426 cm to the SI unit for length.

Solution

$$\text{SI unit} = \text{m}$$
$$426 \text{ cm} = \underline{} \text{ m}$$
$$100 \text{ cm} = 1 \text{ m}$$
$$(426 \text{ cm})\left(\frac{1 \text{ m}}{100 \text{ cm}}\right) = \frac{426 \text{ cm} \times 1 \text{ m}}{100 \text{ cm}}$$
$$426 \text{ cm} = 4.26 \text{ m}$$

Example 2
Convert 1.7 h into SI units.

Solution

$$\text{SI unit} = \text{s}$$
$$1.7 \text{ h} = \underline{} \text{ s}$$
$$1 \text{ h} = 60 \text{ min}$$
$$1 \text{ min} = 60 \text{ s}$$
$$(1.7 \text{ h})\left(\frac{60 \text{ min}}{\text{h}}\right)\left(\frac{60 \text{ s}}{\text{min}}\right) = \frac{1.7 \text{ h} \times 60 \text{ min} \times 60 \text{ s}}{\text{h} \times \text{min}}$$
$$1.7 \text{ h} = 6120 \text{ s}$$

Instant Practice

Convert the following quantities to the SI base unit.

1. 7.02 g
2. 32 min
3. 8.13 km
4. 25 961 mm
5. 223 625 cm
6. 3.25 h

Skill POWER 3

MEASUREMENT

Measuring Length

You can use a metre stick or a ruler to measure short distances. Metre sticks and rulers are usually marked off in millimetres and centimetres. To measure a distance, place the zero mark of the metre stick or ruler at one end of the distance to be measured and read the length at the other. Notice that the zero point is usually *not* at the end of the ruler.

Instant Practice

Use a ruler to measure the distance between the following pairs of points: A and D; C and E; B and F.

A •   • B

    • C   • E   • D

       • F

Measuring Area

Area is the amount of the surface of an object. You might want to determine the area of a piece of paper, a wall, or the surface area of a cube. If the area of an object is made up of common geometric shapes, you can measure the dimensions of the shape and use mathematical formulas to calculate the area. A piece of paper, for example, is often rectangular. You would measure the length and the width and then use this formula to calculate the area of the paper:

$A = l \times w$ (area = length × width)

Note: When you are making calculations, always be sure to use the same units of measurement. If you mix centimeters and metres, your calculations will be incorrect.

Instant Practice

Imagine that you are in charge of an art project that will transform one wall of your classroom into a large mural to show a variety of patterns. You may use as many different patterns and materials as you like. However, each piece must be a 30 cm × 30 cm square. How many squares will you need to make your mural?

1. Choose the unit of measurement that will be most practical for the area of the mural.

2. Measure the height and width of the wall you will cover.

3. Calculate the area of the wall in the unit you have chosen.

4. Calculate the number of 30 cm × 30 cm squares that you will need to fill one square unit. For example, if you chose to use metres, the square unit would be square metres (m^2).

5. Multiply this number by the area of the wall in the square unit that you chose.

These students are using 10 cm × 10 cm squares. You are using 30 cm × 30 cm squares.

Measuring Volume

The **volume** of an object is the amount of space it occupies or the empty space in a hollow object. If an object is made up of common geometric shapes, you can measure the dimensions of the object and use mathematical formulas to calculate the volume:

$V = l \times w \times h$

$V = 3 \text{ cm} \times 3 \text{ cm} \times 3 \text{ cm}$
$= 27 \text{ cm}^3$

$V = l \times w \times h$

$V = 6 \text{ cm} \times 4 \text{ cm} \times 2 \text{ cm}$
$= 48 \text{ cm}^3$

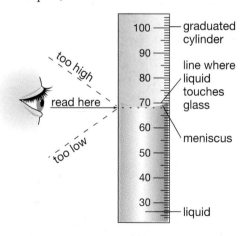

The volume of fluids is usually expressed in litres or millilitres. You can measure the volume of a liquid directly by using a graduated cylinder as shown below. To measure accurately, be sure that your eye is level with the *bottom* of the meniscus. The **meniscus** is the curved surface of the liquid, as shown below.

You can use liquids to help you measure the volume of an irregularly shaped solid. You would first measure a volume of a liquid such as water. Then, carefully slide the object into the water. Determine the total volume in the cylinder. The volume of the solid is equal to the difference between the volume before and after adding the solid.

Volume of object = volume of water with solid − volume of water only

You will practise this method of measuring the volume of a solid in Investigation 4-D on page 114.

Instant Practice

1. Read the volume indicated by each graduated cylinder below.

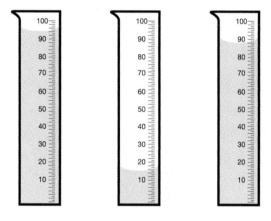

2. Write an explanation of how to find the volume of a solid that is made of common geometric shapes.

Measuring Mass

The **mass** of an object is the amount of matter that makes up the object. Mass is usually measured in milligrams, grams, kilograms, or tonnes. A balance is used for measuring mass. Your school might have triple beam balances like the one shown here. To measure the mass of a solid using a triple beam balance, follow these steps:

1. Set the balance to zero by sliding all three riders back to their zero points. Use the adjusting screw to be sure that the pointer rests at the zero point at the far end of the balance.
2. Place the object on the pan of the balance. The pointer will move up.
3. Slide the largest rider along until the pointer is just below zero. Then move the rider back one notch.
4. Repeat step 3 with the middle rider.
5. Adjust the smallest rider until the pointer swings equally above and below zero.
6. Add the readings on the three scales to find the mass.

To measure the mass of a powdery solid such as sugar, start by measuring the mass of an empty beaker. Next, pour the solid into the beaker. Then measure the mass of the beaker and solid together. To find the mass of the solid, subtract the mass of the empty beaker from the mass of the solid and beaker.

$$\text{Mass of sugar} = \text{Mass of sugar and beaker} - \text{Mass of empty beaker}$$

Instant Practice

1. Which takes more "muscle" to carry, your favourite paperback book or a calculator? Find out by using a balance to compare their masses.
2. Write the steps you would take to find the mass of the contents of a glass of juice.

The mass of the empty beaker is 61.5 g.

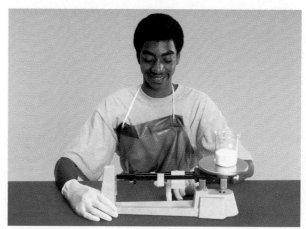

The mass of the sugar and beaker together is 161.5 g. The mass of the sugar equals the mass of the sugar and beaker together minus the mass of the beaker: 161.5 g − 61.5 g = 100 g.

In advanced science courses, you might use an electronic balance like this. Just place the object on the pan and read the mass.

Measuring Temperature

Temperature is a measure of the average energy of motion of the particles in a substance. You can measure the temperature of a material by using a thermometer. You usually express the temperature of a substance in degrees Celsius. Zero degrees Celsius (0°C) is the freezing point of water. One hundred degrees Celsius (100°C) is the boiling point of water. The SI unit of temperature is the kelvin (K). Zero kelvins (0 K) is the coldest possible temperature. It is also called absolute zero. 0 K is equal to −273°C.

When using a thermometer to measure temperature, remember these three important tips.

- Handle the thermometer extremely carefully. It is made of glass and can break easily.
- Do not use the thermometer as a stirring rod.
- Do not let the bulb of the thermometer touch the walls of the container.

In school laboratories, you will probably use a thermometer like this one. The temperature scale is in degrees Celsius.

Instant Practice

Your teacher will supply your class with three large containers of water, each at a different temperature.

1. Twelve students will each be provided with a thermometer. When your teacher says "now," the students will take temperature readings of the water in the different containers. Four students will be asked to take a temperature reading of the water in one container. Four others will take the temperature reading of the water in the second container, and four others will take a reading of the water in the third container. Each student should keep the temperature reading a secret until putting it on a class chart.

2. Make a class chart on the chalkboard to record each of the students' temperature readings. The three columns will be:

Container 1	Container 2	Container 3
Temperature Reading (°C)	Temperature Reading (°C)	Temperature Reading (°C)

3. Each student will record the temperature reading of the water in the container used.

4. Did each person record the same temperature reading of the water in the same container? If the temperature readings were not all the same, explain why you think this might be so.

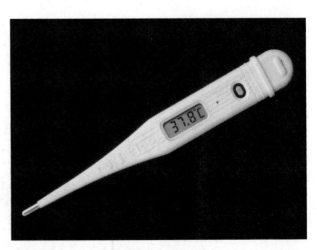

A digital thermometer like this one can be used to take your temperature.

Skill POWER 4

DEVELOPING MODELS IN SCIENCE

A **scientific model** is an idea that helps to explain some part of the natural world. Scientists often develop models by using words, pictures, or objects. Many of the illustrations, diagrams and graphs in textbooks are scientific models. A model can help identify the key parts of a system and how those parts interact. As one example, Investigation 3-G involves building a model of an ecosystem in a bottle. These photographs show a model of the formation of geologic structures called sedimentary rock formations.

The land is once again covered with water and another layer of sediment forms.

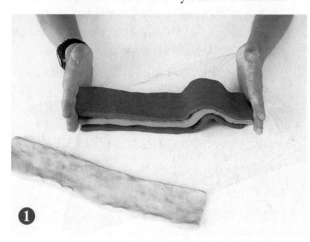

The different colours of modelling clay represent layers of sediment that build up on the ocean floor. Large movements of continents cause the sediments to become dry land. Movements deep beneath Earth's surface push the layers together, creating hills and mountains.

More movement causes one section of the land to push up and over another section of land.

Sometimes important structures are hidden from view. A model allows you to visualize these structures. The following diagram shows you the structure of the inside of Earth. You would call this type of model a *cutaway model*.

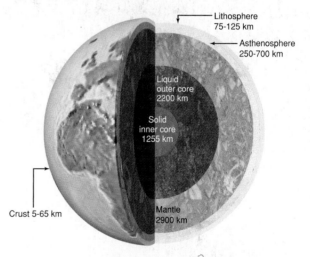

Sometimes distances are so great that you cannot see all of the important structures at the same time. For example, even with a telescope, you cannot see all of the planets in the solar system at the same time. Therefore, you cannot see how the planets are arranged around the Sun. The model on the next page allows you to see the arrangement of the planets. In this case, the model cannot be built to scale. Recall that the Unit 3 Project features a scale model of an earthquake-resistant building. A scale model has

SkillPower 4 • MHR **301**

exactly the same shape as the object that it is modelling but its size is either much larger or much smaller. If the model of the Sun were approximately the size of a baseball, Earth would be nearly one kilometre away and its diameter would be less than one millimetre.

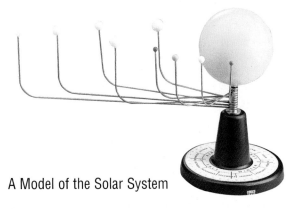

A Model of the Solar System

Sometimes structures are too small to be seen. Models can help you visualize microscopic structures. The diagram below shows you how the inside of a living animal cell might look.

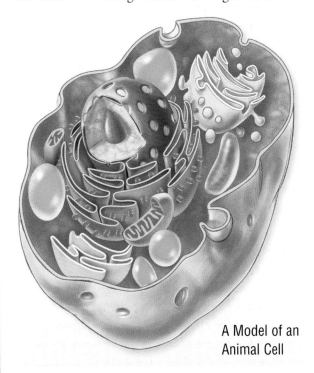

A Model of an Animal Cell

Sometimes the model does not look exactly like the system that it represents. In the following activity, you can develop a model of the greenhouse effect.

Modelling the Greenhouse Effect

You will create models of Earth with and without greenhouse gases. You will need the following materials.

2 1 L clear plastic soft drink bottles with tops cut off and labels removed
2 thermometers (wrap bulbs with masking tape)
  potting soil
  masking tape
  plastic wrap
  elastic band
  lamp with 100 Watt lightbulb
  watch or clock with a second hand

Procedure

1. Put equal amounts of potting soil in the bottom of the two bottles.

2. Tape the thermometers to the inside wall of the bottles with the bulbs about a centimetre above the top of the soil.

3. Seal the top of one bottle with plastic wrap that is held in place with an elastic band.

4. Place the lamp with the exposed 100 W lightbulb between the two bottles and about one centimetre from each bottle.

5. Let the apparatus sit for about 5 min and then record the temperature in each bottle.

6. Turn on the light. Read and record the temperature in each container every 2 min for 16 min.

7. Draw a graph of temperature versus time.

Conclude and Apply

1. What does the lightbulb model?

2. What does the plastic wrap model?

3. (a) Compare the temperatures in the two bottles throughout the experiment. How does the experiment model the greenhouse effect?

   (b) What are some limitations of your model?

Skill POWER 5

ORGANIZING AND COMMUNICATING SCIENTIFIC DATA

Drawing a Data Table

To learn how to set up a data table, read the paragraph below that contains data. Then compare the data in the paragraph to the data in the table below.

White rats have an average mass of 0.15 kg and a resting heart rate of about 350 beats per minute. An average 12.0 kg dog has a resting heart rate of about 100 beats per minute. Adult humans have an average mass of about 70 kg and a heart rate of about 72 beats per minute. An elephant with a mass of 4000 kg has a heart rate of about 30 beats per minute.

Table 1 Heart Rates and Masses of Four Animals

Type of Animal	Mass (kilograms)	Heart Rate (beats per min)
white rat	0.15	350
dog	12.0	100
human	70.0	72
elephant	4000	30

Notice that the data contain three categories:
- type of animal
- mass of animal
- heart rate of animal.

The number of categories tells you how many columns you need in your table.

There is a complete set of data for each of four animals. The number of sets of data tells you how many rows you will need in your table. You need one row for each set of data and one row for headings for each column. As well, you should always give your table a title.

Tips for Drawing Data Tables

1. Use a ruler to draw your table.
2. Print all letters and words so they will be easy to read.
3. Express numbers in your data as numerals (for example, 1, 2, 3) not as words (one, two, three).
4. Give the table a title. If there are small numbers in the title, you can express them in words. The variables in the observations or experiments can become the title of your data table. For example, "The Effects of Light on Plant Growth."
5. Draw a box around the data. Place the title outside the box.
6. Print headings neatly at the top of each column. Include units for the data with the headings, not in the columns.
7. Draw a line separating the headings from the data.
8. Record the data in the columns below the headings.
9. Separate the columns of data with lines.

Instant Practice

Draw a table from the data in the following paragraph.

A serving of cooked pasta has about 15 g of carbohydrate, 3 g of protein, and no fat. A serving of beef has no carbohydrate, 7 g of protein, and about 5 g of fat. A serving of bananas has 15 g of carbohydrate, no protein, and no fat. A serving of milk has about 12 g of carbohydrate, 8 g of protein, and about 3 g of fat. A serving of tomatoes has about 5 g of carbohydrate, 2 g of protein, and no fat.

Graphing

A **graph** is often a line or curve that shows how one quantity depends on changes in another quantity. A graph is a good way to display data so that you can see patterns and relationships among variables.

Graphing Terms You Need to Know

x- and y- axes: The *x*-axis is the horizontal line on a graph. The independent variable is placed on the *x*-axis. The *y*-axis is the vertical line on a graph. The dependent variable is placed on the *y*-axis.

Origin: The point at which the *x* and *y* axes meet. The origin is often, but not always, the point on a graph at which *x* and *y* are equal to zero.

Plotting: Drawing a point or line on a graph that represents data.

Scale of numbers: Equal divisions marked on the *x* and *y* axes so that they can be used for measuring. The magnitude (size) of the numbers on a scale must increase as you move away from the origin. As you go to the right or upward, the numbers become increasingly positive. As you go left or downward, the numbers become increasingly negative.

Interval: The value of the divisions between units marked on a scale (for example, 0, 1, 2, 3, … or 0, 5, 10, 15, …..). The interval between the numbers on the scale must be equal in size.

Range: The difference between the largest and smallest of a series of numbers.

Key: (Legend): A small table that explains or identifies symbols on a graph. The legend should be placed close to the graph.

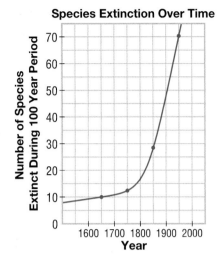

Drawing a Line Graph

You can use a line graph whenever the data you collect depend on each other. Line graphs make it easy to see relationships between two quantities. You can use a line graph to predict values that are not even on the graph.

Example

The following data table gives the number of species that became extinct during each 100 year period between 1600 and 1900. To learn how to draw a line graph from the data, examine the graph above as you read the steps.

Table 2 Species Extinction Over Time

Year	Species Extinct During Century
1600 – 1700	10
1700 – 1800	11
1800 – 1900	28
1900 – 2000	70

1. With a ruler and pencil, draw an *x*-axis and a *y*-axis on a piece of graph paper.

2. Label the axes. Write the years along the *x*-axis and "Number of Species Extinct During 100 Year Period" along the *y*-axis.

3. Decide on a number scale to use. Your *x*-axis will have four intervals of 100 years each. Your *y*-axis will go from 10 to 70. You can start at zero and end at a number above 70. Use a tick mark at major intervals on your scale.

4. Go to the point representing the middle of the century 1600 to 1700. Now move up until you are at a point that represents 10 species. Make a dot at this point. Repeat this procedure for all data points on the table.

5. Do not simply connect the data points when you draw a line graph. Experimental data points usually have some error. If the points are almost in a straight line, draw a straight line as close to most of the points as possible. There should be about as many points above the line as there are below the line. This line is called a **line of best fit**. If the data points do not appear to follow a straight line, then draw a smooth curve that comes as close to the points as possible. A line of best fit shows the trend of the data. It can be extended beyond the first and last points.

6. Give your graph a title.

Drawing a Bar Graph

A bar graph helps you to compare the number of items in one category with the number of items in other categories. The height of the bar represents the number of items in the category. Study the example, and then make your own bar graph.

Example

Make a bar graph from the data in the table below. Use only the data on time in the second column. Do not graph the data (Cost) in the third column. As you read the steps, examine the completed graph on the right to see how the steps were followed.

Table 3 Time and Cost to Boil 900 mL of Water		
Method	Time (min)	Cost (cents)
electric kettle	4.5	0.71
electric stove	8.0	1.25
gas stove	6.0	0.32
camp stove	7.0	11.1

1. Draw an *x*-axis and a *y*-axis on a sheet of graph paper. Label the *x*-axis with the names of the heating methods. Label the *y*-axis with the title "Time (min)." (See the bar graph below.)

2. Look at the data carefully in order to select an appropriate scale. The range of the data is between 4.5 min and 7.0 min. You can number the scale from 0 to 9. Write the numbers on your *y*-axis scale.

3. Decide on a width for the bars in the graph. They should be large and easy to read.

4. Mark the width of the bars on the *x*-axis. Leave the same amount of space between each bar.

5. To draw the bar for the electric kettle, go to the centre of the bar on the *x*-axis, then go up until you are halfway between 4 and 5. Make a mark to represent 4.5 min. Use a pencil and a ruler to draw in the first bar.

6. Repeat the procedure for the other heating methods.

7. When you have drawn all the bars, colour them so that each one is different. If you are comparing two or more independent variables that you have plotted on the *x*-axis, you will need to make a legend or key to explain the meanings of each bar.

8. Give your graph a title.

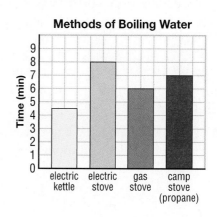

Instant Practice

A science teacher asked her students what kind of pet they had or would like to have. She planned to have her students research the proper care and feeding of each pet. The responses are shown in the following table. Make a bar graph using the data.

Table 4 Pets Chosen by Students	
Pet	Number of students choosing pet
cat	16
gerbil	11
hamster	13
goldfish	3
mice	7
guinea pig	10
dog	14
hedgehog	10

Constructing a Histogram

The graph on the right is called a histogram. It is another type of bar graph. Notice that there is no space between the bars in a histogram but there is space between the bars in a bar graph. The reason for the space is that each bar represents a different item.

In a histogram, the x-axis represents one continuous item, divided into size categories. In the histogram, notice that the x-axis represents temperature, divided into two-degree categories. Thus, the x-axis is quantitative, meaning that it contains numerical values. In a typical bar graph, the x-axis is qualitative and cannot be described with numbers.

Example

A group of students conducted a test on plant growth. They placed plants in rooms with different temperature ranges and observed the effects on the plants. Table 5, called a frequency table, shows how many plants were placed in each location. As you read the steps, examine the completed histogram below.

Table 5 Temperature of Rooms	
Temperature (°C)	Frequency (number of plants)
10–12	1
13–15	3
16–18	3
19–21	2
22–24	3
25–27	5
28–30	5
31–33	3

1. On a piece of graph paper, draw an x-axis and a y-axis. Label the x-axis "Temperature (°C)" and the y-axis, "Frequency (number of plants)."
2. Separate the x-axis into eight equal segments. Label the segments using the temperatures listed in the frequency table.
3. Make a scale on the y-axis that goes to 6.
4. Move your pencil up the y-axis to 1, and make a light mark. Then, using a ruler, make a bar that is the width of the 10–12°C temperature range.
5. Repeat this procedure for each temperature range and corresponding number of plants.

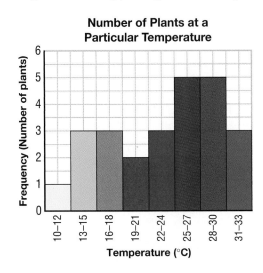

Number of Plants at a Particular Temperature

Instant Practice

The following frequency table gives the number of launches of the space shuttle in two-year intervals. Make a histogram using the data.

Table 6 Launches of the Space Shuttle	
Years	Number of launches
1981–1982	5
1983–1984	9
1985–1986	11
1987–1988	2
1989–1990	11
1991–1992	14
1993–1994	14
1995–1996	14
1997–1998	13
1999–2000 (planned)	10

Drawing a Circle Graph

A circle graph (also called a pie chart) is a very good way to present data as percentages of a total. To learn how to make a circle graph, study the following example.

Example

The table below lists energy use by cows. Make a circle graph from these data.

Table 7 Energy Use by Cows		
Use of Energy	Energy (%)	Degrees in "piece of pie"
Building and repairing body tissues	4	14.4
Breathing, mooing, pumping blood	33	118.8
Urine, feces, gas, heat	63	226.8

1. Copy the first two columns of Table 7 into your notebook. Add a third column to your table. Label it "Degrees in piece of pie."

2. Determine the number of degrees in the "piece of pie" that represents the percentage of energy use, using the following formula:

   Degrees for "piece of pie" =
   $$\frac{\text{Percent for energy use}}{100\%} \times 360°$$

   For example, the calculation for the "piece of pie" representing the building and repairing of body tissues is

   $$\frac{4\%}{100\%} \times 360° = 14.4°$$

   Write the number of degrees in the third column of the table.

continued

SkillPower 5 • MHR 307

3. Use a compass to make a large circle on a sheet of paper. Put a dot in the centre of the circle.

4. Draw a straight line from the centre of the circle to the edge. Place a protractor on this line and make a mark almost halfway between 14° and 15°. Draw a line from the centre of the circle, through the mark, to the circumference of the circle. This is the "slice of pie" that represents the percentage of the total energy that is used to build and repair the cow's tissues.

5. To draw the next "piece of pie," representing breathing, mooing, and pumping blood, use the line that ended the first "piece of pie" as the starting place.

6. Repeat the procedure for the final "piece of pie."

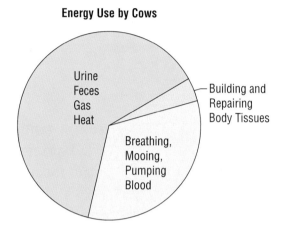

Energy Use by Cows

Instant Practice

1. Chapter 1 discusses the ecoprovinces in British Columbia. Table 8 (below) shows the percentage of the total land and water area in British Columbia that is covered by each ecoprovince. Use the data in Table 8 to make a circle graph.

Table 8 Percentage of Area Covered by Ecoprovince		
Ecoprovince	**Percent of total area**	**Degrees in "piece of pie"**
Boreal Plains	4	
Central Interior	11	
Coast and Mountains	18	
Georgia Depression	2	
Northeast Pacific	25	
Northern Boreal Mountains	12	
Southern Interior	3	
Southern Interior Mountains	11	
Sub-Boreal Interior	9	
Taiga Plains	5	
Total Circle		360°

DESIGNING AND CONDUCTING EXPERIMENTS

A useful experiment usually starts with an observation. The observation might be as simple as watching a child on a playground swing such as the one shown below.

A swing has a rhythmic motion.

Observing and Inferring

If your observation can lead to the development of an experiment, you must be able to ask a question that can be answered by a scientific investigation. You might ask, "What features of a playground swing or any other swinging object determine the rhythm of the swinging motion?" To answer the question, you might think back to when you were younger. You knew that you could make a swing go higher, but it always seemed to swing back and forth at the same rate. You have narrowed the question down to what determines the rate at which a swinging object goes back and forth. You have *inferred* that some characteristic of a swinging object determines the rhythm. You could compare a child on a swing to some other previous experience. Maybe you bumped into a hanging planter at home and watched it swing. It seemed to swing faster than the child on the swing. The hanging planter is a shorter system than the child on the swing. You might infer that the length of the rope influences the rate of swinging. This is an idea that you can test by creating a model of a child on a swing. You can model a child on a swing by building a simple pendulum (see the photograph below).

Hypothesizing and Predicting

When you have developed a model, you can use it to formulate a hypothesis. You might think about the size of the mass at the end of the string on your pendulum. You might hypothesize that a heavier mass would swing faster. If this is the case, you could predict that a larger mass would cause the time for one complete swing to be shorter. The term that describes the length of time for the pendulum to swing from one end to the other and back again is the *period*. You are nearly ready to carry out your experiment. First, you must identify all of the variables involved when a pendulum swings.

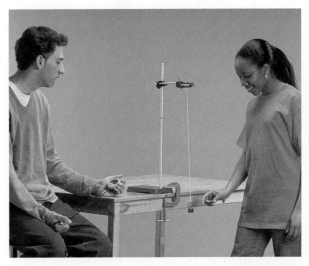

These students are modelling a child on a swing by setting up a pendulum.

Variables

A variable is anything that might affect the outcome of the experiment. The period of the pendulum is a variable. All of the other variables that might affect the period are:

- the mass of the pendulum bob (the object at the end of the string),
- the length of the string, and
- how far you pull the pendulum bob to one side before letting it go.

In order to carry out a fair test, you must test the effect of only one specific variable at a time. You must ensure that all the other variables do not change. Scientists say that you *control* the other variables. This is important because, if you changed both the mass of the bob and the length of the string and you observed a change in the period, you would not know whether the change in the mass or in the length of the string caused the period to change. Study the figure below and see if you can identify the variables in another experiment.

Independent, Dependent, and Controlled Variables

The variable that you choose to test is called the **independent variable.** You can remember which variable is the independent variable by remembering, "**I** choose to test the **I**ndependent variable." You can select which independent variable that you will test. The **dependent variable** is the one that you observe to see if it has been affected by the changes to the independent variable. All of the other variables that you are not testing are **controlled variables.** They remain the same while you change the conditions of the independent variable. These concepts will become clear as you study the pictures of the student doing a pendulum experiment in the photographs on this page and the next page.

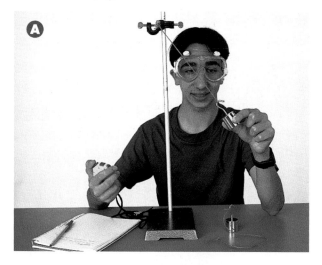

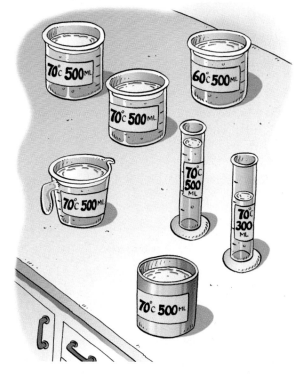

Examine the containers of water. Assume that you want to know what factors affect the length of time it takes for water to evaporate. How many variables can you find in the figure? What are those variables? How would you test each variable individually?

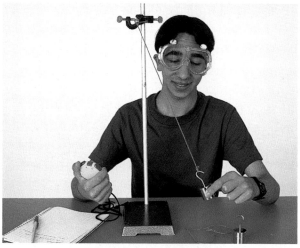

In part (A) of the pendulum experiment, the length of the string is the independent variable.

310 MHR • Science Skills Guide

The student is testing two different lengths to see how they affect the period of the pendulum. He will start the stopwatch when he releases the pendulum bob and measure the time for one period. In this case, time is the dependent variable. Notice that the student is using the same mass and he is holding the mass the same distance from the central position. These are the controlled variables.

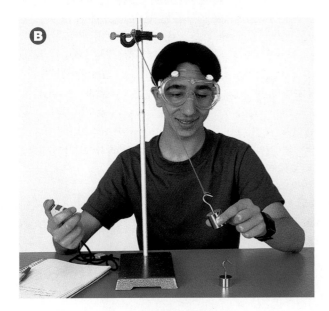

In part (B) of the pendulum experiment, the mass is the independent variable. Notice that the length of the string is the same in both cases. The length of the string is now a controlled variable.

Control Sample

In many experiments, you need to have a control sample in which there is no independent variable at all. Examine the figures below to understand when you need to have a control sample.

(a) Find the best filter for muddy water

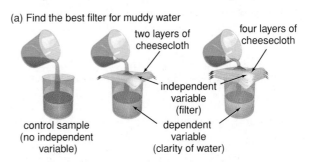

(a) If you compared only the two filters, you would not know whether the filters were doing anything at all. You need to compare both filters to the case in which there is no filter. The sample with no filter is the control sample.

(b) Find the best plant food for plant growth

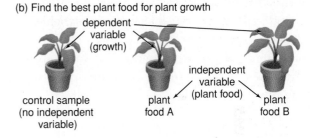

(b) If you tested only the two plant foods and plant (B) grew taller than plant (A), you would not know if plant food A had any effect at all. You need to compare both plants with plant food to a third plant that received no plant food.

(c) Does the amount of acid in water affect seed germination?

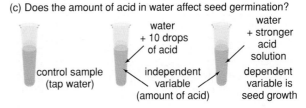

(c) To fully understand the effect of acid on seed germination, you have to compare the effect of two acid strengths to a control sample with water only and no acid.

SkillPower 6 • MHR **311**

Recording Data

Before you conduct an experiment, be sure that you know what tests you are going to perform and what measurements you will take. Prepare tables for your data before you start your experiment. Also, list values of the controlled variables with each data table. If you were going to do the pendulum experiment, you might make a table such as the one on the right.

Mass of Pendulum Bob		Length of string		Angle between string and vertical support	
Mass (g)	Average Period (s)	Length (cm)	Average Period (s)	Angle (°)	Average Period (s)
25		50		5	
50		100		10	
75		150		15	
100		200		20	
Values of Controlled Variables					
Length constant at 100 cm Angle constant at 10°		Mass constant at 50 g Angle constant at 10°		Length constant at 100 cm Mass constant at 50 g	

Note that the controlled variables are listed on the bottom row. Now you are ready to conduct your experiment.

After you have set up your apparatus and received your teacher's permission, you can begin to collect data. Taking one measurement is never enough. It is too easy to make an error in measurements. You need to take at least four measurements for each value of the independent variable. You can then calculate an average value. Write your average value in the table.

Often, it is easier to interpret your results if you graph your data. Your pendulum results might look like those in the table and graphs below. Because there is always some error in the data, the points will not lie on a smooth line. Therefore, draw a smooth line or curve close to the points.

Mass of Pendulum Bob		Length of string		Angle between string and vertical support	
Mass (g)	Average Period (s)	Length (cm)	Average Period (s)	Angle (°)	Average Period (s)
25	2.1	50	1.4	5	2.1
50	2.0	100	2.0	10	1.9
75	2.2	150	2.4	15	2.2
100	1.9	200	2.7	20	2.0
Length constant at 100 cm Angle constant at 10°		Mass constant at 50 g Angle constant at 10°		Length constant at 100 cm Mass constant at 50 g	

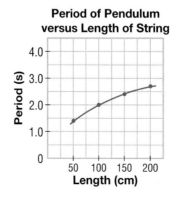

Period of Pendulum versus Mass of Pendulum Bob

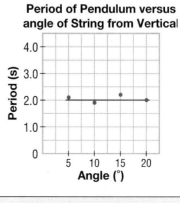

Period of Pendulum versus Length of String

Period of Pendulum versus angle of String from Vertical

Summarize Results

Always write a summary of your results in a descriptive paragraph. Your summary of the pendulum experiment might be similar to this.

> When the mass of the pendulum bob was changed, the period stayed the same. When the length of the string was changed, the period changed. When the string was made longer, the period became longer. When the angle between the string and the vertical support stand was changed, the period stayed the same.

Conclusion

Finally, you will write a conclusion statement. Your conclusion will explain the meaning of the results. If your experiment was based on a hypothesis, your conclusion should state whether the results supported your hypothesis. Imagine that your hypothesis for the pendulum experiment had been:

> "I hypothesize that a larger mass will make the pendulum swing faster than a smaller mass will make it swing. Therefore, the period of the pendulum will be shorter for larger masses."

Your conclusion statement would have to say that the results did not support your hypothesis. Your conclusion statement might read like this:

> "The results did not support my hypothesis that a large mass will produce a shorter period than will a small mass. Mass had no effect on the period of the pendulum. The angle between the string and the vertical support stand also had no effect on the period. The length of the string was the only variable that affected the period of the pendulum. As I made the string longer, the period of the pendulum swing also became longer."

Should you consider an experiment to be a failure if the results did not support your hypothesis? No! Your recorded data is valid. You demonstrated that the mass of the pendulum bob and the angle between the string and the support stand had no effect on the period. That information is just as important as the conclusion that the length of the string *did* determine the period of the pendulum.

Practice

How could you design an experiment to test the question, "What effect does the size of the hole in tubing have on the rate at which water flows through the tube?" or "What factors affect the amount of sugar that will completely dissolve in water?" When you design your experiment, be sure to answer the following questions.

1. What are all of the possible variables that could affect your results?
2. What is your independent variable?
3. What is your dependent variable?
4. What are the controlled variables for this experiment?
5. How are you going to ensure that all of the controlled variables are held constant?
6. Do you need a control sample? If so, what will your control sample be?
7. What will be the form of the data that you are recording?
8. What will you need to put in your data table?
9. Will you need to graph the data? If so, what type of graph will you choose?

Plan your experiment and obtain your teacher's approval before conducting your experiment. **Note:** Be sure to include all necessary safety precautions.

Glossary

How to Use This Glossary

This Glossary provides the definitions of the key terms that are shown in **boldface** type in the textbook. (Instructional boldfaced words, such as "observe" and "predict" used in investigations, are not included.) Other terms that are not critical to your understanding, but that you may wish to know, are also included in the Glossary. The Glossary entries show the section where you can find the boldfaced key terms. In addition, the Glossary entries list related words in brackets after the definition. A pronunciation guide appears in square brackets after selected terms. Use the following pronunciation key to read these terms:

a = mask, back	ih = ice, life	uh = sun, caption
ae = same, day	i = simple, this	uhr = insert, turn
air = stare, where	o = stop, thought	yoo = cute, human
e = met, less	oh = home, loan	
ee = leaf, clean	oo = food, boot	

A

abiotic [AE-bih-o-tik]   non-living; refers to non-living things in the environment, such as water, soil, or air (1.1) (see also *biotic*)

absolute age   the age, in years, of a rock or other object (9.3)

abyssal plain   a large, flat area on the sea floor (8.2)

acid   a compound that dissolves in water to form a solution that has a pH lower than 7; tastes sour, is corrosive, and can cause burns to skin (6.3) (see also *base, compound*)

acid rain   rain that contains higher than normal levels of acid; caused by waste gases released into the environment by industries and automobiles; damaging to the environment (3.2)

alloy   a solution that is made from two or more metals (5.2) (see also *solution*)

annual surveys   yearly counts of species such as birds and other animals (3.1)

aquatic   growing or living in water; having to do with water (3.2)

asthenosphere   the semi-molten zone of Earth's mantle just below the lithosphere (7.1) (see also *mantle, lithosphere*)

atmosphere   the air surrounding Earth

atom   small particles from which all matter is composed; the basic unit of an element, which still retains the properties of that element (5.3)

atomic-molecular theory   a combined theory stating that all matter is composed of atoms, and atoms can combine to form molecules (5.3)

bar graph   a diagram consisting of horizontal or vertical bars that represent (often numerical) data

base   a compound that dissolves in water to form a solution that has a pH greater than 7; tastes bitter, breaks down oil and fats, has a slippery texture, and can cause burns to skin (6.3) (see also *acid, compound*)

baseline data   data gathered by scientists to be used as a starting point to track changes in the environment (3.1)

bedrock   the solid rock that lies beneath the soil (9.2)

bioaccumulation   the accumulation of pollutants in organisms through the levels of a food chain, so that organisms higher up the food chain have higher levels of pollutants (2.2)

biome   a large region on Earth that contains particular types of plant life, and has about the same temperature and amount of precipitation throughout (1.1)

biosphere reserves   regions that are internationally recognized for trying to promote environmental conservation and sustainable development (3.1) (see also *sustainable development*)

biotic [bih-O-tik]   living; refers to living things in the environment, such as humans, trees, or fish (1.1) (see also *abiotic*)

boiling   rapid evaporation of a liquid occurring at a specific temperature called the boiling point

boiling point   the temperature at which a liquid boils to become a gas (4.1) (see also *melting point*)

capacity unit   unit used to measure the volume of liquids; an example is the litre (L)

captive breeding program   a program to boost the population of a species, by mating animals in captivity and then releasing the offspring into the wild (3.3)

314   MHR • Glossary

carbon cycle a model showing how carbon is used and re-used throughout ecosystems (2.2)

carbon dioxide a gas that is involved in the process of photosynthesis, and that is produced by the process of cellular respiration (2.1)

carnivore an animal that eats other animals; examples include wolves and hawks (1.3) (see also *consumer*)

carrying capacity the maximum number of individuals that an environment can support over a long period of time (2.3)

cellular respiration the process by which an organism obtains energy by breaking down high-energy foods, such as sugar and starch (2.1) (see also *photosynthesis*)

cementation the process in which pieces of sediment are held together by another material (7.3)

chemical change a process in which one type of matter changes to produce one or more different types of matter with different identities from those of the original matter (4.3) (see also *physical change*)

chemical property a property of a substance that can be observed only as the substance changes into another substance through a chemical change (4.3)

chemical weathering the breaking down of rocks by chemical changes (8.1) (see also *mechanical weathering*)

chemistry the study of matter and its changes (5.1)

chlorophyll [KLOHR-uh-fil] a green pigment found in producers that traps the energy of the Sun during photosynthesis (2.1)

cinder cone volcano a volcano made of cinder (ashes) with steep sides and a bowl-shaped crater at the top (9.3)

circle graph a circle divided into sections (like pieces of a pie) to represent data; also called a pie chart

cirque a bowl-shaped basin formed by erosion at the start of a mountain glacier (9.1)

cleavage in a mineral, the characteristic of breaking along smooth, flat planes (7.2)

climate the average weather pattern of a region over a long period of time (1.1)

commensalism a symbiotic relationship between two different types of organisms, in which one partner benefits and the other partner does not appear to lose or gain anything (1.3)

community a group made up of all the interacting populations that live in an area (1.2)

compaction the process in which layers of sediment are squeezed together by the weight and pressure of water and other sediment, to form sedimentary rock (7.3)

composite volcano a volcano composed of alternating layers of lava and cinder or ash, also called a stratovolcano (9.3)

compound a pure substance composed of two or more elements, which are chemically combined in fixed ratios (5.3) (see also *element*)

concentrated solution a solution that has a large mass of dissolved solute for a certain quantity of solvent (6.1) (see also *dilute solution*)

concentration the quantity of solute that is dissolved in a certain quantity of solvent (6.1)

concept map a graphic organizer made up of words or phrases in circles or boxes and connecting lines; used to show relationships among ideas

condensation the process in which a gas or vapour changes in state to become a liquid; for example, water vapour condenses into liquid water (2.2, 4.3) (see also *evaporation*)

condense change from a gas or vapour into a liquid

consumer an organism that eats the food made by producers; can be a herbivore, carnivore, or omnivore (1.3) (see also *producer*)

continental shelf a shallow underwater ledge located between a continent and the deep ocean (8.2)

continental slope a deep slope in the seabed between a continental shelf and the sea floor (8.2)

control in a scientific experiment, a standard to which the results are compared; ensures a fair test

controlled variable in an experiment, a condition that is not allowed to change

convection current in geology, a movement of material produced by the rising of a warmer material and the sinking of a cooler and denser material (7.1)

convergent boundary in geology, a plate boundary at which the plates move towards each other and collide (8.3) (see also *divergent boundary*, *transform boundary*)

co-ordinate graph a grid that has data points named as ordered pairs of numbers

core sample a long, cylindrical sample of rock obtained by drilling (8.2)

criteria a set of standards or expectations to be met

cross section a drawing of what you see when you cut through an object (8.2)

crust the thin, outer layer of Earth; made of solid rock (7.1)

crystal in geology, a structure with straight edges, flat sides, and regular angles (7.2)

crystallization the process of evaporating the solvent from a solution, leaving the solute behind in crystal form (6.4)

cubic units the units used to report the volume of a substance, such as cm^3

cycle (concept) map an events chain map in which a series of events does not produce a final outcome; this type of concept map has no beginning and no end

data facts or information

database an organized or sorted list of facts or information, usually generated by computer

decomposer an organism that breaks down (decomposes) dead or waste materials, such as rotting wood, dead animals, or animal waste (1.3)

delta a triangular deposit of sediment at the mouth of a river (9.1)

density the quantity of mass in a certain volume of a material (4.2)

dependent variable in an experiment, a condition that is changed as a result of changes to the independent variable (see also *independent variable*)

deposition the dropping of sediments by an agent of erosion, such as wind or water (9.1)

dike a volcanic feature formed when magma flows across existing layers of rock, and then hardens (9.3) (see also *sill*)

dilute solution a solution that contains a small mass of solute for a certain quantity of solvent (6.1) (see also *concentrated solution*)

dissolve to mix a solute completely with a solvent to form a solution; the physical properties of the solute and solvent combine into one set of properties (5.2)

distillation a process used to separate and collect the components of a solution; the solvent is heated to change it into a gas, then converted back to a liquid through condensation (6.4)

divergent boundary in geology, a plate boundary at which the plates move away from each other (8.3) (see also *convergent boundary*, *transform boundary*)

dormant the term used to describe a volcano that is not active (9.3)

E

earthquake shaking of the ground caused by the sudden release of energy stored in the bedrock (9.2)

ecological reserves areas that are set aside to protect examples of the different habitats in a province, as well as rare and endangered plants and animals (3.3)

ecological succession the gradual change in the make-up of a biological community over time (2.3)

ecologist [ee-KOL-oh-jist] a scientist who studies the interactions among organisms, and/or the interactions between organisms and their environment (1.1)

ecology the study of the interactions among organisms, as well as the interactions between organisms and their environment (1.1)

ecoprovinces ten different natural regions in British Columbia that have been mapped out and described by biologists (1.1)

ecosystem all the interacting organisms that live in an environment, as well as the abiotic (non-living) parts of the environment that affect the organisms (1.1)

ecosystem monitoring a method of checking the condition of an ecosystem by comparing the results of investigations done at different times (3.1)

elasticity the ability of a material to go back to its original shape after it has been stretched or compressed

element a pure substance composed of one type of atom, that has its own unique set of properties, and that cannot be broken down or separated into simpler substances by physical or chemical changes (5.1) (see also *compound*)

emulsion a suspension to which a substance has been added to allow the parts of the suspension to remain distributed throughout the mixture (5.2)

endangered species a species that is in danger of dying out entirely, so that the species no longer exists (3.2) (see also *extinct*)

energy the ability to do work and to cause chemical or physical change

energy flow the transfer of energy that begins with the Sun and passes from one organism to the next in a food chain (2.1) (see also *food chain*)

environmental impact assessment a report that outlines how an activity will affect the environment (3.1)

epicentre [E-pi-sen-tuhr] the location on Earth's surface that is directly above the focus of an earthquake (9.2) (see also *focus*)

erosion the transportation of sediments to another location (8.1)

erupt of volcanoes, to become active (9.3)

evaporation the process in which a liquid changes into a gas or vapour; for example, liquid water evaporates to become water vapour (2.2, 4.3) (see also *condensation*)

events chain map a concept map used to describe a sequence of events, the steps in a procedure, or the stages of a process

extinct no longer existing (3.2)

extrusive rock igneous rock that forms when lava cools on Earth's surface (7.3) (see also *intrusive rock*)

F

fault the surface along which rocks break and move during an earthquake (9.2)

floodplain a flat layer of sediment deposited by a river in flood stage (9.1)

focus the place deep in Earth's crust where an earthquake begins (9.2) (see also *epicentre*)

foliated having thin, leaf-like layers (7.3)

food chain a model that shows how food energy passes from organism to organism (2.1) (see also *food web*)

food web a network of interconnected food chains in an ecosystem (2.1) (see also *food chain*)

fossil fuels coal, oil, and natural gas; natural materials formed in Earth's crust by the compression of plant and animal matter over millions of years; used as fuel and as the basis of many manufactured goods (3.2, 7.2) (see also *raw materials*)

fossils evidence of once-living organisms (7.3)

fracture in a mineral, the characteristic of breaking with rough or jagged edges (7.2)

freezing point the temperature at which a material freezes to become a solid, or melts to become a liquid; also called the melting point

gas a state of matter that has no fixed shape or volume of its own, but fills its container completely, with no surface (4.1) (see also *solid, liquid*)

genus a group of species that are related (see also *species*)

geothermal energy the energy harnessed from Earth's interior (9.3)

glacier a large body of ice that moves slowly downhill (9.1)

global warming the steady increase in the temperature of Earth's atmosphere (2.3, 3.2)

graphic organizer a visual learning tool that helps clarify the relationship between a central concept and related ideas or terms

greenhouse effect the trapping of heat in the atmosphere around Earth by greenhouse gases, such as carbon dioxide and aerosol gases (3.2)

ground water the water contained in Earth's crust (2.2)

habitat the place in which an organism lives (1.2) (see also *niche*)

habitat enhancement project a project that improves an existing habitat (3.3)

habitat fragmentation the separation of one part of a habitat from another (3.2)

habitat restoration project a project that restores or improves a habitat that has been damaged (3.3)

hardness in a mineral, the characteristic of resisting being scratched (7.2)

herbivore an animal that eats only plants (1.3) (see also *carnivore, omnivore*)

heterogeneous [het-uhr-oh-JEEN-ee-uhs] of a mixture, having different parts that retain their own properties and that can be detected quite easily (5.1) (see also *homogeneous*)

histogram a type of bar graph in which each bar represents a range of values and in which the data are continuous

homogeneous [hoh-moh-JEEN-ee-uhs] of a mixture or pure substance, every part of the material the same as every other part (5.1) (see also *heterogeneous*)

horn a sharpened peak formed by the glacial action of three cirques on a mountaintop (9.1)

host the organism that a parasite lives and feeds on (1.3) (see also *parasitism, parasite*)

hot spot an area under Earth's crust where the temperature is much hotter than normal, forcing magma towards the surface (9.3)

hydrothermal vent an opening in the sea floor through which mineral-rich water escapes (8.2)

igneous rock [IG-nee-uhs] the type of rock that forms when hot magma or lava cools and becomes solid (7.3) (see also *sedimentary rock, metamorphic rock*)

independent variable in an experiment, a condition that is selected or adjusted to see what effect the change will have on the dependent variable (see also *dependent variable*)

indicator a compound that turns one colour in an acid and a different colour in a base (6.3)

inner core the deepest layer at the centre of Earth, made of iron, and possibly other elements (7.1)

insoluble unable to dissolve in a particular solvent (6.1) (see also *soluble*)

introduced species new species that are brought into an ecosystem either on purpose or by accident (3.2)

intrusive rock igneous rock that forms when molten rock cools below Earth's surface (7.3) (see also *extrusive rock*)

landslide the rapid movement of large amounts of rock fragments and soil down a slope (9.1)

lava the term used for magma, or molten rock, when it breaks through Earth's crust and reaches the surface, as in a volcanic eruption (7.3)

lichen [LIH-kuhn] a plant-like organism that can grow on rocks

limiting factor any living or non-living factor that controls the number of individuals in a population (2.3)

line graph a diagram that shows how one value depends on or changes according to another value; produced by drawing a line that connects data points plotted in relation to a *y*-axis (vertical axis) and an *x*-axis (horizontal axis)

liquid a fluid state of matter that has a specific volume, but no shape of its own; it takes the shape of its container and forms a surface in the container (4.1) (see also *solid, gas*)

lithosphere the rigid outer layer of Earth composed of the crust and upper mantle (7.1) (see also *crust, mantle*)

long-term monitoring studying organisms for many years (3.1)

lustre a measure of how much light is reflected from the surface of a mineral, that is, how shiny the mineral is (7.2)

magma hot melted rock formed deep below Earth's crust by high temperatures and pressures (7.3) (see also *lava*)

magnetic field a region of force around a magnet; Earth has a magnetic field (7.1)

magnetic reversal the switching of Earth's north and south poles, an event that has occurred many times in the past

magnetometer [maeg-nuh-TOM-uh-tuhr] a sensing device used to detect the direction and strength of magnetic fields (8.2)

malleability of matter, the characteristic of being able to be hammered, stretched, or rolled in all directions, without breaking or cracking

Glossary • MHR **317**

mantle a large and complex layer of rock, both solid and molten, that lies under Earth's crust (7.1)

mass the quantity of matter in a solid, liquid, or gas (4.2)

matter makes up every living thing and material object; anything that takes up space and has mass (4.1)

mechanical mixture a mixture made of more than one type of matter, in which the different parts can be easily seen with the unaided eye (5.2)

mechanical weathering the breaking down of rocks by the action of physical forces such as water freezing and melting (8.1) (see also *chemical weathering*)

melting point the temperature at which a solid melts to become a liquid; also called the freezing point (4.1) (see also *boiling point*)

meniscus the slight curve where a liquid touches the sides of its container

metamorphic rock the type of rock produced when heat, pressure, or fluids change one type of rock into a new form (7.3) (see also *igneous rock, sedimentary rock*)

metric system a system of measurement based on multiples of ten and in which the basic unit of length is the metre

mineral a chemical element or compound that is naturally occurring and has a crystal structure (7.2) (see also *rock, mineral resource*)

mineral resource a rock or mineral that can be mined and used for a specific purpose (7.2)

mixture a combination of two or more different types of matter that can be separated by physical changes (5.1)

model a representation of a scientific structure or process (4.3)

Moho the boundary separating the crust from the mantle (7.1) (see also *crust, mantle*)

Mohs Hardness Scale in geology, a scale of ten minerals with a hardness value of 1 to 10; the higher the number on the scale, the harder the mineral (7.2)

molecule a small particle composed of two or more atoms linked together (5.3)

molten melted; a word used to describe rock that is heated and/or under enough pressure to behave like a liquid

moraine a large ridge of material deposited by a glacier (9.1)

mutualism [MYOO-choo-al-is-uhm] a symbiotic relationship between two different types of organisms, in which each partner benefits from the relationship (1.3)

national park a protected area created by the Canadian government to preserve an ecosystem (3.3) (see also *provincial park*)

native species the organisms that occur naturally in an ecosystem (3.2)

natural resources materials and products found in nature, such as trees, water, oil, fish, and minerals, which people use to meet their basic needs (3.2)

network tree a concept map in which some terms are written in circles or boxes while other terms are written on connecting lines

neutral neither acidic nor basic, with a pH of 7 (6.3)

niche [NEESH] both the location in which an organism lives, and the role it plays within its ecosystem (1.3) (see also *habitat*)

non-foliated not having layers (7.3)

non-renewable resources resources that take millions of years to form and that cannot be replaced, such as coal (3.2) (see also *natural resources, renewable resources*)

ocean ridge a raised part of the sea floor, which can become large enough to be considered an underwater mountain range (8.2)

omnivore an animal that eats both plants and animals (1.3) (see also *herbivore, carnivore*)

ore a rock in the ground that contains one or more valuable substances

organism any living thing, such as a plant or animal

outer core the deep layer close to the centre of Earth that is made of liquid iron and nickel (7.1)

oxygen a gas that is produced by the process of photosynthesis, and that is used in the process of cellular respiration (2.1)

Pangaea [pan-JEE-uh] a theoretical, huge land mass that once, millions of years ago, included all the continents (8.2)

parasite an organism that lives on or in another organism (the host) and feeds on it (1.3) (see also *parasitism, host*)

parasitism a symbiotic relationship between two different types of organisms, in which one partner benefits and the other partner is harmed (1.3) (see also *parasite, host*)

parent rock the original rock that has been changed by pressure and heat into a metamorphic rock (7.3) (see also *metamorphic rock*)

particle model of matter a scientific description of many different features of matter; according to the particle model, all matter is made up of extremely tiny particles, and each pure substance has its own kind of particle, different from the particles of other pure substances (4.3)

periodic table a table in which the elements are arranged into groups and periods according to their properties (5.3)

permanent plots study sites that scientists monitor year after year to gather information about an ecosystem (3.1)

pesticide a substance used to control insects or other organisms that are harmful to plants or animals

pH scale a numerical scale used to describe the acidity or basicity of a solution, ranging from 0 (very acidic) to 14 (very basic) (6.3)

photosynthesis [foh-toh-SIN-thuh-sis] the process by which producers use trapped energy from the Sun to make food (2.1) (see also *cellular respiration*)

physical change a change in which a material changes its shape or state but keeps its identity (4.3) (see also *chemical change*)

phytoplankton [fih-toh-PLANK-tuhn] microscopic organisms found in the ocean, that trap energy from the Sun and convert it into food; they form a basis for the ocean food chain (see also *zooplankton*)

plain a large, flat area often found in the interior region of a continent (9.1)

plasticity the ability of a material to be stretched or bent into another shape, and to keep that shape without breaking or cracking

plateau a flat, raised area of land made up of nearly horizontal rocks that have been uplifted by forces within Earth (9.1)

plates very large pieces of the outer layer of Earth, the lithosphere, that fit together like the pieces of a jigsaw puzzle according to the theory of plate tectonics (7.1) (see also *lithosphere*)

pollutant a material that causes pollution (2.2) (see also *pollution*)

pollution a collective term for the different types of harmful materials that are released into the environment through human activities (2.2) (see also *pollutant*)

population a group of organisms of the same species, living together in one ecosystem (1.2) (see also *ecosystem*)

precipitation the process in which the tiny droplets inside clouds combine to form large drops which fall to Earth as rain, sleet, snow, or hail; also, the water (in liquid or solid state) that falls to Earth as rain, sleet, snow, or hail (2.2)

predator an organism that catches and eats other organisms of a different species

prey an organism that is caught and eaten by another organism of a different species

primary (P) wave the fastest moving of the three types of seismic waves that are produced by an earthquake, originating from its focus; can pass through solids, liquids, and gases (9.2) (see also *secondary wave, surface wave*)

principle of superposition the principle stating that for undisturbed layers of rock, the oldest rocks are on the bottom and the youngest rocks are on the top (9.3)

producer an organism that creates its own food rather than eating other organisms to obtain food; for example, a plant (1.3) (see also *consumer*)

property a characteristic or feature of matter; every type of matter has its own set of properties, such as colour and density (4.1)

provincial park a protected area created by the government of a province to preserve an ecosystem (3.3) (see also *national park*)

pure substance a material that is the same throughout and is composed of only one type of particle; examples include water and gold (5.1)

pyramid of numbers a model used by ecologists to show how many organisms are at each level in a food chain; there are many producers at the bottom of a pyramid, and fewer organisms at the top (2.1)

quadrat a square that marks off a specific area for study (1.2)

qualitative observation an observation in which numbers are not used (see also *quantitative observation*)

qualitative property a characteristic of a substance that is observed and described, usually with words, but not with a numerical value (4.1) (see also *quantitative property*)

quantitative observation an observation that uses numbers (see also *qualitative observation*)

quantitative property a characteristic of matter that can be measured or described with a numerical value (4.1) (see also *qualitative property*)

rate of dissolving a measurement of how fast a solute dissolves in a solvent (6.2)

raw material a material that must be processed to obtain useful products; examples are crude oil and iron ore (5.4)

recycling the process of using the same item over again; recycling can either use the item as it was originally used, or find new uses for it, perhaps by changing its composition

relative age age in comparison to other things (9.3)

renewable resources resources that can be recycled or replaced by natural processes in less than 100 years, such as trees (3.2) (see also *natural resources, non-renewable resources*)

Richter scale [RIK-tuhr] a scale used to describe the magnitude (strength) of an earthquake (9.2)

rift an opening in the oceanic crust as plates move away from each other, where molten materials from Earth's mantle can escape (8.2)

rock a natural material composed of one or more minerals (7.2) (see also *mineral*)

rock cycle the naturally occurring processes by which rocks are continually changed over long periods of time (8.1)

run-off water that runs along the surface of the ground into lakes and rivers (2.2)

sampling in population studies, a method used by ecologists to estimate population sizes in ecosystems (1.2)

saturated solution a solution in which no more of a solute will dissolve at a particular temperature (6.2) (see also *unsaturated solution*)

scavenger an organism that eats decaying plant or animal matter (1.3)

science a body of facts or knowledge about the natural world, but also a way of thinking and asking questions about nature and the universe

science inquiry the orderly process of asking well-focussed questions and designing experiments or conducting research that will give clear answers to those questions

scientific investigation an investigation that involves the systematic application of concepts and procedures (e.g., experimental design and research, observation and measurement, analysis and sharing of data)

sea floor spreading the process in which the sea floor slowly increases in size because of the formation of new crust (8.2)

secondary (S) wave the second fastest moving of the three types of seismic waves that are produced by an earthquake, originating from its focus; can pass through solids but not liquids or gases (9.2) (see also *primary wave, surface wave*)

sediment loose material, such as bits of rocks, minerals, and plant and animal remains (7.3)

sedimentary rock a type of rock formed by compacting and/or cementing sediment (loose material) and/or chemical reactions (7.3) (see also *igneous rock, sediment, metamorphic rock*)

seismic waves [SIHZ-mik] the energy waves that are released by an earthquake and that travel outward from its focus (8.2, 9.2)

seismogram [SIHZ-moh-gram] the record of seismic waves detected by a seismograph (9.2) (see also *seismic waves, seismograph*)

seismograph [SIHZ-moh-graf] a machine used by scientists to measure the strength of earthquakes (9.2) (see also *seismogram*)

selective harvesting taking only some of the resources available, and leaving the rest, so that resources are not used up (3.3)

shield volcano a volcano made of gently sloping layers composed entirely of cooled lava (9.3)

SI (from the French *Le Système international d'unités*) the international system of measurement units, including such terms as kilogram, metre, and second

sill a volcanic feature formed when magma flows parallel to existing layers of rock, and then hardens (9.3) (see also *dike*)

solid a state of matter that has a fixed shape and volume (4.1) (see also *liquid, gas*)

solid waste garbage, such as plastic, paper, cans, bottles, metals, food, etc. (3.2)

solubility the mass of a certain solute that can dissolve in a certain volume or mass of solvent, at a certain temperature (6.2)

soluble able to dissolve in a particular solvent (6.1) (see also *insoluble*)

solute a substance that can be dissolved in a solvent, to form a solution (5.2) (see also *solvent*)

solution a homogeneous mixture of two or more substances that combine so that the mixture is the same throughout and the properties of the substances blend (5.2)

solvent a substance into which a solute dissolves, to form a solution (5.2) (see also *solute*)

sonar (**so**und **n**avigation **a**nd **r**anging) a technology used to gather information about the sea floor, in which the time that the sound waves take to bounce back from the ocean floor is recorded (8.2)

species a group of organisms that can successfully mate with each other and reproduce (1.1)

spider map a concept map used to organize a central idea and a set of associated ideas that are not necessarily related to each other

states of matter the different forms in which matter is found: solid, liquid, or gas (4.1)

stewardship the careful and responsible management of something for which you are responsible (3.3)

streak the colour of the powdered form of a mineral (7.2)

strength the ability of a material to resist forces that squeeze or press on it

subduction zone a place on Earth's crust where high pressure pushes an oceanic plate under another plate into the mantle (8.3)

submersible a vehicle designed for use under the surface of the ocean

substance in science, this term means "pure substance" (see also *pure substance*)

succession the slow process in which a natural ecosystem gradually develops and changes over time; also called ecological succession (2.3)

surface wave the slowest moving of the three types of seismic waves that are produced by an earthquake, originating from its epicentre; surface waves do the most damage of the three types of waves (9.2) (see also *primary wave, secondary wave*)

suspension a heterogeneous mixture in which the particles settle slowly after mixing (5.2)

sustainability the practice in which natural resources are being renewed at least as quickly as they are being used, and wastes can be completely absorbed (3.2)

sustainable development human activity that can continue for a long time period without damaging the long-term health of the environment (3.1)

symbiosis [sim-bih-OH-sis] a biological relationship in which two species live closely together in a relationship that lasts over time (1.3) (see also *symbiotic*)

symbiotic a word used to describe a situation in which two species live closely together in a relationship that lasts over time (1.3) (see also *symbiosis*)

table an orderly arrangement of facts set out for easy reference; for example, an arrangement of numerical values in rows or columns; also called a data table

technology the design and construction of devices, processes, and materials to solve practical problems and to satisfy human needs and wants

tectonics [tek-TON-iks] the study of the movement of large-scale structural features on Earth's crust (8.3)

temperature a measurement of how hot or cold a material is (4.1)

theory an explanation of an event or events that has been supported by consistent, repeated experimental results and has therefore been accepted by many scientists (5.3)

theory of continental drift a theory proposing that the continents change position very slowly, moving over the surface of Earth a few centimetres every year (8.2)

theory of plate tectonics a theory proposing that the plates of Earth's lithosphere interact with each other (8.3) (see also *tectonics*)

threatened species a species that could become endangered if the factors limiting its population are not reversed (3.2) (see also *endangered species*)

top consumer the carnivore at the end of a food chain (2.1) (see also *carnivore, consumer, food chain*)

traditional ecological knowledge knowledge that Aboriginal peoples have gathered about their home environments over a long period of time (3.1)

transform boundary in geology, a plate boundary at which plates slide past each other (8.3) (see also *convergent boundary, divergent boundary*)

transpiration the process in which water that is taken in through a plant's roots evaporates from the plant's leaves, stem, and flowers (2.2)

trench a long, narrow, deep depression in the sea floor (8.3)

trial and error process of solving a problem by trying several ways and learning from errors made

unsaturated solution a solution in which more solute can be dissolved in the solvent at a particular temperature (6.2) (see also *saturated solution*)

variable a condition or factor that can influence the outcome of an experiment

Venn diagram a graphic organizer consisting of overlapping circles or ovals; used to compare and contrast two concepts or objects

vent an opening in Earth's crust through which magma can escape, forming lava (9.3) (see also *magma, lava*)

viscosity a description of how thick or thin a liquid is

volcanic neck the core of a volcano left behind after the softer cone has been eroded (9.3)

volcano an opening in Earth's crust that can release lava, smoke, and ash when it erupts (9.3) (see also *lava*)

volume the amount of space occupied by an object (4.1)

water cycle the continuous movement of water through the environment; the water cycle consists of evaporation, transpiration, condensation, and precipitation (2.2)

water vapour the gaseous form of water

weathering the process in which rocks are broken down into sediment (8.1)

WHMIS an acronym that stands for Workplace Hazardous Materials Information System

zooplankton [zoh-oh-PLANK-tuhn] microscopic organisms found in the ocean that consume other organisms for food; they form a basis for the ocean food chain (see also *phytoplankton*)

Index

The page numbers in **boldface** type indicate the pages where the terms are defined.
Terms that occur in investigations (*inv*) and activities (*act*) are also indicated.

Abiotic, **6**, 12
Absolute age, **281**
Abyssal plain, **236**, 240*act*
Acid, **174**, 176–177*inv*, 178
Acid rain, **79**, 178, 230
Air, 7
Air pollution, 78, 230
Alien species, **76**, 77*act*
Alloys, **139**
Annual surveys, **71**
Ants, observation of, 35*act*
Aquatic, **79**
Archimedes, 113, 143
Area, **295**, **297**
Asthenosphere, **200**, 201*inv*
Atomic-molecular theory, **149**
Atoms, **148**
Avalanches, 54

Bacteria, 29, 38
Badger, 87
Bar graph, **305**
Barnacle, 28, 29
Base, **174**, 176–177*inv*, 178
Baseline data, **71**
Bats, 17
Beaver, 12
Bedrock, **268**
Beetles, 54
Bioaccumulation, **48**, 49
Biomes, **13**
Biotic, **6**, **9**, 12
Bluebirds, 69
Boiling point, **108**
Boreal forest, 13
Brittleness, 109
Burgess Shale, 217

Calcite, 230
Captive breeding program, **85**
Carbon cycle, **45**
Carbon dioxide, **9**, 37, 45, 78, 230
Career Connect, 11, 73, 152, 183, 211, 264
Caribou, 17, 68–69*inv*
Carnivores, 23
Carrying capacity, **51**
Carson, Rachel, 48
Cellular respiration, **37**, 38
Cementation, **216**
Chain of events, 5*act*

Chemical change/property, **125**
Chemical weathering, **230**
Chemistry, **102**
Chemosynthesis, 38
Chlorophyll, **37**
Cinder cone volcano, **274**
Circle graph, **307**
Cirque, **262**
Clarity, 109
Cleavage, **208**
Climate, 7
Coal, 75, 204
Colour, **207**
Commensalism, **29**
Community, **17**
Compaction, **216**
Competition, 52
Composite volcano, **276**
Composting, 26–27*inv*
Compound, **145**, 146*act*
Concentrated solution, **163**
Concentration, **163**, 164
Condensation, **46**
Consumer, **23**, 37, 40
Continental crust, 199
Continental shelf, **236**
Continental slope, **236**
Controlled variable, **310**
Convection currents, **202**
water, 202*act*
Convergent boundary, **245**, 246
Copepods, 41–42*inv*
Core sample, **241**
Corundum, 207
Coyotes, 52
Cross section, **236**
Crust, **199**
Crystal, **206**
Crystallization, **180**
Crystals, formation of, 215*inv*
Cycle map, **292**

Dalton, John, 148, 149
Data table, **303**
Decomposers, **25**
Deep-sea trench, 246
Delta, **258**
Density, **116**
building a density tower, 117*inv*
comparing, 118
formula for, 119–120

Dependent variable, **310**
Deposition, **259**
Diamond, 204
Dichloro-diphenyl-trichloroethane (DDT), 48–49
Dike, **279**
Dilute solution, **163**
Direct evidence, 198
Diseases, 52
Displacement, 113
Dissolves, **140**
Dissolving, 164–173
Distillation, **182**
Divergent boundary, **245**, 247
Dormant, **274**
Ducks, 62
Ductility, 109

Eagle, 40
Earth, model/structure, 198, 200*act*
Earthquake, **265**
arrival time of waves, 269*act*
epicentre, 267, 269*act*
focus, 267
measuring, 268
preparing for, 271*inv*
Ecological reserves, **82**
Ecological succession, **56**, 57*inv*
Ecologists, **12**, 13
Ecology, **12**
Aboriginal peoples, 14, 65
traditional knowledge, xviii–xxiii, 64, 65
Ecoprovinces, **14**
Ecosystem monitoring, **67**
Ecosystems, 2–11, **12**, 13–30
Aboriginal peoples, 70
air pollution, 78
carrying capacity, 51
change in, 66*act*
conserving, 82–88
counting caribou, 68–69*inv*
cycles in, 34–58
greenhouse effect, 78
human impact on, 75–80
learning about, 63*act*, 64–74
limiting factors, 51

mapping, 72*act*
model in a bottle, 84*inv*
monitoring, long-term, 69
monitoring, methods of, 70
organisms, 16–22, 23–30
organization, 16–22
population, 51
population sampling, 20–21*inv*
predicting changes in, 73
protecting, 82–88
temperature, 51
Elasticity, 109
Electrons, 149
Element, **145**, 146*act*, 147
Elk, 16, 52
Emulsifying agent, 138
Emulsion, **138**
Endangered species, **79**, 80*act*, 85
Energy flow, 36, **43**
Environment, 2–30
abiotic. See Abiotic
adapting to, 14
air, 7
biotic. See Biotic
chain of events, 5*act*
interaction, 4
light, 7
organisms, 4
relationship, 4
soil, 7
temperature, 7
water, 7
Environmental impact assessment, 73
Epicentre, 267, 269*act*, 270
Eras, **281**
Erosion, **232**, 258, 259, 260, 261, 262
Erupt, **274**
Eruptions, 279*act*
Estuary, 158
Evaporation, **46**
Events chain, **292**
Evidence, 197*act*
Exotic species, 76
Experiments, 309–313
Extinct, **79**
Extrusive rock, **214**

322 MHR • Index

Fault, **265**
Fireweed, 56
Flexibility, 100
Floodplain, **260**
Floods, 54
Flowchart, **292**
Focus, 267
Foliated, **221**
Food, **9**
Food chain, **36**
  consumer, 37
  mealworm, 39*inv*
  organisms, 43
  producer, 37
Food web, **36**, **40**, 43
Force, 257*act*
Forest fires, 54
Fossil fuels, 75, **78**, **204**
Fossils, **217**, 218*inv*
Foxes, 52
Fracture, **208**
Frogs, 83
Fungus, 25, 54

Gas, **107**
Gas chromatography (GC), 130
Geologic time scale, 281, 282*act*
Geothermal energy, **280**
Glaciers, **261**, 262, 263*act*
Global warming, **55**, 78
Gneiss, 220, 221
Gold, 106, 133, 143, 152, 153, 160, 204, 207
  panning for, 153*act*
Grams, 111
Granite, 213
Graph, **304**
Graphic organizer, **292**
Grassland, 13
Greenhouse effect, **78**
Grizzly bears, 52
Ground water, **46**
Groups, 147

Habitat, **17**
  designer, 18*act*
  enhancement, 83
  loss of, 62, 76
  preserving, 82
  restoration, 83
Habitat enhancement projects, 83
Habitat fragmentation, **76**
Habitat restoration projects, 83
Hardness, 100, 109, **206**
Hares, 6, 24*inv*, 52
Herbivores, **23**

Heterogeneous, **132**
Heterogeneous mixture, 141*inv*
Histogram, 306
Hoary marmot, 40
Homogeneous, 133
Horn, **262**
Host, **28**
Hot spot, **275**
Hot-water vents, 226
Hydrogen peroxide, 103
Hydrothermal vents, 38, **243**

Igneous rocks, 213, **214**, 232
Independent variable, **310**
Indicator, **175**, 178*act*
Indirect evidence, 198
Inner core, **199**
Insoluble, **160**, 161
Intertidal marine ecosystem, 19
Interval, **304**
Introduced species, **76**, 77
Intrusion, 281*act*
Intrusive rock, **214**

Key, **304**
Kilograms, 111

Landforms, 258
Landslides, 54, **260**
Lava, **214**
Lice, 28
Light, **7**
Limestone, 167, 213, 230
Limiting factors, **51**, 53*inv*
Limpet, 29
Line graph, 304
Liquefaction, 273
Liquid, **107**
Lithosphere, **200**, 226
Litmus, 178*act*
Litres, 113
Long-term monitoring, **69**
Lustre, 109, **207**
Lynx, 6, 24*inv*, 52

Magma, **214**
Magnetic field, **203**
Magnetic reversals, **242**
Magnetometer, 242
Malleability, 109
Mantle, **199**, 201*inv*, 202
Marble, 220, 221
Marmot, 85
Mass, **111**, 112*act*, 295, 299

Matter, 45, 100, **102**, 111
  changes in, 122, 123*inv*
  chemical changes in, 125
  description, 102–110
  identification, 101*act*
  measuring, 111–121
  particle model. See Particle model of matter
  physical changes in, 124
  properties, 100, 106–110
  quantitative properties, 108
  recording description, 105*act*
  state of, 107
  volume, 113
Mealworm, food chain, 39*inv*
Measurement, 297
Mechanical mixture, **137**, 140
Mechanical weathering, **228**, 229
Melting point, **108**
Metamorphic rocks, 213, **220**, 232
Metric system, **294**
Milligrams, 111
Millilitres, 113
Mineral, **204**, 206, 210*inv*
Mineral resource, **204**, 209*inv*
Mistletoe, 28
Mixture, 131*act*, **132**, 134*act*
  classification, 137–142
  heterogeneous, 132, 135*inv*
  homogeneous, 133, 134, 135*inv*
  mechanical, 137
  pure substance, 151
  separating, 151*act*
Model, **124**
Moho, **199**
Mohs hardness scale, 207, 210*inv*
Mohs, Friedrich, 206
Molecule, **149**
Moraine, **262**
Mountain goat, 2
Mountains, 258
Mutualism, **29**

National parks, 82
Native species, **76**, 77
Natural disturbances, 54
Natural resources, **75**, 87
Neptune Project, 252
Network tree, **292**
Neutral, **175**, 176–177*inv*
Neutrons, 149
Newton, 295
Newts, 17
Niches, **23**, 24*inv*
Non-foliated, **221**
Non-renewable resources, **76**
Non-reversible, 125
Normal fault, 265

Oak, 64
Ocean ridges, **236**, 243
Oceanic crust, 199
Oil, 75, 204
Omnivores, **23**
Opaque, 208
Ore, 152
Organisms
  ecosystems, 16–22, 23–30
  environment, 4
  food chain/web, 43
  limiting factors, 53*inv*
  pyramid of numbers, 43
Organization, ecosystems, 16–22
Origin, **304**
Original remains, 218
Outer core, **199**
Owl, 80
Oxygen, **9**, 37

Pangaea, **237**
Parasites, **28**, 52
Parasitism, **28**
Parent rock, **220**
Particle model of matter, **124**, 144, 148, 162
  using, 124*act*
Pelican, 80
Peregrine falcons, 49
Periodic table, **146**, 147
Periods, 147
Permanent plots, **71**
Petroleum, 204
pH paper, 176–177*inv*
pH scale, **175**
Photosynthesis, **37**, 38
Physical change, **122**
Phytoplankton, 34, 36, 48
Pie chart, 307
Pine trees, 54
Plains, **258**

Plasma, 107
Plasticity, 109
Plate tectonics, 245,
    248*inv*, 249, 250–251*inv*
Plateaus, **258**
Plates, **200**, 226
Plotting, **304**
Pollutants, **48**, 49
Pollution, **48**, 78, 79,
    230
Polychlorinated biphenyls
    (PCBs), 48–49
Pond, 12
Population, **17**
    ecosystems, 51
    sampling, 19, 20–21*inv*
Porcupine, 4
Precipitation, **46**
Predator/prey, 40, 52
Primary waves, **266**
Principle of superposition,
    280
Producers, **23**, 37, 40
Property, **106**
Protons, 149
Protozoa, 29
Provincial parks, 82
Pure substance, 131*act*,
    133, 143
    classifying, 144
    identifying, 143
    mixture, 151
    particle model of matter,
    144
Purple martins, 83
Pyramid of numbers, **43**
Pyrite, 106

Quadrat, **19**
Qualitative property, **106**
Quantitative property,
    106, 108
Quartzite, 221

Raccoons, 52
Range, **304**
Rate of dissolving, **169**
    changing, 170–171*inv*
    size of solute, 172
    temperature, 172
Rattlesnakes, 76
Raw material, **152**
Recycling, 26–27*inv*
Relative age, **280**
Renewable resources, **75**
Resources, 52, 75, 205*act*
Reverse fault, 265
Richter scale, **268**
Rift, **236**
Ring of fire, 276

Rock cycle, 228, **232**,
    233–234*inv*, 235*act*
Rocks, **204**, 213
    chewable, 232*act*
    classification, 221*act*
    reaction with acid,
    231*inv*
Run-off, 46

Safety, viii–xi
Salmon, 6
Salty seeds, 8*act*
Sampling, **19**, 20–21*inv*
Saturated solution,
    164*act*, **165**
Scale, **304**
Scale worm, 29
Scavengers, 25
Schist, 220
Science, *xiii*
Scientific model, **301**
Scotch broom, 77
Sea floor spreading, **240**
Secondary waves, **266**
Sediment, **216**, 227*act*
Sedimentary rocks, 213,
    216, 232
Seismic waves, **241**,
    266–272
    modelling, 272*act*
Seismograph/seismogram,
    268
Selective harvesting, **87**
Separation, 180, 181*inv*,
    182
Shape, 109
Sheep, 52
Shield volcano, **275**
SI units, 294, 296
Sill, **279**
Silver, 207
Slate, 220
Snakes, 75
Soil, 7, 229
    environment, 7
    sleuths, 10–11*inv*
Solid, **107**
Solid waste, 79
Solubility, 161*act*, **166**
    temperature, 168*inv*
Soluble, **160**, 162
Solutes, **140**, 160–164
Solution, **139**, 141*inv*,
    159*act*
    identifying, 140
    parts of, 140
    processing, 180–184
Solvents, **140**, 160–164
Sonar, **241**
Species, **9**

Spider map, **293**
Spruce, 52
States of matter, **107**
Stewardship, **85**–87
Streak, **208**
Strength, 109
Strike slip fault, 265
Subduction zone, **246**,
    276
Sugar, 167, 172
Sulfur, 204
Supersaturated solution,
    165
Surface waves, **266**
Suspension, **137**, 138*act*
Sustainability, **76**
Symbiotic/symbiosis, **28**,
    29

Tapeworm, 29
Technology, *xiv*–xvii
Tectonics, 245
Temperate forest, 13
Temperature, 7, **108**,
    300
    ecosystems, 51
    environment, 7
    rate of dissolving, 172
    solubility, 168*inv*
Termite, 29
Texture, 109
Theory, **148**
Theory of continental
    drift, **237**, 239*inv*
Theory of plate tectonics,
    245
Threatened species, **79**,
    80*act*
Top consumer, **37**
Traditional ecological
    knowledge, xviii–***xxi***,
    xxii–xxiii, **65**
    science and, 67
Transform boundary,
    245, 247
Transform fault, 265
Translucence, 208
Transparency, **208**
Transpiration, **46**
Tree line, 51
Trench, **246**
Trilobite, 217
Tsunamis, **272**
Tundra, 13
Turkey vulture, 25

Universal solvent, 160
Unsaturated solution,
    165

Variables, 310
Venn diagram, 293
Vent, **274**
Viscosity, 109
Volcanic ash, 277*act*
Volcanic neck, **279**
Volcano, **274**
    impact on humans,
    278
Volume, **111**, 113,
    114–115*inv*, 295, **298**

Water, **7**, **9**, 160
    convention currents,
    202*act*
    environment, 7
    purifying, 183
Water cycle, **46**, 47*act*
Water lily, 14
Weasels, 52
Weather, 54
Weathering, **228**
Wegener, Alfred, 237
Whales, 28, 29, 34, 48,
    49
Whimbrel, 14
Wilson, J. Tuzo, 245
Wolf, 52
Wolverine, 25
Workplace Hazardous
    Materials Information
    System, xi

x-axis, **304**

y-axis, **304**

Zooplankton, 34, 48

Photo Credits

iv centre left George McCallum Photography/Alamy; **iv bottom left** Barry L. Runk/Schoenberger/Grant Heilman Photography Inc.; **v top right** Jacqueline Windh/windhphotos.com; **v centre right** Dick Hemingway; **vi top left** Maximilian Weinzierl/Alamy; **vi centre left** Gerald Newlands, High River AB; **vi bottom** Peter M. Wilson/CORBIS/Magma; **vii centre right** Larry Neubauer/CORBIS/Magma; **viii centre left** Ian Crysler; **x** Ian Crysler; **xii top centre** Bob Rowan; Progressive Image/CORBIS/Magma; **xii** Roger Ressmeyer/CORBIS/Magma; **xii bottom centre** Ed Young/CORBIS/Magma; **xviii centre left** Marilyn "Angel" Wynn/nativestock.com; **xviii bottom** George McCallum Photography/Alamy; **xx bottom right** Stuart Dee Photography; **xxi bottom right** archivberlin Fotoagentur GmbH/Alamy; **xxi inset** PHOTOTAKE Inc./Alamy; **xxii bottom left** Hans Blohm/Masterfile; **xxii bottom right** Jacqueline Windh/windhphotos.com; **xxii top left** Marilyn "Angel" Wynn/Native Stock; **xxiii top** Chris Cheadle/Cheadle Photography; **2–3** Shin Yoshino/Minden Pictures; **4 top** Thomas& Pat Leeson/Photo Researchers Inc.; **5 bottom left** Mike Grandmaison Photography; **6 top centre** Thomas&Pat Leeson/Photo Researchers Inc.; **6 bottom left** Michael Quinton/Minden Pictures; **7 top right** Janis Burger/Bruce Coleman Inc.; **7 bottom right** J.M. Labat/Jacana/Photo Researchers, Inc.; **8 top left** Linda J. Moore; **9 right** Chris Newbert/Minden Pictures; **11 bottom** J. David Andrews; **12 top** Pam Hickman/Valan Photos; **17 centre right** Tom McHugh/Photo Researchers, Inc.; **17 bottom right** Mike Grandmaison Photography; **18** Ian Crysler; **19 top right** Joel W. Rogers; **19 bottom right** Bamfield Marine Sciences Centre; **19 bottom left** Ian Crysler; **25 centre left** Dave Taylor; **25 centre right** Gordon&Cathy Illg/Maxximages.com; **25 bottom right** Val Wilinkson/Valan Photos; **26** Ian Crysler; **27** Ian Crysler; **28 centre** Francois Gohier/Photo Researchers, Inc.; **28 bottom left** Gibson Stock Photography; **29 top left** Robert Calentine/Visuals Unlimited; **29 top right** Ken Lucas/Visuals Unlimited; **29 centre right** Mark Conlin/Seapics.com; **34 top centre** Barry L. Runk, Schoenberger/Grant Heilman Photography Inc.; **34 inset** Francois Gohier/Photo Researchers, Inc.; **35 bottom left** Peter Ward/Bruce Coleman Inc.; **38 bottom left** Science VU/Visuals Unlimited; **40 top left** Andrew A. Bryant; **41 centre left** Ian Crysler; **44 bottom right** Ian Crysler; **46 centre** John Fowler/Valan Photos; **48 bottom centre** Dave Taylor; **49 bottom** Dave Taylor; **51 centre** Al Harvey/slidefarm.com; **52 top left** Joel Sartore/www.joelsartore.com; **52 centre right** Natural Resources Canada, Canadian Forest Service; **52 bottom left** Dave Taylor; **54 top left** Lenard Sanders; **54 centre left** 2004 Province of British Columbia All Rights Reserved; **54 centre right** 2004 Province of British Columbia All Rights Reserved; **55 centre left** T. Kitchin/Tom Stack & Associates; **56 bottom left** Parks Canada/R. Grey; **62 top** Steve Greer Photography; **63 bottom left** Call Kullman/Beyond Books; **64 centre** Al Harvey/slidefarm.com; **64 bottom left** David M. Schleser/Photo Researchers, Inc.; **65 top right** www.guntermarx-stockphotos.com; **66 centre left** Ian Crysler; **68 top left** Dave Taylor; **69 bottom** Wolfgang Kaehler/CORBIS/Magma; **70 top left** Call Kullman/Beyond Books; **70 bottom left** www.guntermarx-stockphotos.com; **71 bottom right** Gary Will/Visuals Unlimited; **72 bottom left** Artbase Inc.; **73 bottom right** Bowles&Sober: Career Connections: Careers for People Who Love Being Outdoors, published by Trifolium Books Inc./Weigl Educational Publisher Limited; **74 bottom left** Steve Greer Photography; **75 top left** Barrett & MacKay Photography; **75 top right** Vancover Public Library; **76 centre left** Mary M. Steinbacher/PhotoEdit Inc.; **77 top right** Al Harvey/slidefarm.com; **79 centre right** Inga Spence/Visuals Unlimited; **80 top left** Tim Davis/Photo Researchers, Inc.; **80 inset** Dave Taylor; **82 top left** Gary Fiegehen/Maxximages.com; **82 bottom right** Russ Heinl/Maxximages.com; **83 top right** John Lowman; **83 centre** Dave Taylor; **84 centre left** Ian Crysler; **85 centre** Andrew A. Bryant; **85 bottom** Chris Cheadle/Cheadle Photography Inc.; **86 top centre** Jacqueline Windh/windhphotos.com; **86 bottom** Marilyn "Angel" Wynn/Nativestock.com; **87 centre** Artbase Inc.; **88 top left** Ian McAllister/Raincoast.org; **89 bottom right** Ian Crysler; **92 top** 1,2,3) Chris Cheadle/Cheadle Photography Inc.; 4) W. Wayne Lockwood, M.D./CORBIS/MAGMA; **92 centre left** Courtesy of Gertrude Frank; **93 bottom right** Canadian Museum of Civilization; **94 top left** Courtesy of Hilda Ching; **96 top** David Young Wolf/PhotoEdit Inc.; **98–99** Mark Tomalty/Masterfile; **100 top** Artbase Inc.; **101 top right** Ian Crysler; **101 bottom left** Brooks Kraft/CORBIS/Magma; **103 top** Ian Crysler; **103 bottom right** Dick Hemingway; **104 top** Ian Crysler; **104 top centre**, Ian Crysler; **105 bottom** Photodisc/Artbase Inc.; **105 top left** Ian Crysler; **106 centre left** George Bernard/Science Photo Library/Photo Researchers, Inc.; **106 centre right** Geoff Tompkinson/Science Photo Library/Photo Researchers, Inc.; **109 top left** Clive Webster/Ivy Images; **109 top right** Tom Kitchin/firstlight.ca; **109 centre left** Harold &Esther Edgerton Foundation 1999, Courtesy of Palm Press Inc.; **109 centre right** Comstock/Alamy; **109 bottom left** Bill Reid/Canadian Museum of Civilization; **109 bottom right** Werner Forman/Art Resource, NY; **111 centre right** Stamp: Canada Post, reproduced by permission, all others, Artbase Inc.; **112, 114, 117** Ian Crysler; **118 bottom left** Richard Megna/Fundamental Photographs; **118 bottom right** Ian Crysler; **120 top right** Brian Pieters/Masterfile; **121 centre right** Marco Cristofori/CORBIS/Magma; **122 top left** Stephen Saks/Photo Researchers, Inc.; **122 centre** Ian Crysler; **122 bottom** Ken Davies/Masterfile; **125 top right** Charles D. Winters/Science Source/Photo Researchers, Inc.; **127 bottom right** Ian Crysler; **130 top** www.guntermarx-stockphotos.com; **131 bottom left** Kit Kittle/CORBIS/Magma; **132 top left** Art Directors/Helene Rogers; **132 top left** Dick Hemingway; **133 bottom left** 2000 Michael Dalton, Fundamental Photographs, NYC; **133 bottom right** Japack Company/CORBIS/Magma; **136 bottom left** Al Harvey/slidefarm.com; **137 centre**, Antonio Luiz Hamdan/Getty Images; **137 bottom right** NASA/JPL/Cornell; **138 top left** Ian Crysler; **139 top right** Phil Degginger; **140 bottom left** Kip Peticolas/Fundamental Photographs; **143 bottom left** Dick Hemingway; **144 centre**, Ricardo Funari/agefotostock/firstlight.ca; **144 bottom left** Charles D. Winters/Photo Researchers, Inc.; **145 bottom left** Ronnie Kaufman/CORBIS/Magma; **145 bottom centre**, Anthony Masterson/Getty Images; **146 bottom right** Getty Images/Artbase; **148 top left** Bettman/CORBIS/Magma; **149 centre left** J. A. Kraulis/Masterfile; **149 centre right** Judd Pilossof/picturearts/firstlight.ca; **149 bottom left** Tom Feiler/Masterfile; **149 bottom right** Melanie Acevedo/Getty Images; **151 top right** Getty Images/Artbase; **152 top right** Don Haddy/Career Connections: Great Careers for People Interested in How Things Work, published by Trifolium Books Inc./Weigl Educational Publisher Limited; **152 bottom left** Ken Lucas/Visuals Unlimited; **153 top right** British Columbia Archives; **153 bottom left** H. Armstrong Roberts/CORBIS/Magma; **154 bottom left** Gregg Andersen/Gallery 19; **155 centre right** Photodisc Green/ArtBase Inc.; **155 bottom right** Winston Fraser/www.stockphotoimages.com; **157 bottom right** Tony Freeman/PhotoEdit Inc.; **158 top centre**, Geoff Tomlins/Pacific Geomatics Ltd.; **159 bottom left** Miles Ertman/Masterfile; **160 centre left** Photodisk/Artbase Inc.; **160 bottom** Photodisk/Artbase Inc.; **163 right** Artbase Inc.; **165 centre right** Royal Geographical Society/Alamy; **166 centre**, Richard Megna/Fundamental Photos; **166 centre right** Richard Megna/Fundamental Photos; **166 top left** Ian Crysler; **167 bottom left** Lauree Feldman/Maxximages.com; **169 top right** Ian Crysler; **171, 172, 173, 175, 176, 177,** Ian Crysler; **178 top left** Townsend P. Dickinson/The Image Works; **180 bottom** www.guntermarx-stockphotos.com; **180 top left** Megapress/Naiman; **183 right** Maximilian Weinzierl/Alamy; **183 bottom right** Vancouver Aquarium; **188 top left** Michael Keefer; **188 top** Glenbow Museum; **188 top** Michael Keefer; **188 top** Erin McNeil; **188 top** Margaret Teneese; **189 centre right** Bev Hills; **190 top left** Courtesy of Hesham Nabih; **191 bottom right** Bettmann/CORBIS/Magma; **192 centre right** Gary Braasch/CORBIS/Magma; **192 top right** Digital Vision; **193** Ian Crysler; **194–195** Alberto Garcia/SABA; **196 top** AP Wide World; **197 bottom left** Tom Bean/CORBIS/Magma; **205 bottom right** R. Ian MacDonald; **206 centre** Barry L. Runk, Grant Heilman Photography; **206** Doug Martin; **207 bottom left** Charles D. Winters/Photo researchers Inc.; **207 bottom** Arthur R. Hill/Visuals Unlimited; **208 all** Doug Martin; **210 centre right** Ian Crysler; **213 centre** ER Degginger/Color-Pic, Inc.; **214 bottom** Douglas Peebles; **216 top left** Val Wilkinson/Valan Photos; **216 both** Joyce Photos/Valan Photos; **217 top right** Gerald Newlands, High River

AB; **217 bottom left** Natural History Museum, London; **217 bottom right** Gerald Newlands, High River AB; **218 top** Marty Snyderman/Visuals Unlimited; **218 centre** Runk/Schoenberger, Grant Heilman Photography; **218 bottom** Frank Staub/Maxximages.com; **220 top** Doug Martin; **220 bottom left** Dr. Morley Reid/Science Photo Library/Photo Researchers Inc.; **220 bottom** Chuck Keeler/Getty Images; **220 bottom** Jeannie R. Kemp/Valan Photos; **220 bottom right** WestLight/FirstLight; **222 bottom** CORBIS/Magma; **226 top** Chris McLaughlin/CORBIS/Magma; **228 top** Peter M. Wilson/CORBIS/Magma; **229 centre left** Pam E. Hickman/Valan Photos; **229 bottom right** Anne-Marie Weber/Getty Images; **230 top left** Susan Van Etten/PhotoEdit; **230 bottom** Hans Strand/Getty Images; **233 bottom left** Ian Crysler; **234 bottom** CORBIS/Magma; **238 bottom right** Martin Land/SPL/Photo Researchers, Inc.; **241 bottom centre** Scripps Institution of Oceanography, UCSD; **243 centre right** Canadian Scientific Submersible Facility; **243 bottom left** B. Murton/Southampton Oceanography Centre/Photo Researchers, Inc.; **243 bottom** Ralph White/CORBIS/Magma; **245 bottom right** Ontario Science Centre; **246 top left** Brad Wrobleski/Masterfile; **256 top** NOAA/Artbase Inc.; **259 centre left** Rod Planck/Photo Researchers, Inc.; **259 centre right** Jim Sugar/CORBIS/Magma; **260 centre right** Lenard Sanders; **260 centre left** www.guntermarx-stockphotos.com; **260 bottom** CORBIS/Magma; **261 centre left** Larry Neubauer/CORBIS/Magma; **261 centre right** John&Barbara Gerlach/Visuals Unlimited; **261 bottom right** MacDonald/PhotoEdit; **261 top right** Steve McCutcheon/Visuals Unlimited; **262 bottom right** Csiro/Simon Fraser/Science Photo Library/Photo Researchers Inc.; **263 centre right** Ian Crysler; **264 top right** Courtesy Dr. Charles Yonge; **265 top right** Natural History of Museum London; **268 top left** David R. Frazier/Photo Researchers Inc.; **271 top right** Mark Richards/PhotoEdit; **271 bottom left** Dolores Baswick/Courtesy Westcoast Energy Inc.; **272 bottom** Bridgeman Art Library/CORBIS/Magma; **273 top right** Francois Gohier/Photo researchers Inc.; **274 top left** R.E. Wilcox/U.S. Geological Survey; **275 top right** Dr. Catherine Hickson/Geological Survey of Canada; **277 bottom right** Joel W. Rogers/CORBIS/Magma; **278 bottom left** by Sommer©Gardiner/Thompson Collection/Photo Researchers Inc.; **280 bottom left** Corel Corporation; **281 bottom right** Musee de l'Histoire Naturelle/Archives Charmet/Bridgeman Art Library; **286 #1,#2,#4**, Chris Cheadle/Cheadle Photography Inc.; **286 top #3**, Tom Collicott/Masterfile; **286 top left** Courtesy of Herb Morven; **287 centre right** Chris Cheadle/Cheadle Photography Inc.; **287 bottom right** Jeff Vinnick Photography; **288 top left** Courtesy of Herb Dragert; **290 centre left** Llloyd Cluff/CORBIS/Magma; **299–302** Ian Crysler; **308 top left** Ian Crysler; **308 bottom left** Ian Crysler; **309 top left** Dennis Kunkel/Phototake; **309 bottom right** Gunter Marx Photography/CORBIS/Magma; **310–311** Ian Crysler

Illustration Credits

XIX Jun Park; **10** Theresa Sakno; **13** Jun Park; **14** Theresa Sakno; **36** Jane Whitney; **59** Steve Attoe; **45** Jane Whitney; **49** Jane Whitney; **111** Steve Attoe; **113** Jun Park; **113** Jun Park; **113** Jun Park; **102** Steve Attoe; **136** Steve Attoe; **141** Jun Park; **143** Steve Attoe; **144** Theresa Sakno; **148** Theresa Sakno; **148** Theresa Sakno; **148** Theresa Sakno; **148** Theresa Sakno; **148** Theresa Sakno; **156** Theresa Sakno; **174** Steve Attoe; **174** Steve Attoe; **175** Jane Whitney; **162** Theresa Sakno; **164** Theresa Sakno; **200** Steve Attoe; **201** Theresa Sakno; **203** Jun Park; **204** Steve Attoe; **209** Jun Park; **241** Jun Park; **249** Jun Park; **258** Jun Park; **270** Jun Park; **277** Jun Park; **278** Jun Park; **301** Theresa Sakno; **289** Jun Park